Knowledge and Space

Volume 8

Series editor
Peter Meusburger, Department of Geography, Heidelberg University, Germany

Knowledge and Space

This book series entitled "Knowledge and Space" is dedicated to topics dealing with the production, dissemination, spatial distribution, and application of knowledge. Recent work on the spatial dimension of knowledge, education, and science; learning organizations; and creative milieus has underlined the importance of spatial disparities and local contexts in the creation, legitimation, diffusion, and application of new knowledge. These studies have shown that spatial disparities in knowledge and creativity are not short-term transitional events but rather a fundamental structural element of society and the economy.

The volumes in the series on Knowledge and Space cover a broad range of topics relevant to all disciplines in the humanities and social sciences focusing on knowledge, intellectual capital, and human capital: clashes of knowledge; milieus of creativity; geographies of science; cultural memories; knowledge and the economy; learning organizations; knowledge and power; ethnic and cultural dimensions of knowledge; knowledge and action; and the spatial mobility of knowledge. These topics are analyzed and discussed by scholars from a range of disciplines, schools of thought, and academic cultures.

Knowledge and Space is the outcome of an agreement concluded by the Klaus Tschira Foundation and Springer in 2006.

More information about this series at http://www.springer.com/series/7568

Peter Meusburger • Tim Freytag
Laura Suarsana

Editors

Ethnic and Cultural Dimensions of Knowledge

 Springer

Editors
Peter Meusburger
Department of Geography
Heidelberg University
Heidelberg, Germany

Tim Freytag
Faculty of Environment and Natural
  Resources
Freiburg University
Freiburg, Germany

Laura Suarsana
Department of Geography
Heidelberg University
Heidelberg, Germany

ISSN 1877-9220
Knowledge and Space
ISBN 978-3-319-21899-1   ISBN 978-3-319-21900-4   (eBook)
DOI 10.1007/978-3-319-21900-4

Library of Congress Control Number: 2015954956

Springer Cham Heidelberg New York Dordrecht London

Printed on acid-free paper

Springer International Publishing AG Switzerland is part of Springer Science+Business Media (www.springer.com)

Acknowledgments

The editors thank the Klaus Tschira Stiftung for funding the symposia and book series on Knowledge and Space. The staff of the Klaus Tschira Stiftung and the Studio Villa Bosch always contribute a great deal to the success of the symposia.

Together with all the authors in this volume, we are especially grateful to David Antal for his tireless dedication to quality as technical editor of the chapters and as translator for some of them. Volker Schniepp at the Department of Geography at Heidelberg University and Erwin Vogl at the Department of Geography at Passau University have been an enormous support in getting the figures and maps to meet the high standards of publication. We also thank the students of Heidelberg University's Department of Geography, who helped organize the eighth symposium and prepare this publication, especially Julia Brasche, Helen Dorn, Maike Frank, Laura Krauß, Julia Lekander, Martina Ries, Veronika Walz, and Angela Zissmann.

Contents

1   Ethnic and Cultural Dimensions of Knowledge and
    Education: An Introduction 1
    Peter Meusburger, Tim Freytag, and Laura Suarsana

2   The School System as an Arena of Ethnic Conflicts 23
    Peter Meusburger

3   Race, Politics, and Geography in the Development
    of Public Schools in the Southern United States 55
    Adam Fairclough

4   Spatial Traditions of Knowledge and Education:
    Ethnic Groups in the United States Reconsidered 69
    Werner Gamerith

5   Educational Inequalities Reflecting Sociocultural
    and Geographical Embeddedness? Exploring the Place
    of Hispanics and Hispanic Cultures in Higher Education
    and Research Institutions in New Mexico 93
    Tim Freytag

6   Local Cultural Resource Knowledge, Identity, Representation,
    Schooling, and Education in Euro-Canadian Contexts 109
    George J. Sefa Dei

7   The Knowing in Indigenous Knowledge: Alternative
    Ways to View Development, Largely from a New Guinea
    Highlands' Perspective 129
    Paul Sillitoe

8   Local Knowledge as a Universal Social Product:
     A General Model and a Case from Southeast Asia 165
     Christoph Antweiler

9   Local Knowledge and Global Concerns: Artificial
     Glaciers as a Focus of Environmental Knowledge
     and Development Interventions 191
     Marcus Nüsser and Ravi Baghel

10   Political Economy, Power, and the Erasure of Pastoralist
     Indigenous Knowledge in the Maghreb and Afghanistan 211
     Diana K. Davis

11   "Masawa—bogeokwa si tuta!": Cultural and Cognitive
     Implications of the Trobriand Islanders' Gradual Loss
     of Their Knowledge of How to Make a Masawa Canoe 229
     Gunter Senft

12   Beyond Merry-Making: Customs of Indigenous Peoples
     and the Normative Functions of Ceremonies in Precolonial
     Igbo Societies 257
     Ikechi Mgbeoji

13   Knowledge, Behavior, and Culture:
     HIV/AIDS in Sub-Saharan Africa 275
     William T.S. Gould

The Klaus Tschira Stiftung 293

In Memory of Klaus Tschira 295

Index 297

Contributors

Christoph Antweiler Department of Southeast Asian Studies, Institute of Oriental and Asian Studies (IOA), University of Bonn, Bonn, Germany

Ravi Baghel Cluster of Excellence: Asia & Europe, Heidelberg University, Heidelberg, Germany

Department of Geography, South Asia Institute, Heidelberg University, Heidelberg, Germany

Diana K. Davis Department of History, The University of California, Davis, CA, USA

George J. Sefa Dei Department of Social Justice Education, University of Toronto, Toronto, ON, Canada

Adam Fairclough Institute for History, Leiden University, Leiden, The Netherlands

Tim Freytag Faculty of Environment and Natural Resources, Freiburg University, Freiburg, Germany

Werner Gamerith Department of Geography, University of Passau, Passau, Germany

William T.S. Gould Department of Geography and Planning, University of Liverpool, Prenton, Merseyside, UK

Peter Meusburger Department of Geography, Heidelberg University, Heidelberg, Germany

Ikechi Mgbeoji Osgoode Hall Law School, Ignat Kaneff Building, York University, Toronto, ON, Canada

Marcus Nüsser South Asia Institute, Department of Geography, Heidelberg University, Heidelberg, Germany

Gunter Senft Max Planck Institute for Psycholinguistics, Nijmegen, The Netherlands

Paul Sillitoe Department of Anthropology, Durham University, Durham, UK

Laura Suarsana Department of Geography, Heidelberg University, Heidelberg, Germany

Chapter 1
Ethnic and Cultural Dimensions of Knowledge and Education: An Introduction

Peter Meusburger, Tim Freytag, and Laura Suarsana

Universally Accepted Knowledge versus Particular Knowledge

It is well known that some forms of knowledge are regarded as true or useful only in particular cultures; by certain ethnic groups[1]; or by people embedded and acting in specific physical, social, and cultural environments, whereas other categories of knowledge are universally accepted. The dichotomy between factual knowledge and orientation knowledge has already been discussed in other volumes of this series (e.g., Meusburger 2015). As a particular form of situated knowledge, orientation knowledge consists chiefly of belief systems, values, cultural traditions, worldviews, ideologies, religions, moral positions, mindsets, action-guiding norms, and reflection about the ethical conduct of one's life (for details see Meusburger 2015; Mittelstraß 1982, 2001, 2010; Stegmaier 2008; Tanner 1999).

---

[1] Culture and ethnicity have much in common, but ethnicity should not be taken as synonym of culture. The concept of ethnicity embraces cultural distinctiveness, deliberate or forced demarcation from other groups, self-denomination, and often social discrimination. The term *culture* is much broader; culture can be practiced unknowingly and without any intention. "[A] cultural group becomes an ethnic group when it decides or is forced, for one reason or another, to live in close contact with a different cultural group, and their differences—not necessarily all of them— are used to mark a social boundary between them" (Melville 1994, p. 87). In many states, "ethnicity" is a personal attribute registered in censuses, whereas culture is not.

P. Meusburger (✉) • L. Suarsana
Department of Geography, Heidelberg University,
Berliner Straße 48, 69120 Heidelberg, Germany
e-mail: peter.meusburger@geog.uni-heidelberg.de; suarsana@uni-heidelberg.de

T. Freytag
Faculty of Environment and Natural Resources, Freiburg University,
79085 Freiburg, Germany
e-mail: tim.freytag@geographie.uni-freiburg.de

© Springer International Publishing Switzerland 2016   1
P. Meusburger et al. (eds.), *Ethnic and Cultural Dimensions of Knowledge*,
Knowledge and Space 8, DOI 10.1007/978-3-319-21900-4_1

Orientation knowledge lays a basis for making moral valuations; providing actors and societal systems with a moral compass, ideologies, goals, values, a cultural memory, and a collective identity; strengthening the motivation and internal cohesion of societal systems; and offering rituals to their members and meeting their spiritual needs. (Meusburger 2015, p. 27)

Factual knowledge can be regarded as widely shared, canonized knowledge that is generated by experts and taken as true on the basis of the prevailing state of the art in research. It is needed in order to achieve a realistic description and analysis of a given situation, to master the complexity of the "real" world, to cope with new challenges, and to manage risks under uncertainty. Factual knowledge can be subdivided into three types:

1. Indisputable matters (e.g., $4 \times 5 = 20$; the distance between A and B is 12,678 miles; the sum of the angles in a triangle equals 180°). This category of knowledge is the only one that is universal and unimpaired by local or cultural influences.
2. Contestable matters provable as true or false only through lengthy empirical examination or theoretical research (e.g., human impact on climate change, viruses can trigger cancer).[2]
3. Knowledge whose validity and practicability have been tested in specific contexts but which is known, accepted, or utilized only by particular people in certain physical, social, or cultural environments. This category of knowledge goes by several names: *local knowledge*, *place-based knowledge*, *situated knowledge*, *traditional knowledge*, and *indigenous knowledge* (for details see the Chaps. 7 and 8 by Sillitoe and Antweiler in this volume). However, indigenous or traditional knowledge may also contain elements of orientation knowledge.

The dichotomy between so-called universal knowledge and local, place-based, or situated knowledge[3] has two facets. First, it is the result of problem-solving, learning processes, and practices provoked by the challenges, needs, stimuli, or incitements that people experience in a particular environment or context in the sense meant by Grossberg (2010, pp. 20–43). This knowledge-generating process is affected by social, cultural, historical, and physical factors and power relations bound to a certain place, region, or context. It is not necessarily restricted to one place or one region but to a specific physical and cultural context (see the Chap. 9 by Nüsser and Baghel in this volume) and can be replicated in environments with comparable conditions and potentials. Geographers as well as the prominent scholar in science and technology studies, Donna Haraway (1988), and other social constructivists are "arguing for politics and epistemologies of location, positioning, and situating, where partiality and not universality is the condition of being heard to make rational knowledge claims…. Situated knowledges are about communities, not about isolated individuals" (pp. 589–590).

---

[2] Types 1 and 2 are also discussed by Felder (2013, p. 14).

[3] Some authors prefer the dichotomy between modern knowledge and traditional knowledge (see the Chaps. 10 and 12 by Davis and Mgbeoji in this volume).

All new knowledge starts as local knowledge. The diffusion of knowledge depends on the need to know and on the prior knowledge that potential recipients require in order to understand the new information (Douglas 2001; Meusburger 2009). If new knowledge is immediately accepted as useful, beneficial, or true by many people, it may become universally accepted knowledge or an indisputable matter after a certain period of time. If such knowledge is valued only as useful or true by people in particular environments with specific conditions, challenges, and traditions and is rejected by people who must cope with other challenges and need different competencies, it will remain local, situated, site-specific, or place-based[4] knowledge.

The second facet of this dichotomy is asymmetric power relations between center and periphery. Centers of state bureaucracies, centers of large social systems, centers of calculation as meant by Latour (1987), imperialistic states, and dominant ethnic majorities often declare their own epistemic position—their knowledge, competence, technology, and interpretation of the world—to be objective, scientific, modern, progressive, or forward-looking. Simultaneously, they declare the epistemic position of their opponents, peripheries, colonies, or nondominant ethnic minorities to be traditional, outmoded, unscientific, or indigenous. Such dichotomies have found use as a political weapon in many ethnic and ideological conflicts, political and cultural imperialism, colonialism, and suppression in totalitarian states[5] to exclude or marginalize other forms of knowledge. Nationalists, imperialists, and racists in many cases have claimed to possess a superior form of knowledge and a historic mission to bring the knowledge and blessings of their civilization to less developed countries (the "White man's burden")[6] or to nondominant minorities within their own state (Ara 1991, p. 276; Eriksen 1991, p. 65; Tomiak 1991, pp. 187–188). The concept of local or indigenous knowledge is therefore not just an analytical construct but a political one as well (see the Chap. 9 by Nüsser and Baghel in this volume).

It would be wrong to apply the concept of local, place-based, or situated knowledge only to traditional (premodern) societies or in the framework of development studies. As disciplines, the geography of science and the history of science provide ample evidence that local knowledge milieus play a notable part in advanced knowledge societies and academia alike (Crang 1998, pp. 182–186). The gaining of expertise is usually understood "as a process of enhancing one's competence in a target domain by accumulating experience of problem solving, understanding, and task performance in that domain" (Hatano and Oura 2001, pp. 3173–3174). The possibility of achieving knowledge and experience in a certain domain is in many cases restricted to a specific place, geographical setting, or knowledge environment (e.g., a specific

---

[4] For definitions and discussion of the terms *place* and *space*, see Harvey (2005), Massey (1999, 2005), and Meusburger (2008).

[5] Examples from the former Soviet bloc are presented by Gyuris (2014) and Győri and Gyuris (2015).

[6] "The White Man's Burden" is a poem by the English poet Rudyard Kipling (1899). This phrase was later used to justify American imperialism in the Philippines as a noble enterprise.

research institute). Some universities provide a unique knowledge environment, creative milieu, culture of science, and scientific potential; others do not (Livingstone 2003; Meusburger and Schuch 2012). Some faculties have produced a large number of distinguished scholars (including Nobel Prize winners); others have not.

Interrelations Between Knowledge and Culture

Numerous authors in various disciplines have documented the inseparable link between knowledge and culture, between culture and education, and between power and education (e.g., Ara 1991; Berry 1979; Ciborowski 1979; Crang 1998; Dinges and Duffy 1979; Eriksen 1991; Freytag 2003; Gamerith 1998a, b, 2006; Havránek 1991; Heinemann 1991; Jordan and Tharp 1979; Kuikka 1991; Meusburger 1996, 1998; Miąso 1991; Mitter 1991; Shweder 2001a, b; Strohmayer 2003; Tomiak 1991; Vroede 1991). Political power tries to influence or control the educational system and the way cultural practices are produced (see the Chap. 2 by Meusburger, in this volume). Cultural values and identities shape learning styles, cognitive styles, comprehension, attitudes toward school and education, and the schemata of interpretation and understanding that individuals use to explain the world. Likewise, learning processes can have an impact on cultural traditions, ethnic identities, and processes of acculturation. Because cultural value systems evolve primarily through processes of communication, learning, imitating, appropriation, and adaptation, it is hard to conceive of investigating knowledge or education apart from culture's influence or of studying culture by omitting learning processes in particular institutions and settings. "[E]very form of knowledge is somehow 'cultural'" (Strohmayer 2003, p. 521). "Knowledge, academic or popular, is about cultural systems of belief and validation—and cultural geography does not escape that" (Crang 1998, p. 185).

It is thus no surprise that many definitions of culture contain terms such as *knowledge*, *learning*, *education*, and *cognitive skills*. British social anthropologist Radcliffe-Brown (1952) defined culture as "the process by which a person acquires, from contact with other persons or from such things as books or works of art, knowledge, skill, ideas, beliefs, tastes, sentiments" (pp. 4–5). "The transmission of learnt ways of thinking, feeling and acting constitutes the cultural process, which is a specific feature of human social life" (p. 5). D'Andrade (1984) specified culture as "learned systems of meanings, communicated by means of natural language and other symbol systems, having representational, directive and affective functions, and capable of creating cultural entities and particular senses of reality" (p. 116). In 1977 the Canadian Commission for UNESCO defined culture as a "dynamic value system of learned elements, with assumptions, conventions, beliefs and rules permitting members of a group to relate to each other and to the world, to communicate and to develop their creative potential" (p. 83). Some anthropologists explicate culture as "patterns of behavior that are learned and passed on from generation to generation" (Shweder 2001a, p. 3106). According to Kögler (1999), culture refers to "the systems of knowledge or sense-making within which the self-understanding of socially situated subjects is articulated" (p. 221). Hörning (1999) takes culture to mean "social practices grounded in knowledge" (p. 88). Hatano and Takahashi (2001) point out that "humans organize learning opportunities for their young to socialize or enculturate them in particular ways" (p. 3041). Knowledge is "one form of culture, which resides alongside other forms of existence" (Strohmayer 2003, p. 522). The educational system and media are the most important institutions "involved in internalizing the ingredients of culture and in making conformity to them a matter of individual will" (Smelser 1992, p. 7). According to Münch (1992),

> social structures of inequality…are produced, reproduced, and transformed in interaction with the cultural code of a society, which entails the language, values, and norms used in discourses on questions of equality and inequality. This interaction is where culture meets social structure, exerts its influence on it, and is itself influenced by it. (p. 244)

Given the close connection between culture, education, social status, and power, it is astounding that some representatives of new cultural geography more or less omitted terms such as *knowledge, education, school, teacher,* and *educational system.* Adherents of new cultural geography rightfully accused traditional cultural geography of having a strong historical orientation and of having concerned itself mainly with traditional issues such as local dialects, music, and material aspects of culture, such as the physiognomy of barns, houses, fences, or gravesites, rather than with the paramount questions and transformations of urban society (Barnes 2003; Cosgrove and Jackson 1987; Mikesell 1978). However, leading proponents of new cultural geography have ignored some of the key issues of a knowledge society— the school system, learning processes, educational attainment, the role of teachers in ethnic conflicts, the cultural causes, and consequences of ethnic disparities in educational attainment, and the suppression of minorities through the educational system. Terms and concepts such as *education, knowledge, school system,* and *learning* do not appear in Mitchell's *Cultural Geography* (2000) and are at best marginal in the *Handbook of Cultural Geography* (Anderson et al. 2003)[7] and in *Kulturgeographie* (edited by Gebhardt et al. 2003). Some of the few exceptions are *Cultural Geography* by M. Crang (1998), which contains a chapter entitled "Cultures of Science: Translations and Knowledge"; *Culture and Space* by J. Bonnemaison (2005, pp. 77–82), and *Schlüsselbegriffe der Kultur- und Sozialgeographie* by Lossau et al. (2014) with its overview on the potential of education and knowledge to be explored in social and cultural geographies (Jahnke 2014).

This disregard of educational systems and learning processes by some proponents of new cultural geography is even more astonishing considering that books seminal to the tradition and style of cultural studies reflected keen interest in inequality in educational trajectories (Winter 1999, p. 157). Literacy was a topic from the outset in Birmingham, England, the cradle of cultural studies (Hoggart 1957). Cultural studies as pursued at the Birmingham School encompassed a broad

---

[7] The handbook covers topics such as cultures of consumption, cultures of money, cultures of labor, geographies of racialization, colonial geographies, and queer cultural geographies. It contains a chapter about epistemology (Strohmayer 2003), but *education, educational attainment, literacy, school, university, research,* and *teacher* do not appear in the index.

range of knowledge forms, including many types of everyday knowledge and ability rooted in popular culture (Hörning 1999, p. 99).

There is much debate and incertitude about how to define cultural studies,[8] but according to Grossberg (2010) cultural studies

> is concerned with describing and intervening in the ways cultural practices are produced within, inserted into, and operate in the everyday life of human beings and social formations, so as to produce, struggle against, and perhaps transform the existing structures of powers....
>
> Cultural studies describes how people's everyday lives are articulated by and with culture. It investigates how people are empowered and disempowered by the particular structures and forces that organize their everyday lives in contradictory ways, and how their (everyday) lives are themselves articulated to and by the trajectories of economic, social, cultural, and political power.... Cultural studies is concerned with the construction of the contexts of life as matrices of power, understanding that discursive practices are inextricably involved in the organization of relations of power. (p. 8)

According to Hörning (1999), a new cultural science must also be a type of "entirely new sociology of knowledge that is not solely concerned with how knowledge is generated as a cultural phenomenon but mainly with how, in all its manifestations and notations as cultural knowledge and expertise, it underlies social practice and thereby animates cultural life" (p. 113). A theory of culture should concern itself with the "implicit and informal aspects of the creation, representation, transfer, practice, materialization, storage, and practical application of knowledge and ability in which the social power of culture is expressed" (p. 89).

Culture and Ethnicity as a Power-Sensitive Field

The question of how to treat difference or otherness is not only an academic issue but a serious political and ideological matter as well. "Culture is politics by another name" (Mitchell 2000, p. 3). Cultural issues are often used and abused for political reasons, to mobilize people, to establish a border between people, to stabilize social class inequalities and power relations, to constitute or enforce a territorial claim, and to consolidate or destabilize imagined communities and identities. "[C]ulture is a contested, conflictual set of practices bound up with the meanings of identity and community" (Denzin 2001, p. 3124). Culture is central to identity politics. It is a pivotal issue in the discussion of racism, xenophobia, colonialism, nation-building, neocolonialism, and the emancipation of and discrimination against nondominant ethnic minorities. Processes of assimilation and acculturation[9] are characterized by clear asymmetries of power relations between a majority and a minority, or between a dominant culture and a less dominant one, or between a political center and

---

[8]"Any definition is likely to disown at least some people who want to locate themselves within cultural studies" (Grossberg 2010, p. 7).

[9]"Acculturation is the process whereby one cultural system conquers the minds of an individual or group" (Boyer 2001, p. 3032).

gradually assimilated peripheral entities (Boyer 2001). Many authors have pointed out that knowledge is conditioned by power, rhetorical or textual construction, and social solidarity (Fernandez 2001; Foucault 1980, 1990; Meusburger 2015). This viewpoint applies not only to orientation knowledge but also to the first two categories of factual knowledge listed in the first section of this introductory chapter.

The great interest that agents of power have in culture lies partly in the fact that culture is

> not stable, homogeneous, and fixed but marked by receptiveness, contradiction, negotiation, conflict, innovation, and resistance. Culture is viewed as a process of social inequality involving a struggle for power.... It is not the integrative function but rather the struggle over meanings, the conflict about the sense and value of cultural traditions, experiences, and practices, that determines analyses [of culture].... Culture is a polyphonic, ever controversial, and complex process of constructing sociocultural meanings and identities. (Hörning and Winter 1999, p. 9)

As Grossberg (2010) formulates it, "while power operates in institutions and in the state, it also operates where people live their daily lives, and in the spaces where these fields interact" (p. 29).

If culture is understood not only as a medium of integration but as a criterion of distinction and as a discourse about values, norms, meaning, validity, and identity (Miller 2001; Spencer 1983; Spencer and Markstrom-Adams 1990), then which institutions of knowledge production and knowledge dissemination serve as an instrument or forum for these processes and power struggles? In states with a history of internal colonialism, in newly founded nation-states, and in nations struggling with ethnic or racial conflicts, it is little wonder that cultural diversity, ethnicity, and the cultural rights of nondominant minorities are highly explosive political topics (see Ara 1991; Eriksen 1991; Frantz 1999; Havránek 1991; Heinemann 1991; Kuikka 1991; Miąso 1991; Tomiak 1991). Nationalist politicians such as Giovanni Gentile in fascist Italy regarded the nation "as a unitary and organic community which could not allow any particularist deviations" (Ara 1991, p. 276). Political elites of many nation-states emphasize the importance of national unity and cultural homogeneity and often advocate an anticulturalist position.

> Anti-culturalists worry that any description of cultural difference merely sows the seed of invidious comparison and conflict, and thus should be disavowed. For the…pluralists, however, the recognition and appreciation of cultural differences is one of the major aims of ethnography in particular and cultural anthropology in general. (Shweder 2001b, p. 3154)

Some intellectuals worry that culture is "an excuse for the maintenance of authoritarian power structures and permits despots and patriarchs around the world to deflect criticism of their practices by saying 'that is our custom' or 'that is the way we do things in our culture'" (Shweder 2001b, p. 3153). Others point out that "cultural rights are significant because culture is an intrinsic part of individual and group identities. Not only does culture exert a crucial influence on the formation of identity of individuals, but it also helps perpetuate ethnic groups" (Renteln 2001, p. 3116; see also Kymlicka 1995). Another reason why culture is often misused as a vehicle for political aims is that people seem much easier to mobilize if political conflicts are presented (disguised) as religious, ethnic, or cultural conflicts.

Goals, Structure, and Contents of This Volume

The chapters of this volume are not intended as fuel for the intensive debate about multiculturalism, cultural pluralism, cultural assimilation and acculturation, cultural rights, or minority rights (see Kymlicka 1992, 1995; Renteln 2001). They focus instead on three other issues. The first five texts deal with the role of the educational system in multiethnic states, with the impact of ethnic identity and race[10] on schooling, and with the reasons and consequences of ethnic disparities of educational achievement. The subsequent four chapters study the relevance of indigenous, native, traditional, and local knowledge compared to universal, scientific, or so-called *objective* knowledge. The final four chapters present case studies on the social and cultural function of indigenous knowledge in nonwestern societies and on the influence that culture may have on action.

Ethnic Minorities and Schooling

Peter Meusburger presents an overview on some of the reasons why the school system in multiethnic or multilingual states is a contested field and an arena of political, racial, and cultural conflicts. In states with compulsory education, it is crucial to ask whether and under which circumstances nondominant ethnic minorities are allowed to use their mother tongue in elementary and secondary schools as the language of instruction, whether the teachers are sympathetic to the minority cultures, whether minorities figure in determining the location and size of elementary schools and the content of the textbooks, and whether those minorities have a say in the "memory industry." The school can support and reinforce the cultural learning process that students have already undergone in their families and neighborhoods but can also interrupt or reverse it, eventually instilling them with serious doubt about their identity. In multiethnic states minority students entering the school system frequently

---

[10] In the academic community it is widely agreed that there is no such biological thing as *race* (for an overview on the literature, see Bonnett and Nayak 2003; Diamond 1994, pp. 85–87; Mitchell 2000, pp. 233–241). In most European countries this term is no longer used by scholars or in official statistics. Nevertheless, in some states (e.g., the United States) race is still one of the personal attributes registered in population censuses. The U.S. population census of 2010 declares: "Our population statistics cover age, sex, race, Hispanic origin, migration, ancestry, language use,…" (http://www.census.gov/programs-surveys/decennial-census.html). However, the connotation of the term *race* in the United States differs from that in Europe. Even if the scientific validity of race as an attribute is denied, one cannot ignore the racialization of many spheres, the power of racial differentiation, and the existence of racism in everyday life (Bonnett and Nayak 2003; Kobayashi 2003). For example, a journal entitled *Ethnic and Racial Studies* has existed since 1978. In the United States scholars have developed the academic discipline called "critical race theory" (Gillborn and Ladson-Billings 2010; Stovall 2006) and discuss "racing-language" (Gutiérrez et al. 2010, p. 359). And in a speech delivered in Philadelphia on March 18, 2008, U.S. President Obama declared: "Race is an issue that I believe this nation cannot afford to ignore right now." See par. 43 of the transcript of President Obama's speech, retrieved from http://edition.cnn.com/2008/POLITICS/03/18/obama.transcript/

find that the values, historical experiences, and cultural practices that their parents have passed on to them are called into question, resisted, or portrayed as backward.

To study the collective social status and the vertical social mobility of nondominant ethnic groups; the ethnic awareness, social cohesion, and ethnic self-esteem of minorities; the power relations and conflicts between ethnic groups; or the discrimination against ethnic minorities, one can scarcely find a better approach than to inquire into a minority's position in the school system and to analyze the large disparities in educational achievement among ethnic groups. The public school system is one of the multiethnic state's best sounding boards for societal problems and conflicts. Since the mid-nineteenth century, the school system has become a battlefield of ethnic strife and conflict where existing power relations have been maintained, reinforced, yet also contested. Nationalist movements in Europe of the nineteenth century as well as the civil rights movement in the United States of the twentieth century originated in conflicts over schooling issues.

Adam Fairclough studies the development of public education in the southern United States, focusing on interrelations between education, race, politics, and geography. The southern states have posted the lowest levels of educational achievement since the first U.S. statistics in this field appeared. In this region public schools have long suffered from the legacy of slavery, racism, and the political consequences of the Civil War. White hostility to the education of Blacks stunted the development of public education after 1865 and produced a racially segregated school system that perpetuated inequality. In the nineteenth and early twentieth century, most Whites wished to restrict Black education to basic literacy, and many of them continued to believe that schooling for Blacks damaged the interests of Whites. Rural schools of the south were mainly primitive one-room, single-teacher elementary schools. Some schools for Blacks suffered the additional handicap of a constantly shifting population. In the cotton South, Black sharecroppers often moved every year in search of a better landlord, an annual migration that continually disrupted the already meager and intermittent education of their children. Another problem was that many teachers in Black schools had never attended a secondary school, let alone a college or university.

The post-1945 civil rights movement attacked and eventually destroyed the segregated school system. By 1970 the South's public schools were substantially integrated. Many Blacks, however, regarded integration as a dubious gain. In implementing integration, White politicians and administrators usually saved public money by closing Black schools, demoting Black principals, and dismissing Black teachers. Black children found it difficult to feel at home in integrated schools where the principal and most of the teachers were White. By the 1990s, however, "White flight" had made the South's big-city public school systems overwhelmingly Black. According to Fairclough, the current nostalgia for the Black schools of the pre-Brown era[11] reflects profound disappointment that the predicted academic and

---

[11] In May 1954 the U.S. Supreme Court handed down a landmark ruling in a case known as Brown v. Board of Education. In effect, the court declared that "separate but equal" public schools for Blacks and Whites were unconstitutional. The decision meant that "separation itself was inherently unequal and a violation of the equal protection clause of the 14th amendment." Retrieved April 14, 2015, from http://www.civilrights.org/education/brown/

social benefits of integration have not been not forthcoming. In his view the all-Black schools provided a safe haven within which they were shielded from White condescension and White bullying.

Werner Gamerith calls attention to the fact that public education in the United States was originally regarded as an instrument for seeding American society with political attitudes and ideological concepts such as the egalitarian society and the melting pot. The American public school was supposed to function as a cradle of democracy offering equal opportunity for any citizen and across all social and ethnic lines. The public school was vital to the process of assimilation as legions of European immigrants poured into the United States in the late nineteenth and early twentieth century. However, the ideology of public education as an incubator of American civic values and as a vehicle for social advancement has never fully embraced African Americans, Native Americans, and Hispanics. The official rhetoric of schooling as an anchor for equal opportunity, democratic values, and social change has always collided with the day-to-day reality of failure, racism, suppression, and low achievement endured by various ethnic groups.

Although some ethnic differences in educational attainment have been reduced in the United States in recent decades, equality still lies far in the future. Equalization in financing the public school system is at best a political slogan, but will not come about as long as school funds mainly depend on the local property taxes. Fourth- and eighth-graders from the Old South but also from parts of the Southwest and the West fail to achieve national standards in reading skills and mathematics. As these individuals proceed through their educational careers, they come to account for a high percentage of school dropouts and only a low percentage of university graduates. The risk of leaving school before graduation depends considerably on the ethnic background of the students, even if one controls for other variables like parents' socioeconomic status, income, or occupation. Even when schools solely for African Americans existed, they were nowhere near equal to the educational institutions for Whites. Differences in financial endowment, the duration of the school year, and the competence of the teachers could vary greatly between White and African Americans schools. Native Americans suffered even more discrimination. White-Anglo America was convinced that the best way to assimilate Native American children was to put them into boarding schools hundreds of miles away from their families and local cultures on the reservations and not allow them to use their mother tongue.

Tim Freytag explores the impact of cultural identity and sociocultural embeddedness on the production of educational inequalities by examining the example of Hispanic students and academic staff at U.S. universities and research institutions. He first analyzes the comparatively low educational performance of Hispanics and the extent to which they are still underrepresented in U.S. higher education, particularly at the University of New Mexico in Albuquerque. The second part of his chapter focuses on the importance of family values in Hispanic culture and explains why it appears rather difficult for some Hispanics to combine university studies with the expectations and duties of the family. Hispanics who grew up in New Mexico tend to feel particularly attached to the cultural and physical environment of that U.S. state. They may experience the period of studying at an out-of-state university as cultural transformation and alienation. This observation is also underlined by Hispanic professors who try to find a place for or improve the integration of Hispanic culture in their professional environment.

Freytag argues that educational success and educational inequalities should be regarded neither as merely resulting from one's ethnic or cultural belonging nor as simply depending on the meaning and significance attributed to educational attainment in an overarching concept, such as social class, socioeconomic status, lifestyle, or gender. In fact, educational attainment arises from a complex interplay of parameters operating at the individual, collective, or structural levels.

George J. Sefa Dei theorizes the link between identity, knowledge, representation, and schooling in the particular case of Black and minority youth education in Euro-Canadian contexts. He adopts an anticolonial perspective to argue that the epistemologies of Black and marginalized youth, particularly the ways in which these learners come to know and act within schools, offer interesting insights into why and how the questions of identity, representation, and social difference are critical for educational success. He focuses on narratives of Canadian youth, parents, and educators from diverse racial and ethnic backgrounds as they speak about the school system. The relevance and implications of their voices are highlighted as legitimate sources of cultural resource knowledge that inform teaching, learning, and the administration of education. The local ways of knowing among young learners, minority parents, and educators stand at the center of Dei's theorizing and search for ways to improve schools in response to the needs and concerns of a diverse body politic. He affirms there is much to learn from the ways in which oppressed bodies relegated to the status of racial minorities eventually claim a sense of intellectual and discursive agency as well as ownership and responsibility for their knowledge about everyday schooling. Dei discusses ways in which African people's local knowledge of cultural resources can help enhance the schooling and education of young African learners.

An anticolonial reading offers a critique of how conventional schooling provides education that fails to help young learners develop a strong sense of identity, self, collective agency, and empowerment for community-building. The anticolonial prism affirms the role and power that local cultural knowledge possesses to subvert such internalized colonial hierarchies of schooling by placing values such as social justice, equity, fairness, resistance, and collective responsibility at the center of the learner's education. Dei argues that a school system should be capable of tapping into youth identities and identifications as valuable sources of knowledge. He asserts that there is something fundamentally and morally wrong when students go through the educational system without being taught by educators who share their cultural, racial, and ethnic backgrounds. He does not mean that only teachers of a particular background can provide the much-needed nurturing environments but rather that it is essential for all learners to have teachers with whom they can identify.

Academic Debates About Indigenous, Traditional, Native, and Local Knowledge

The vast majority of chapters in the first seven volumes of the series *Knowledge and Space* refer to the context of literate, economically advanced societies, where concepts such as meritocracy, efficiency, competitiveness, professionalization, and bureaucracy give science, professional expertise, technology, high educational attainment, and rationality a function as cornerstones of society. However, there exist other societies that challenge the western view of knowledge and expertise. In preindustrial societies, illiterate societies, or acephalous[12] communities, educational achievement, science, and technology do not enjoy the same status as in advanced knowledge societies.

Paul Sillitoe demonstrates in his chapter that people of acephalous communities are not so much inclined to trust experts, especially external ones, as people in western societies are. The perception of what constitutes knowledge differs from one spatial and cultural context to the next. The terms *indigenous knowledge*, *traditional knowledge*, *native knowledge*, and *local knowledge* have been used to describe categories of knowledge that are more or less counter to western science, to so-called objective knowledge, the knowledge of former colonial powers, or the knowledge of the centers of calculation (Latour 1987). Approaches marked by preference for these terms stress the subjective nature of experience, understanding, and knowing; mirror an interest in oral tradition, illiterate societies, embodied knowing, and individual knowledge; reflect an attempt to increase the prominence of local voices and practices (e.g., local knowledge about medical plants or farming) in development contexts, and point out that these terms are often used by minorities in their resistance against hegemonic powers.

The terms *indigenous knowledge*, *native knowledge*, and *traditional knowledge* have been criticized by some scholars for several reasons. According to Sillitoe, there is anxiety from a liberal perspective that use of the words *indigenous* or *native* may encourage xenophobic or racial emotions. The use of these terms may also be unwelcome because it reveals asymmetric power relations, inequality, exploitation, and histories of colonialism. He indicates that some states that began as colonies, such as the United States of America and Australia, given their history of domination, ethnocide, and even genocide, do not wish to draw attention to indigenous issues and rights. Despite warnings from some academics about using the word *indigenous*, many autochthonous minorities do so deliberately to position themselves in society and to remind others of their traditions, way of life, culture, and painful history.

The situation and the research question determine which of these adjectives are appropriate or useful. The term *indigenous knowledge* may be problematic in development studies, but quite useful in political debates about identities, minority rights,

---

[12] An acephalous society or community is one that lacks political leaders or hierarchies. Most foraging or hunter-gatherer societies are acephalous.

colonialism, and postcolonialism. The adjectives *traditional*, *indigenous*, and *native* express the need to see knowledge in a wider cultural context, not just from the position of western science. Dei and Sillitoe argue that the notion of static traditionalism sometimes exists only in the imagination of the critics of indigenous knowledge. They hold that criticisms of these adjectives seemingly have more to do with reservations of western intellectuals than with those of people to whom the words are regularly applied and who happily use these labels for themselves because—unlike western scientific, rationalist epistemology—they characterize unique attributes of the native epistemologies.

In political discourses or power struggles the adjectives *indigenous* and *native* entail a certain moral stance. They are meant to recall the history of colonialism, aggression, and genocide and are prominent in the native people's resistance to dominant political forces. Sillitoe argues that attempts to discourage the use of the adjective *indigenous* could induce new tensions because some people fear there will lose their identity or be disadvantaged in their fights for their rights and interests, of which national governments seek to dispossess them.

Diana K. Davis employs the term *indigenous knowledge* in her chapter to refer to "knowledge that the nomads and other pastoralists have garnered, often over centuries, from working with animals in an arid, stochastic environment" (see p. 27 in this volume). She also underscores the point that indigenous knowledge is very dynamic, not static as some critics assume. In her view indigenous knowledge is knowledge and practice executed at the local level as opposed to a variety of expert knowledge coming from elsewhere, such as those of international development agencies and national agricultural institutions. She uses the terms *indigenous knowledge* and *local knowledge* synonymously. Some people prefer the term *traditional* knowledge because they wish to keep their cultural tradition and cultural heritage intact and want to protect them from extinction due to pressures of current globalization. In a bid to revive their traditional culture, others seek to recover traditions that had been suppressed by colonial powers and that are now lost.

Two chapters in this volume focus on the category of local knowledge. Christoph Antweiler deals with the specific ontic character of local knowledge, the epistemic implications of that character, and the proper methods for studying such knowledge. In his view, research on local knowledge is often idiosyncratic, and the possibility of a truly comparative analysis of local knowledge from different local settings is often dismissed prematurely. He claims that research about local knowledge needs well-defined methodologies, which require a clear theoretical understanding of the phenomenon of local knowledge. Using an empirical study of urban knowledge from Indonesia, he outlines ten universal features of local knowledge, which he presents in a general model. According to Antweiler, local knowledge is usually gained through intense personal experience in a specific environment. "Local knowledge comprises skills and acquired intelligence, which are action-oriented, culturally situated, and responsive to constantly changing social and natural environments" (see p. 27 in this volume). In his view local knowledge consists of more than technological and environmental knowledge (e.g., how to survive in a desert). It entails not only cognitive but also emotive and corporeal aspects and may be best understood as a cultural or social product.

The term *local knowledge*, too, has its critics. Some of them hold that external relations, migration, and the impact of globalization make definition of the local problematic. However, Marcus Nüsser and Ravi Baghel see such critique as a misunderstanding of the term *local*. Local does not mean stable, fixed, or isolated from external influences but site-specific or geographically situated. Nüsser and Baghel's concept of local or situated knowledge is that certain types of knowledge are geographically bounded and, even more important, engendered in specific localities that affect both its form and its content. Site-specific knowledge emerges through people's practical engagement with their environmental setting, including its material resource base. Such site-specific knowledge is not timeless or static but part of the ever-changing constellation of human and environmental factors.

Nüsser and Baghel explore in their chapter so-called artificial glaciers in Ladakh (the Himalayas), structures that people build to store frozen water to help them through seasonal water scarcity. These ice reservoirs meet local irrigation needs in the agricultural season by melting before natural glacial meltwater becomes available. At the same time, they are also framed as an adaptation to climate change, a perspective suggesting that, contrary to romanticized notions, local knowledge is not disconnected from globally circulating knowledge about climate change. Nüsser and Baghel state that the dynamics of the transformation or persistence of land-use practices in peripheral high mountain regions strongly depend on the interplay of local environmental knowledge, social cooperation, livelihood strategies, political organization, and external relations. Nüsser and Baghel argue that the emphasis on local knowledge as a central pillar of development is one of the most significant changes to emerge from the turn from top-down to bottom-up approaches and has the ostensible objective of integrating local perspectives and perceptions. Local knowledge and external interventions tend to interact in a dynamic and fluid relation that changes the local setting through local observation, external concerns, and strategies to sustain local livelihoods. Local knowledge and scientific knowledge are thus not necessarily incompatible or in contrast to each other; both forms of knowledge can easily be combined.

Knowledge, Culture, and Action: Case Studies on Indigenous and Local Knowledge

Four chapters present case studies on the social and cultural role of indigenous knowledge in nonwestern societies and on the influence that culture has on action. Diana K. Davis studies how colonialism and the influence of western development projects had deleterious effects on the knowledge system of pastoralists in the Maghreb and in Afghanistan. She explores the ways that local knowledge became suppressed and western expert knowledge became privileged and how that processes had far-reaching consequences for power and gender relations. In the nineteenth century the nomads in the Maghreb and Afghanistan possessed an impressive body of ecological knowledge about herding and range management complemented by sophisticated indigenous veterinary knowledge.

The French state appropriated land and forests, banned common management techniques like burning for pasture regeneration, and, later, criminalized traditional veterinary medicine. These changes and those that were wrought in trading systems by the imposition of western veterinary medicine and land-management techniques reduced indigenous knowledge and practice. The erosion of pastoralist indigenous knowledge persisted under the postcolonial government, which retained many colonial laws and policies as well as mainstream development projects that privileged "expert" knowledge. In the colonial Maghreb changes in property laws, restrictions on natural resource use, the imposition of western veterinary medicine, and the spread of capitalist social relations all had profound, mostly adverse impacts on pastoralists' ways of life and knowledge systems.

Similar developments threaten to destroy the sophisticated and highly valued indigenous veterinary knowledge of women among Koochi nomads in Afghanistan. Afghan Koochi women have a rich array of knowledge of animal diseases and their treatments. They help with difficult births, regularly care for sick and newborn animals, and often have a more accurate understanding of livestock diseases than men do. In many cases it is the ignorance of western development agencies that devalues indigenous knowledge and undermines the knowledge and skills of Afghani women. Despite the sophisticated knowledge of Koochi women, western development programs trained only men to treat livestock diseases that had traditionally been dealt with by women. The NGO and USAID staff had assumed that Koochi women do no work with animals other than milk them and that they had little or no knowledge of animal health and disease. Biased, ignorant views of Muslim women by western development "experts," in conjunction with the extremely conservative religious government of the country, is curtailing women's work with livestock in ways that endanger their knowledge, practice, social status, and well-being.

Gunter Senft discusses the interrelations between indigenous knowledge, culture, and ceremonies by studying the process of by which large seagoing *masawa* canoes used to be constructed by the Trobriand Islanders of Papua New Guinea. The joint communal construction of such canoes was a highly complex enterprise that required different forms of technical and magical knowledge. The knowledge needed was distributed among various craftsmen and magicians within a community. The magicians had to know more than ten forms of magical rites and the respective formulae. It was simply inconceivable to a Trobriand Islander that a *masawa* could be constructed without the appropriate magical rites being performed. The dispersed knowledge of craftsmen and magicians had to be integrated and coordinated at the communal meals arranged by the *toliwaga* (an owner of a *masawa*) during or after specific phases of the building process. The process of building a canoe continuously tested and monitored the security and stability of a village community's social network. They all had to cooperate in good faith to ensure the success of the canoe under construction, and they all had to be paid in the form of adequate food-distribution ceremonies after certain stages in the construction process.

Senft explains why most of the village communities on the Trobriand Islands have gradually lost the knowledge of how to make a *masawa* and what social and cultural consequences this loss of knowledge has had for the communities. Because the construction process of a canoe was intertwined with social events and had a stabilizing influence on the communities' social network, a loss of technologies entailed losses of social activity and cultural identity and had an impact on the population's social construction of reality and on their cognitive capacities.

Ikechi Mgbeoji argues in his chapter that ceremonies in precolonial Igbo societies of Southeast Nigeria had not only been greatly significant normative activities and relevant instruments of social governance but also crucial vehicles for the communication of values and beliefs and the diffusion of information and knowledge. Some of the ceremonies had had major ecological and environmental implications and had transferred key knowledge to the young generation. He examines the ways in which those ceremonies had often been deployed to educate the young and prepare them for life in the traditional settings. Ceremonies were veritable instruments that reflected, transmitted, shared, and modified the society's sense of social justice, legitimacy of law, public participation in governance, integrity of the human person, and protection of the family. Ceremonies in the traditional knowledge framework accomplished norm-bearing and norm-iterating functions. They marked certain events that affected social bonds and complex interrelationships.

According to Mgbeoji, Igbo ecological and traditional knowledge was often enmeshed in a body of law that asserted the multiple linkage of mankind in a complex chain and interrelation of other parts. The elaborate ceremonies on adulthood and maturation ceremonies were equally seminal in teaching teenagers the importance of self-reliance and respect for constituted authority. These beliefs were narrated and validated in Igbo law and life through ceremonies and religious observations. Since the age of colonialism, however, Igbo cultural practices have been subjected to Eurocentric sanitization and cleansing processes designed to remake native law in the image of English common law. In the colonial encounter, Igbo culture, though somewhat displaced, has not been totally vanquished.

People experience a gap between knowledge and action every day. Very few cigarette smokers respond positively or immediately, if at all, to medical evidence that smoking affects health and reduces life expectancy. William T. S. Gould explores the nature and extent of the disconnect between knowledge and behavior in connection with the prevalence of HIV/AIDS in sub-Saharan Africa. Analyzing surveys in Kenya, Tanzania, and Uganda, he tries to answer two questions. First, why does widely distributed knowledge of this disease's social biology and its prevention coincide with high and even rising infection rates. Second, why do knowledge-based policies and HIV/AIDS programs often fail to match local cultural practices and socioeconomic conditions? Available knowledge has not been sufficiently internalized to induce substantial change in the sexual behavior of those people most at risk of contracting the disease. Cultural practices affecting sexual behavior and exposure to HIV are telling determinants of the knowledge/behavior gap.

Obviously, access to the most recent biomedical knowledge or a high level of education are not sufficient to influence sexual behavior and to change attitudes.

People do not always react directly or immediately to new knowledge. The path from knowledge to behavior is filtered through attitudes and cultures. Behaviors rooted in traditional values, such as multiple partnerships, are not based on ignorance. Nor are they irrational in the social and cultural contexts in which they are embedded. Gould mentions, for example, that sexual intercourse has a symbolic function among the Luo in Kenya. It plays an important part in festivals for the fertility of people and land, for the harvest, and for prosperity in general. Both resistance to male circumcision and reluctance to abandon sexual cleansing rituals (a widow has to marry the brother of her deceased husband) are two specific examples of cultural attitudes that Gould and many other people regard as barriers to reducing HIV prevalence and AIDS deaths. One of the main conclusions of Gould's research is that effective national HIV/AIDS policies cannot be driven only by national public information programs or the formal school system but need cultural sensitivity as well. He calls for new awareness of the need to understand and build on existing indigenous knowledge systems.

Conclusion

Knowledge and culture are a research topic of at least a dozen disciplines. Each of them can offer something different to the discussion. Ideally, cultural studies, science studies, and educational studies should be interdisciplinary in practice; they should cover various scales of analysis (individuals, communities, social systems, spatial contexts, spatial disparities), and cross the boundaries between disciplines. We hope that the variety of chapters in this volume will extend the research field on the relationship between knowledge and culture and that it will substantiate the importance of spatiality and contextuality for the production and diffusion of knowledge. We also hope that readers interested in traditional knowledge, indigenous knowledge, local knowledge, the cultural dimension of knowledge, and ethnic disparities of educational attainment will be motivated to read beyond their own discipline. May this collection of perspectives foster the discussion on how fruitful and challenging the research and concepts involving situated, local, or indigenous knowledge can be—especially when they bring in postcolonial perspectives and power relations.

References

Anderson, K., Domosh, M., Pile, S., & Thrift, N. (Eds.). (2003). *Handbook of cultural geography*. London: Sage.
Ara, A. (1991). Italian educational policy towards national minorities, 1860–1940. In J. Tomiak, K. Eriksen, A. Kazamias, & R. Okey (Eds.), *Schooling, educational policy and ethnic identity: Vol. 1. Comparative studies on governments and non-dominant ethnic groups in Europe, 1850–1940* (pp. 263–290). New York: New York University Press.

Barnes, T. J. (2003). Vom Bauernhof zum Großstadtdschungel: "Kultur" in der anglo-amerikanischen Stadtgeographie der 1990er Jahre [From farm to urban jungle: "Culture" in American urban geography in the 1990s]. *Berichte zur deutschen Landeskunde, 77,* 91–104.

Berry, J. W. (1979). Culture and cognitive style. In A. J. Marsella, R. G. Tharp, & T. J. Ciborowski (Eds.), *Perspectives on cross-cultural psychology* (pp. 300–312). New York: Academic.

Bonnemaison, J. (2005). *Culture and space: Conceiving a new cultural geography.* First published in 2000 as *La géographie culturelle* (trans: Pénot-Demetry, J.). London: Tauris.

Bonnett, A., & Nayak, A. (2003). Cultural geographies of racialization: The territory of race. In K. Anderson, M. Domosh, S. Pile, & N. Thrift (Eds.), *Handbook of cultural geography* (pp. 300–312). London: Sage.

Boyer, P. (2001). Cultural assimilation. In N. J. Smelser & P. B. Baltes (Eds.), *International encyclopedia of the social & behavioral sciences* (Vol. 5, pp. 3032–3035). Amsterdam: Elsevier. doi:10.1016/B0-08-043076-7/00364-8.

Canadian Commission for UNESCO. (1977). A working definition of 'culture'. *Cultures, 4*(4), 78–85.

Ciborowski, T. J. (1979). Cross-cultural aspects of cognitive functioning: Culture and knowledge. In A. J. Marsella, R. G. Tharp, & T. J. Ciborowski (Eds.), *Perspectives on cross-cultural psychology* (pp. 101–116). New York: Academic.

Cosgrove, D. E., & Jackson, P. (1987). New directions in cultural geography. *Area, 19,* 95–101. doi:10.2307/20002425.

Crang, M. (1998). *Cultural geography.* London: Routledge.

D'Andrade, R. (1984). Cultural meaning systems. In R. A. Shweder, R. A. LeVine, & Social Science Research Council (U.S.) Committee on Social and Affective Development During Childhood (Eds.), *Cultural theory: Essays on mind, self and emotion* (pp. 88–119). Cambridge: Cambridge University Press.

de Vroede, M. (1991). Language in education in Belgium up to 1940. In J. Tomiak, K. Eriksen, A. Kazamias, & R. Okey (Eds.), *Schooling, educational policy and ethnic identity: Vol. 1. Comparative studies on governments and non-dominant ethnic groups in Europe, 1850–1940* (pp. 111–131). New York: New York University Press.

Denzin, N. K. (2001). Cultural studies: Cultural concerns. In N. J. Smelser & P. B. Baltes (Eds.), *International encyclopedia of the social & behavioral sciences* (Vol. 5, pp. 3121–3125). Amsterdam: Elsevier. doi:10.1016/B0-08-043076-7/04583-6.

Diamond, J. (1994). Race without color. *Discover, 15*(11), 83–89.

Dinges, N., & Duffy, L. (1979). Culture and competence. In A. J. Marsella, R. G. Tharp, & T. J. Ciborowski (Eds.), *Perspectives on cross-cultural psychology* (pp. 209–232). New York: Academic.

Douglas, M. (2001). Culture as an explanation: Cultural concerns. In N. J. Smelser & P. B. Baltes (Eds.), *International encyclopedia of the social & behavioral sciences* (Vol. 5, pp. 3147–3151). Amsterdam: Elsevier. doi:10.1016/B0-08-043076-7/04657-X.

Eriksen, K. (1991). Norwegian and Swedish educational policies vis-à-vis non-dominant ethnic groups, 1850–1940. In J. Tomiak, K. Eriksen, A. Kazamias, & R. Okey (Eds.), *Schooling, educational policy and ethnic identity: Vol. 1. Comparative studies on governments and non-dominant ethnic groups in Europe, 1850–1940* (pp. 63–85). New York: New York University Press.

Felder, E. (2013). Faktizitätsherstellung mittels handlungsleitender Konzepte und agonaler Zentren. Der diskursive Wettkampf um Geltungsansprüche [Creating facticity by means of discourses through concepts-in-use and agonal centers: The discursive competition over claims of validity]. In E. Felder (Ed.), *Faktizitätsherstellung in Diskursen: Die Macht des Deklarativen* (pp. 13–28). Berlin: De Gruyter. doi:10.1515/9783110289954.13.

Fernandez, J. W. (2001). Anthropology of cultural relativism. In N. J. Smelser & P. B. Baltes (Eds.), *International encyclopedia of the social & behavioral sciences* (Vol. 5, pp. 3110–3113). Amsterdam: Elsevier. doi:10.1016/B0-08-043076-7/00839-1.

Foucault, M. (1980). *Power/knowledge: Selected interviews and other writings, 1972–1977.* (trans: Gordon, C., Marschall, L., Mepham, J. & Soper, K.). New York: Pantheon Books.

Foucault, M. (1990). *The will to knowledge: Vol. 1. The history of sexuality.* London: Penguin.

Frantz, K. (1999). *Indian reservations in the United States: Territory, sovereignty, and socioeconomic change* (University of Chicago geography research paper, Vol. 242). Chicago: The University of Chicago Press.

Freytag, T. (2003). *Bildungswesen, Bildungsverhalten und kulturelle Identität: Ursachen für das unterdurchschnittliche Ausbildungsniveau der hispanischen Bevölkerung in New Mexico [Education, educational behavior, and cultural identity: Causes of the below-average level of education among the Hispanic population in New Mexico]* (Heidelberger Geographische Arbeiten, Vol. 118). Heidelberg: Selbstverlag des Geographischen Instituts der Universität Heidelberg.

Gamerith, W. (1998a). Education in the United States: How ethnic minorities are faring. In F. J. Kemper & P. Gans (Eds.), *Ethnische Minoritäten in Europa und Amerika. Geographische Perspektiven und empirische Fallstudien* (Berliner Geographische Arbeiten, Vol. 86, pp. 89–104). Berlin: Department of Geography, Humboldt University.

Gamerith, W. (1998b). Das US-amerikanische Bildungswesen—Räumlich-soziale Disparitäten im Spannungsfeld zwischen egalitären und elitären Prinzipien [The U.S. education system—Spatiosocial disparities between egalitarian and elitist principles]. *Mitteilungen der Österreichischen Geographischen Gesellschaft, 140,* 161–196.

Gamerith, W. (2006). Ethnizität und Bildungsverhalten. Ein kritisches Plädoyer für eine "Neue" Kulturgeographie [Ethnicity and educational behavior: A critical call for a "new" cultural geography]. In P. Meusburger & K. Kempter (Eds.), *Bildung und Wissensgesellschaf* (Heidelberger Jahrbücher, Vol. 49, pp. 309–332). Heidelberg: Springer. doi:10.1007/3-540-29517-8_13.

Gebhardt, H., Reuber, P., & Wolkersdorfer, G. (Eds.). (2003). *Kulturgeographie. Aktuelle Ansätze und Entwicklungen [Cultural geography: Current approaches and developments].* Heidelberg: Spektrum Akademischer Verlag.

Gillborn, D., & Ladson-Billings, G. (2010). Education and critical race theory. In M. W. Apple, S. J. Ball, & L. A. Gandin (Eds.), *The Routledge international handbook of the sociology of education* (pp. 15–26). London: Routledge.

Grossberg, L. (2010). *Cultural studies in the future tense.* Durham: Duke University Press.

Gutiérrez, K. D., Ali, A., & Henríquez, C. (2010). Syncretism and hybridity. Schooling, language, and race and students from non-dominant communities. In M. W. Apple, S. J. Ball, & L. A. Gandin (Eds.), *The Routledge international handbook of the sociology of education* (pp. 358–369). London: Routledge.

Győri, R., & Gyuris, F. (2015). Knowledge and power in Sovietized Hungarian geography. In P. Meusburger, D. Gregory, & L. Suarsana (Eds.), *Geographies of knowledge and power* (Knowledge and space, Vol. 7, pp. 203–233). Dordrecht: Springer. doi:10.1007/978-94-017-9960-7_10.

Gyuris, F. (2014). *The political discourse of spatial disparities: Geographical inequalities between science and propaganda.* Cham: Springer. doi:10.1007/978-3-319-01508-8.

Haraway, D. (1988). Situated knowledges: The science question in feminism and the privilege of partial perspective. *Feminist Studies, 14,* 575–599. doi:10.2307/3178066.

Harvey, D. (2005). Space as a key word. In D. Harvey (Ed.), *Spaces of neoliberalization: Towards a theory of uneven geographical development* (Hettner-lecture, Vol. 8, pp. 93–115). Stuttgart: Franz Steiner.

Hatano, G., & Oura, Y. (2001). Culture-rooted expertise: Psychological and educational aspects. In N. J. Smelser & P. B. Baltes (Eds.), *International encyclopedia of the social & behavioral sciences* (Vol. 5, pp. 3173–3176). Amsterdam: Elsevier. doi:10.1016/B0-08-043076-7/02351-2.

Hatano, G., & Takahashi, K. (2001). Cultural diversity, human development and education. In N. J. Smelser & P. B. Baltes (Eds.), *International encyclopedia of the social & behavioral sciences* (Vol. 5, pp. 3041–3045). Amsterdam: Elsevier. doi:10.1016/B0-08-043076-7/02322-6.

Havránek, J. (1991). The education of Czechs and Slovaks under foreign domination, 1850–1918. In J. Tomiak, K. Eriksen, A. Kazamias, & R. Okey (Eds.), *Schooling, educational policy and ethnic identity: Vol. 1. Comparative studies on governments and non-dominant ethnic groups in Europe, 1850–1940* (pp. 235–261). New York: New York University Press.

Heinemann, M. (1991). State, school and ethnic minorities in Prussia, 1860–1914. In J. Tomiak, K. Eriksen, A. Kazamias, & R. Okey (Eds.), *Schooling, educational policy and ethnic identity: Vol. 1. Comparative studies on governments and non-dominant ethnic groups in Europe, 1850–1940* (pp. 133–161). New York: New York University Press.

Hoggart, R. (1957). *The uses of literacy: Aspects of working-class life with special reference to publications and entertainments.* London: Chatto and Windus.

Hörning, K.-H. (1999). Kulturelle Kollisionen. Die Soziologie vor neuen Aufgaben [Cultural collisions: New challenges in sociology]. In K.-H. Hörning & R. Winter (Eds.), *Widerspenstige Kulturen: Cultural Studies als Herausforderung* (pp. 84–115). Frankfurt am Main: Suhrkamp.

Hörning, K. H., & Winter, R. (1999). Widerspenstige Kulturen. Cultural Studies als Herausforderung [Intractable cultures: Cultural studies as a challenge]. In K.-H. Hörning & R. Winter (Eds.), *Widerspenstige Kulturen: Cultural Studies als Herausforderung* (pp. 7–12). Frankfurt am Main: Suhrkamp.

Jahnke, H. (2014). Bildung und Wissen [Education and knowledge]. In J. Lossau, T. Freytag, & R. Lippuner (Eds.), *Schlüsselbegriffe der Kultur- und Sozialgeographie* (pp. 153–166). Stuttgart: Ulmer UTB.

Jordan, C., & Tharp, R. G. (1979). Culture and education. In A. J. Marsella, R. G. Tharp, & T. J. Ciborowski (Eds.), *Perspectives on cross-cultural psychology* (pp. 265–285). New York: Academic.

Kipling, R. (1899). The White man's burden. *McClure's Magazine, 12,* 290–291.

Kobayashi, A. (2003). The construction of geographical knowledge: Racialization, spatialization. In K. Anderson, M. Domosh, S. Pile, & N. Thrift (Eds.), *Handbook of cultural geography* (pp. 544–556). London: Sage.

Kögler, H.-H. (1999). Kritische Hermeneutik des Subjekts: Cultural Studies als Erbe der Kritischen Theorie [Critical hermeneutics of the subject: Cultural studies as a legacy of critical theory]. In K. Hörning & R. Winter (Eds.), *Widerspenstige Kulturen: Cultural Studies als Herausforderung* (pp. 196–237). Frankfurt am Main: Suhrkamp.

Kuikka, M. T. (1991). Educational policy in Finland under Russian domination, 1850–1917. In J. Tomiak, K. Eriksen, A. Kazamias, & R. Okey (Eds.), *Schooling, educational policy and ethnic identity: Vol. 1. Comparative studies on governments and non-dominant ethnic groups in Europe, 1850–1940* (pp. 87–110). New York: New York University Press.

Kymlicka, W. (1992). The rights of minority cultures: Reply to Kukathas. *Political Theory, 20,* 140–146. doi:10.2307/191782.

Kymlicka, W. (1995). *Multicultural citizenship: A liberal theory of minority rights.* Oxford: Clarendon.

Latour, B. (1987). *Science in action: How to follow scientists and engineers through society.* Milton Keynes: Open University Press.

Livingstone, D. N. (2003). *Putting science in its place: Geographies of scientific knowledge.* Chicago: The University of Chicago Press.

Lossau, J., Freytag, T., & Lippuner, R. (Eds.). (2014). *Schlüsselbegriffe der Kultur- und Sozialgeographie [Key terms of cultural and social geography].* Stuttgart: Ulmer UTB.

Massey, D. (1999). Philosophy and politics of spatiality: Some considerations. In D. Massey (Ed.), *Power-geometries and the politics of space-time* (Hettner-lecture, Vol. 2, pp. 27–42). Heidelberg: University Department of Geography.

Massey, D. (2005). *For space.* London: Sage.

Melville, M. B. (1994). "Hispanic" ethnicity, race and class. In T. Weaver (Ed.), *Handbook of Hispanic culture in the United States: Anthropology* (pp. 85–106). Houston: Arte Público Press.

Meusburger, P. (1996). Educational achievement, language of instruction, and school system as key elements of minority research. In K. Frantz & R. A. Sauder (Eds.), *Ethnic persistence and change in Europe and America* (Veröffentlichungen der Universität Innsbruck, Vol. 213, pp. 187–222). Innsbruck: University of Innsbruck.

Meusburger, P. (1998). *Bildungsgeographie. Wissen und Ausbildung in der räumlichen Dimension [Geography of education: Knowledge and education in the spatial dimension]*. Heidelberg: Spektrum Akademischer Verlag.

Meusburger, P. (2008). The nexus of knowledge and space. In P. Meusburger, M. Welker, & E. Wunder (Eds.), *Clashes of knowledge: Orthodoxies and heterodoxies in science and religion* (Knowledge and space, Vol. 1, pp. 35–90). Dordrecht: Springer. doi:10.1007/978-1-4020-5555-3_2.

Meusburger, P. (2009). Spatial mobility of knowledge: A proposal for a more realistic communication model. *disP—The Planning Review, 177*(2), 29–39. doi:10.1080/02513625.2009.10557033.

Meusburger, P. (2015). Relations between knowledge and power: An overview of research questions and concepts. In P. Meusburger, D. Gregory, & L. Suarsana (Eds.), *Geographies of knowledge and power* (Knowledge and space, Vol. 7, pp. 19–74). Dordrecht: Springer. doi:10.1007/978-94-017-9960-7_2.

Meusburger, P., & Schuch, T. (Eds.). (2012). *Wissenschaftsatlas of Heidelberg University. Spatiotemporal relations of academic knowledge production*. Knittlingen: Bibliotheca Palatina.

Miąso, J. (1991). Educational policy and educational development in the Polish territories under Austrian, Russian and German rule, 1850–1918. In J. Tomiak, K. Eriksen, A. Kazamias, & R. Okey (Eds.), *Schooling, educational policy and ethnic identity: Vol. 1. Comparative studies on governments and non-dominant ethnic groups in Europe, 1850–1940* (pp. 163–184). New York: New York University Press.

Mikesell, M. (1978). Tradition and innovation in cultural geography. *Annals of the Association of American Geographers, 68*, 1–16. doi:10.1111/j.1467-8306.1978.tb01176.x.

Miller, J. G. (2001). Culture and the self (implications for psychological theory): Cultural concerns. In N. J. Smelser & P. B. Baltes (Eds.), *International encyclopedia of the social & behavioral sciences* (Vol. 5, pp. 3139–3143). Amsterdam: Elsevier. doi:10.1016/B0-08-043076-7/04644-1.

Mitchell, D. (2000). *Cultural geography: A critical introduction*. Oxford: Blackwell.

Mittelstraß, J. (1982). *Wissenschaft als Lebensform. Reden über philosophische Orientierungen in Wissenschaft und Universität [Science as way of life: Addresses on philosophical orientations in science and the university]*. Frankfurt am Main: Suhrkamp.

Mittelstraß, J. (2001). *Wissen und Grenzen. Philosophische Studien [Knowledge and borders: Philosophical studies]*. Frankfurt am Main: Suhrkamp.

Mittelstraß, J. (2010). The loss of knowledge in the information age. In E. De Corte & J. E. Fenstad (Eds.), *From information to knowledge; from knowledge to wisdom* (Wenner–Gren international series, Vol. 85, pp. 19–28). London: Portland Press.

Mitter, W. (1991). German schools in Czechoslovakia, 1918–1938. In J. Tomiak, K. Eriksen, A. Kazamias, & R. Okey (Eds.), *Schooling, educational policy and ethnic identity: Vol. 1. Comparative studies on governments and non-dominant ethnic groups in Europe, 1850–1940* (pp. 211–233). New York: New York University Press.

Münch, R. (1992). The production and reproduction of inequality: A theoretical cultural analysis. In R. Münch & N. J. Smelser (Eds.), *Theory of culture* (pp. 243–264). Berkeley: University of California Press.

Radcliffe-Brown, A. R. (1952). *Structure and function in primitive society: Essays and addresses*. London: Cohen and West.

Renteln, A. D. (2001). Cultural rights and culture defense: Cultural concerns. In N. J. Smelser & P. B. Baltes (Eds.), *International encyclopedia of the social & behavioral sciences* (Vol. 5, pp. 3116–3121). Amsterdam: Elsevier. doi:10.1016/B0-08-043076-7/04582-4.

Shweder, R. A. (2001a). Cultural psychology. In N. J. Smelser & P. B. Baltes (Eds.), *International encyclopedia of the social & behavioral sciences* (Vol. 5, pp. 3104–3110). Amsterdam: Elsevier. doi:10.1016/B0-08-043076-7/01666-1.

Shweder, R. A. (2001b). Culture: Contemporary views. In N. J. Smelser & P. B. Baltes (Eds.), *International encyclopedia of the social & behavioral sciences* (Vol. 5, pp. 3151–3158). Amsterdam: Elsevier. doi:10.1016/B0-08-043076-7/00840-8.

Smelser, N. J. (1992). Culture: Coherent or incoherent. In R. Münch & N. J. Smelser (Eds.), *Theory of culture* (pp. 3–28). Berkeley: University of California Press.

Spencer, M. B. (1983). Children's cultural values and parental childrearing strategies. *Developmental Review, 3,* 351–370. doi:10.1016/0273-2297(83)90020-5.

Spencer, M. B., & Markstrom-Adams, C. (1990). Identity processes among racial and ethnic minority children in America. *Child Development, 61,* 290–310. doi:10.1111/1467-8624. ep5878983.

Stegmaier, W. (2008). *Philosophie der Orientierung [Philosophy of orientation].* Berlin: Walter de Gruyter.

Stovall, D. (2006). Forging community in race and class: Critical race theory and the quest for social justice in education. *Race, Ethnicity & Education, 9,* 243–259. doi:10.1080/13613320600807550.

Strohmayer, U. (2003). The culture of epistemology. In K. Anderson, M. Domosh, S. Pile, & N. Thrift (Eds.), *Handbook of cultural geography* (pp. 520–531). London: Sage.

Tanner, K. (1999). Ethik und Religion [Ethics and religion]. In R. Anselm, S. Schleissing, & K. Tanner (Eds.), *Die Kunst des Auslegens. Zur Hermeneutik des Christentums in der Kultur der Gegenwart* (pp. 225–241). Frankfurt am Main: Peter Lang.

Tomiak, J. (1991). Education of the non-dominant ethnic groups in the Polish Republic, 1918–1939. In J. Tomiak, K. Eriksen, A. Kazamias, & R. Okey (Eds.), *Schooling, educational policy and ethnic identity: Vol. 1. Comparative studies on governments and non-dominant ethnic groups in Europe, 1850–1940* (pp. 185–209). New York: New York University Press.

Winter, R. (1999). Die Zentralität von Kultur. Zum Verhältnis von Kultursoziologie und Cultural Studies [The centrality of culture: On the relation between cultural sociology and cultural studies]. In K.-H. Hörning & R. Winter (Eds.), *Widerspenstige Kulturen: Cultural Studies als Herausforderung* (pp. 146–195). Frankfurt am Main: Suhrkamp.

Chapter 2
The School System as an Arena of Ethnic Conflicts

Peter Meusburger

> *A nation is not only a political entity, but also something that produces meaning—a system of cultural representations.*
>
> Hall (1996, p. 612)

Nation-Building, Schooling, and Ethnic Conflicts

In the nation-states of nineteenth and early twentieth century, compulsory schooling[1] was introduced to eliminate illiteracy and create a well-educated labor force prepared for economic, technical, and societal innovations.[2] This decision had other effects as well, however. According to Hobsbawm (1989), Smith (1979), Tomiak and Kazamias (1991), and other authors, secular education—especially compulsory primary education—played a paramount role in the formation of nation-states, the functioning of democracy, and the genesis of nationalism. One of the first measures taken in territories annexed after wars was to replace teachers and textbooks of the former system and to change the language of instruction. After all, teachers and priests have in many cases been spokesmen of national movements. The educational system—especially subjects such as history, geography, social studies, civics, literature, and the arts—have served as a means of creating a heroic national

---

[1] In most European countries compulsory schooling was introduced in late eighteenth or nineteenth century. One of the first countries to introduce compulsory education was Austria in 1774, though it took decades to enforce this law throughout the country.

[2] Literacy was a precondition for employment in large and complex bureaucratic organizations with a high degree of division of labor.

P. Meusburger (✉)
Department of Geography, Heidelberg University,
Berliner Straße 48, 69120 Heidelberg, Germany
e-mail: peter.meusburger@geog.uni-heidelberg.de

© Springer International Publishing Switzerland 2016
P. Meusburger et al. (eds.), *Ethnic and Cultural Dimensions of Knowledge*,
Knowledge and Space 8, DOI 10.1007/978-3-319-21900-4_2

historiography, constructing a collective cultural memory, promoting cultural homogeneity, shaping stable interpretations of the world, enforcing an official language of instruction, and infusing the masses with new loyalties to the state and its institutions. "From the state's point of view the school had [an] essential advantage: [I]t could teach all children how to be good subjects and citizens" (Hobsbawm 1989, p. 155).

Culture and ethnic identities are not fixed. It is possible to exit an ethnic group and adopt a new culture. Culture is a continuous and "complex process of constructing sociocultural meanings and identities" (Hörning and Winter 1999, p. 9). In the words of Lewis (2002),

> Culture begins with an imagining of the world about us;…[these imaginings] are formed in discourse, language, symbols, signs and texts—all concepts applied to meaning systems. These imaginings and meanings, however, can never be fixed or solidified, but remain assemblages that can be dismantled through time, space, and human action…. Culture is always transitional, transformative, open, and unstable… [It] can never be closed since it is made up of competing interests and many different individuals and groups. (pp. 22–23)

Processes of assimilation, acculturation, and transculturation occur continuously. This dynamism and fluidity of culture and identity may be one of the reasons why governments and educational administrations repeatedly try to regulate and interfere in the formal education of minorities. The key issue is whether an ethnic group is forced by a government and its institutions to give up parts of its culture and identity or whether a group voluntarily and gradually acquires some elements of a "larger society's repertoire of concepts and norms" (Boyer 2001, p. 3033). Highly educated members of an ethnic minority striving to climb the social ladder will seek to achieve a high proficiency not only in their mother tongue but in other languages as well and will be able to accept some cultural elements of the majority without forfeiting their own cultural identity and practices.

Such voluntary and omnipresent processes of assimilation, acculturation, and multiculturalism are not the topic of this chapter, nor are minority rights[3] or normative judgments on the educational policies of diverse states. This chapter starts instead from the finding that the educational system is a sensitive sounding board of societal problems, social inequalities, asymmetric power relations, racial tensions, and political conflicts in multiethnic states (Meusburger 1996, 1998).

My aims in this chapter are twofold. The first is to provide a set of ideas and tools for analyzing the schooling situation of nondominant minorities[4] and to discuss various practices of political and cultural hegemony as exercised through the school system in order to assimilate nondominant ethnic minorities in multiethnic or mul-

---

[3] There has been intense discussion between liberals denying the collective rights of minorities and authors supporting the cultural rights of minorities (see Brett 1991; Jacobs 1991; Kukathas 1992; Kymlicka 1989, 1991, 1992; Lenihan 1991; Sigler 1983). It makes a difference whether minorities have been forcibly integrated by internal colonialism or by expansion of territory (e.g., during or after a war or through colonial "boundary-making") or whether members of a foreign culture immigrated voluntarily in search of better economic opportunities.

[4] This chapter deals predominantly with autochthonous (native) minorities and not with voluntary immigrants.

tilingual states. The second aim is to show that the achievement, treatment, and opportunities of ethnic minorities in the educational system reflect their collective social status; the degree of discrimination, nationalism (racism), and cultural imperialism to which ethnic groups have been exposed; and the development of ethnic awareness and self-assessment.[5] In multilingual (multiethnic) regions, the educational achievements and vertical social mobility of minorities, the spatial distribution pattern and size of schools, the language of instruction, the ethnic and social origin and qualification of teachers, the extracurricular activities and turnover rates of teachers, and the policies of the school administration offer deep insight into a state's support, toleration, or suppression of its nondominant ethnic groups and into the reactions of minorities to political pressure.

The first part of the chapter focuses on some of the reasons why the educational system is a politically contested field in multiethnic or multilingual states. It is followed by an overview of methods used since the early nineteenth century to suppress, acculturate, or emancipate ethnic minorities through the public educational system. The emphasis is on ethnic disparities of educational achievement and attainment,[6] the language of instruction, the selection of teachers, and the minimum size and location of primary schools in rural areas.

The Educational System as a Sounding Board of Societal Problems and Conflicts in Multiethnic States

Why have state authorities, churches, interest groups, and leaders of ethnic minorities had a genuine interest in exerting influence or administrative control on public education, at least on compulsory schools?[7] Orientation knowledge (Meusburger 2015b, pp. 26–28), cultural practices, and ethnic identities do not simply come from the blue like manna; they are acquired through processes of learning, appropriation, and imitation and are imparted to the next generation by institutions of education and knowledge transfer. In a multiethnic state institutions of education and knowledge transfer (e.g., parents, schools, museums, memorials, and other institutions) compete with each other to give children specific identities, cultural memories, worldviews, and interpretations of history. These stores of knowledge, values, and narratives accumulated during socialization in childhood function later as a kind of

---

[5] The fact that schools serve as an early-warning system for societal problems was discovered in the early nineteenth century by members of the social survey movement in England (for an overview see Marsden 1982, 1987; Meusburger 1998, pp. 191–197) and in the 1920s by members of the Chicago school of social ecology, who had analyzed the relationship between truancy and juvenile delinquency (Shaw 1929).

[6] Indicators of educational achievement describe the final result of an educational process. Indicators of educational attainment are used to describe a current status in an ongoing (unfinished) process of education (for details see Meusburger 1998, 2015a).

[7] This section does not deal with informal education or noninstitutional formal education as defined by Jordan and Tharp (1979, p. 266).

filter for the selective perception and evaluation of information. In multiethnic states, minority students entering the school system frequently experience how the values, historical experiences, and cultural practices that their parents and neighborhoods have passed on to them are called into question, resisted, portrayed as backward, or ridiculed. The school can support and reinforce the family-based cultural learning process that students have already undergone, but it can also interrupt or reverse it, eventually instilling them with serious doubt about their identity.

These considerations raise several crucial questions. Are autochthonous minorities (native people) allowed to use their native language as the language of instruction (at least in the first years of schooling)? Are teachers sympathetic to cultures of native people or ethnic minorities? Do native people have a part in determining the location of schools and the content of textbooks?[8] Do those minorities have a say in the "memory industry" (e.g., the selection of commemoration days, the establishment of museums, or the erection of memorials) in their regions? Nationalist movements in nineteenth- and early twentieth-century Europe started directly from conflicts about schooling issues, as did the U.S. civil rights movement in the twentieth century.

Many national movements, liberation movements, and the development of the standard form of a given language started in universities. These origins are especially evident in Eastern Europe. "In the 1830s, the Ukrainian students of Lvov University began working on the development of their native language…These activities contributed to the awakening of a feeling of Ukrainian national identity" (Miąso 1991, p. 168). Universities contributed a great deal to the education of the functional elites of nation-states, but in some countries they also legitimated imperialism and colonialism (Brown 1993; Craig 1984; Driver 2001; Heffernan 1994; Jarausch 1983; Välimaa 2004).

For more than 150 years, schools have served as appropriate means to assimilate ethnic minorities and—in some cases—to convince them of their cultural inferiority (Boyer 1997; Churchill and Hill 1979; Frantz 1999). As pointed out by Eriksen (1991), school authorities in a number of states have pursued policies designed to assimilate "so-called 'primitive groups' that had not yet reached a 'civilised' level" (p. 65). The Norwegian nationalism of the late nineteenth and early twentieth century "was directed towards the destruction of separate Sami and Finnish culture and identity" (p. 65). "Together with Social Darwinist thinking, nationalism provided clear grounds for ousting the Sami and Finnish languages and cultures so as to make way for the homogenous national state. Nationalism expressed by ethnic groups excited alarm" (p. 65).

Since the mid-nineteenth century, nationalization by schooling has become a salient policy of many states. Whenever a nation achieved independence or a territory was annexed by another state, some of the first measures were to dismiss or relocate the teachers of the former system and replace them with teachers and text-

---

[8] During field work in the 1980s, Frantz (1999) found that American Indian children at some schools run by the Bureau of Indian Affairs were told "that the history of the American continent began in the sixteenth century" (p. 145)—that is, after America was discovered by Columbus.

books sympathetic to the new system; to change the language of instruction; to reorganize the location and size of primary schools, and—in some cases—to found at least one university teaching in the new language of instruction (for details see Ara 1991; Eriksen 1991; Frantz 1999; Meusburger 1979, 1996; Tomiak 1991). It goes beyond the scope of this chapter to analyze the influence of politics on the content of textbooks (for details see Dierkes 2010). In annexed territories, history, geography, arts, literature, and other subjects have been taught in many cases without any reference to the cultural heritage of the nondominant minorities formerly belonging to the majority of the neighboring state. Officials in fascist Italy regarded this neglect "as a means of inculcating an Italian national consciousness into the alloglot children, and to persuade them of the superior value of Italian culture and civilization" (Ara 1991, p. 276).

In some states, ethnic minorities living near a border were regarded as a security problem or a potential fifth column, especially when the minority could rely on the support of an external mother nation. In such cases, elementary schools had an important political function, and the duty of teachers was not only to educate pupils but also to control national movements and political activities among ethnic groups. In many states cultural policy became part of security policy. Eriksen (1991) reported that in Norway at the end of nineteenth century "schools and church buildings near the border were regarded as national symbols and just as important to the security of the country as the establishment of garrisons" (p. 71). In Norway "the teacher had to prove to his superiors that his Norwegian instruction bore fruit [among the minority children] or risk severe sanctions. If he failed, the teacher lost a salary supplement amounting to about a quarter of his total income" (p. 69). Mussolini, too, assigned schools an essential task in his national policy in the alloglot or mixed territories (Ara 1991, p. 279).

Methodological Problems of Minority Research

The utility of any universal categorization of ethnic groups suffers from idiosyncrasies and peculiarities in the history, cultural heritage, and political struggles of minorities as well as from differences in the history of immigration. It is also reduced by dissimilar modes of incorporating minorities into dominant society; differences in the collective social status, cultural level, economic performance, ethnic awareness, and ethnic solidarity of ethnic groups; and the importance of the local (regional) context.[9] Almost every ethnic minority has a unique history and a specific

---

[9] Minorities in urban agglomerations have problems different from those of minorities in sparsely inhabited peripheral areas. Economically well-off minorities living in prosperous areas have interests different from those of minorities struggling with poverty in decayed urban areas. A minority in a border region with strong connections to its former "motherland" is confronted with more suspicion and mistrust from the central state than is a minority not involved in border claims. A law having positive effects in country A (e.g., desegregation of schooling) may be inimical to a minority in country B.

political and economic position that preclude generalizing assertions about how minorities should be treated by the relevant state administration. Because categories of ethnic minorities widely vary (e.g., natives, autochthonous minorities, people in occupied territories, immigrants, refugees, asylum seekers), it is almost impossible to make valid normative statements on the entitlement of ethnic minorities to cultural rights and, therefore, to protection in the educational system.

There is also methodological concern about categories such as mother-tongue, race and ethnicity, and the way in which certain ethnic groups are counted. Melville (1988, 1994) and Weaver (1994) demonstrated the extraordinary complexity and internal fragmentation of the categories *Hispanic* and *Latino*. Frantz (1999, pp. 70–84) discussed the problems and inadequacies of counting "American Indians" in the United States.[10] Until the U.S. census of 1950, the individual census takers decided the race to which a person belonged. Self-denomination—the principle that members of the population could decide on their own to which ethnic group they belonged—was introduced in the census of 1960. Unsurprisingly, the Bureau of Indian Affairs (BIA) has its own data, and its figures on "American Indians" are much lower than those of the official population census. "In order to be recognized as an American Indian by the BIA and thus to have this legal status and title, one had to either live on or near a reservation and be an enrolled member of a federally recognized tribe or be able to prove at least one-quarter Indian ancestry" (Frantz 1999, p. 73). The result of the BIA's procedure is that "thousands of people who can give no evidence of Indian ancestry are officially recognized as American Indian, while, on the other hand, tens of thousands American Indians, some of them full-blooded, are deprived of this status" (p. 73). As Frantz (1999) stated in his summary,

> The data on American Indians are imprecise, and in some cases, extremely variable from one data source to another…First, since the beginning of the U.S. census, there has never been general agreement on the question of who is an Indian. Furthermore, the census takers who collect data in the field did not have an adequate knowledge of the Indian way of life. Not until 1980 were investigations carried out by people with sufficient geographical and linguistic knowledge to locate and enumerate tribal members accurately. (pp. 70–71)

In a number of bilingual areas, the informative value of using ethnic identity as a category in research becomes even more complicated because of "floating identities." Individual members or groups of an ethnic minority can change their self-denomination from one census to the next in response to external pressure, internal conflicts, the offer of minority rights or privileges, and an increasing political fragmentation or intraethnic antagonism.[11]

---

[10] Analogous problems of definition and counting exist with some minorities on other continents (e.g., gypsies in Europe).

[11] Such floating identities were evident, for example, among some of the Slovenes in South Carinthia in the nineteenth and twentieth centuries (Mannhardt 1965, pp. 113–114; Meusburger 1979, pp. 229–230), Hispanics (Freytag 2003) in the United States, and the Germans in post-1945 eastern Europe.

Because ethnicity is difficult to record, it has its limits as an analytical category. However, race and ethnicity "feature as part of the power structure of a society" (Bulmer and Solomos 1998, p. 823), and belonging to a particular ethnic or racial group has distinct social, educational, material and political consequences. Indicators such as ethnicity, race, and mother-tongue therefore prove useful, given the methods by which the data have been collected. As categories, race and ethnicity "are used by both dominant and subordinate groups for the purposes of legitimizing and furthering their own social identities and interests" (p. 823). In multiethnic nations almost every aspect of contemporary social and political relations is deeply infected with an ethnic dimension (p. 824). The importance of ethnicity and race fluctuates over time, surging in political debate predominantly during periods of struggle, oppression, political transformation, resistance, and power asymmetries.

Despite these methodological problems, indicators such as ethnicity (race) or mother-tongue prove to be very useful in view of the methods by which the data have been collected. Personal attributes such as mother tongue, nationality, race or ethnicity have additional advantage that they are registered in censuses and other official statistics of many states. Such census data facilitate research about some issues of social structures, processes, and relations whose study would not be possible with small samples or qualitative data. Indicators such as ethnicity, mother tongue, or national identity make a large difference in educational achievement and are important heuristic devices for uncovering various types of inequalities and asymmetries of power relations.

Educational Achievement and Attainment as a Measure of the Collective Social Status of Ethnic Minorities

The collective social status[12] and vertical social mobility of ethnic minorities; ethnic awareness, social cohesion, and ethnic self-esteem of minorities; the power relations and conflicts between one ethnic group and the majority; and discrimination against ethnic minorities are all worthy topics of analysis. There is hardly a better way to start that research process than to study the disparities of educational achievement and attainment among ethnic groups. Spatial and social disparities of educational attainment (achievement) are influenced by a large range of factors (for details see Meusburger 2015a). Educational achievement strongly affects a person's social status, income, and his or her chances on the labor market. Schooling ramifies into other sectors of the social, political, economic, and ideological milieus (Tomiak and Kazamias 1991, p. 3). And well-chosen indicators[13] of educational attainment or

---

[12] It goes without saying that the collective social status of an ethnic group is just a rough concept. Within each ethnic group, too, there are major regional and social inequalities.

[13] The explanatory power of indicators of educational achievement varies in space and time. Whether an indicator has explanatory power depends first on the selectivity of the school system and on the indicator's life cycle. Many indicators eventually lose their value. In states with 50 %

achievement correlate highly with all other indicators of social status. For all these reasons, indicators of educational achievement have frequently been used in human geography as a means to reveal social, economic, and spatial inequalities and to point out asymmetric power relations.

Vast ethnic disparities of educational achievement within a state can signal potential social and political conflicts. An ethnic group with a high percentage of illiterates or a very low level of educational achievement is caught in a vicious circle. If its value system does not meet the expectations and attitudes of a modern meritocratic knowledge society, the students of this group will develop strategies of resistance to the rules and goals of the school. This response will create an anti-school culture marked by truancy, refusal to study, and other behavior that all but guarantees academic failure (Achor and Morales 1990, p. 270).

Table 2.1 ranks the ethnic groups in former Yugoslavia according to their degree of illiteracy shortly after World War II. At that time illiteracy rates of ethnic groups

Table 2.1 The percentage of illiterates in the Yugoslavian population in 1948, by Nations (age groups above 15 years)

Ethnic group (Nation)	Percentage of total population	Percentage of illiterates in the ethnic group		
		Total	Male	Female
Slovenian	9.0	2.3	2.3[a]	2.3
Czech	0.2	3.0	2.4	3.5
Slovak	0.5	4.8	4.2	5.3
German	0.4	6.2	5.9	6.4
Italian	0.5	6.9	4.9	8.7
Hungarian	3.1	8.0	6.0	9.6
Romanian	0.4	17.6	11.2	23.2
Croatian	24.0	18.1	11.2	24.0
Bulgarian	0.4	19.9	6.8	33.1
Montenegrin	2.7	24.1	11.0	35.5
Serbian	41.5	27.7	14.1	40.2
Macedonian	5.1	30.2	18.7	42.0
Wallachian	0.7	38.2	22.7	51.6
Autonomous Moslem	5.1	54.6	36.6	70.6
Turkish	0.6	63.9	47.4	80.7
Albanian	4.8	73.7	55.9	92.5
Gypsy	0.5	74.0	61.9	85.7
Total	100[b]	25.4	15.4	34.4

From UNESCO (1953, p. 166). Reprinted with permission

[a]That is, 2.3 % of the male Slovenian population in Yugoslavia in 1948 was recorded as being illiterate

[b]Rounding accounts for the 0.5 % of the population belonging to other ethnic groups

illiteracy or more, the indicator "percentage of illiterates" has a very high informative value. In states with 2 % illiteracy it is irrelevant (for details see Meusburger 1998, pp. 303–339).

such as the Gypsies, Albanians, Turks, and autonomous Moslems were still exceptionally high, ranging between 54 % and 74 %, as compared to less than 10 % among the Slovenes, Czechs, Slovaks, Germans, Italians, and Hungarians in Yugoslavia. In other words, the ethnic groups at the bottom of the ranking had literacy rates that those at the top had already surpassed 150–200 years earlier (for details see Meusburger 1998, pp. 261–270). Such gaping disparities in literacy within a nation almost inevitably result in corresponding social and economic inequalities, social distance, prejudice, and political tensions, especially if the social inequalities are coupled with religion or the mother tongue or if the politically dominating group (the Serbs) has a higher percentage of illiterates than the groups who feel politically disadvantaged (Slovenes, Croats, Hungarians). Table 2.1 also affords insight into gender relations and the social status of women in the ethnic groups. Not only do substantial gender disparities in educational attainment indicate socioeconomic underdevelopment, high illiteracy rates of mothers lead to low educational attainment of children and thus perpetuate educational inequalities between ethnic groups.

Similar disparities between the educational achievement of ethnic groups exist in many other countries, too (for details see Freytag 2003; Frantz 1999; Gamerith 1998a, b, 2002; Meusburger 1979, 1980, 1996, 1998, 2001; Tomiak 1991). Apart from educational achievement and educational attainment, a number of other indicators can convey the experience that minorities have with schooling. The most frequently used are school enrollment quota, dropout rates, truancy rates, suspension rates, expulsion rates, number of girls per 100 students, persistence of school attendance, school life expectancy, and scores on general certificates and achievement tests, such as the National Assessment of Educational Progress (the United States) and the Pupil Level Annual School Census (the United Kingdom) (for details see Gamerith 2002, pp. 50–105; Meusburger 1998, pp. 303–339).

Measures Used to Suppress, Assimilate, Disadvantage, or Emancipate Minority Children in the School System[14]

In many multiethnic or multilingual regions, the educational system has intermittently become a battlefield of ethnic strife, national movements, and racial conflicts. In such periods, a wide range of measures have been taken to suppress, assimilate, or support minorities in the educational system and to sustain, reinforce, or contest existing power relations. The package of measures has ranged from brutal racism or racializing practices (see Gillborn 2005, 2008; Gillborn and Ladson-Billings 2010; Gutiérrez et al. 2010; Stovall 2006; Tomlinson 1991, 2008) to "soft" democratic arrangements that are nevertheless very efficient. Some administrative authorities have simply denied or ignored the existence of minorities on the territory of their state.

---

[14] This chapter draws on material from my earlier studies (Meusburger 1996, 1998).

[After World War I] Norway and Sweden were members of the League of Nations, and often championed the cause of minorities in this forum. Yet, neither of them admitted that they had minorities within their own boundaries and that the minority rights of Samis and Finns were set aside. (Eriksen 1991, p. 80)

Other governments have taken an anticulturalist position, worrying "that any description of cultural difference merely sows the seed of invidious comparison and conflict, and thus should be disavowed" (Shweder 2001, p. 3154). A third group of politicians, administrators, and authors has rejected the view that minority cultures have any special rights vis-à-vis the larger community, maintaining that only individuals are entitled to human rights (e.g., Kukathas 1992).

A number of states have resorted to purely bureaucratic measures in a bid to end or avoid inequalities and discrimination in their educational systems. They have thereby neglected the deep roots of inequalities and minority problems and have ignored the impact of the spatial and cultural context on educational behavior. In the United States the tendency in the 1950s and 1960s was to see the desegregation of schools as the solution to racial problems. In the 1954 *Brown vs Board of Education* decision, the U.S. Supreme Court declared that racial segregation "generates a feeling of inferiority" among Blacks (347 U.S. 483, 494; cited in Gutiérrez et al. 2010, p. 360). The Court ruled that segregated schools were "inherently unequal" and indicated that these schools must be desegregated (Pratte 1982, p. 398). By contrast, the utmost goal of minorities in many European States had been to have their own schools, that is, schools in which the language of instruction was that of the minority population they served and in which the teachers were sympathetic to that population's culture and identity (Ara 1991; Eriksen et al. 1991; Havranek 1991; Heinemann 1991; Höfle and Meusburger 1983; Kuikka 1991; Meusburger 1979, 1998; Miąso 1991; Mitter 1991; Vroede 1991). In the 1970s it became clear that desegregation, busing and other bureaucratic measures of school administrations had not eliminated racial inequalities at American schools and that it had been naïve to ignore the importance of culture, ethnicity, spatial contexts, and all the other factors influencing educational attainment and causing social and spatial inequalities (see Meusburger 1998, pp. 273–302). Even with a focus on desegregation, the interesting research question is not whether busing or desegregated schools have a positive or negative outcome for minorities in general but rather which contexts, milieus, and societal and political conditions lead to the assessment that (de)segregated schools are an advantage or disadvantage for an ethnic group and which ideology lies behind the attendant policies.

After the 1960s the focus of U.S. educational policies shifted to bilingual education (see also Gunderson 1982; Gutiérrez et al. 2010; Pratte 1982). In *Lau v. Nichols* (1974)—a landmark ruling on bilingual education—the issue arose "whether non-English-speaking students who constitute national-origin minority groups can receive an education free from unlawful discrimination when instructed in English, a language they did not understand" (Pratte 1982, p. 398). The court reasoned that

[u]nder these state-imposed standards there is no equality of treatment merely by providing students with the same facilities, textbooks, teachers, and curriculum; for students who do not understand English are effectively foreclosed from any meaningful education. (*Lau v. Nichols* 1974, p. 566)

This insight was a critical step forward to accepting the fact that educational inequalities stem from a large number of factors whose local interaction constitutes a milieu and to recognizing that a milieu is more than just infrastructure and staff.

Compulsory Ignorance

Probably the harshest measure in the history of education was "compulsory ignorance"—the prohibition against teaching a certain group of people how to read and write or against giving them access to schools. As applied to ethnic groups, this ban is largely a thing of the past,[15] but it serves as a reminder that keeping people ignorant has been one of the major control instruments of political power. Slave owners in the Deep South viewed education as a menace to the slave system (Whiteaker 1990, pp. 7–8). In the eighteenth century, a number of southern British colonies in the New World introduced compulsory ignorance laws that made it illegal to teach blacks to read and write. In 1740 South Carolina adopted the first compulsory ignorance law. It declared that every person who teaches a slave to write[16] "shall, for each offence, forfeit the sum of one hundred pounds current money" (Whiteaker 1990, p. 7; see also Erickson 1997, p. 207; Miller 1982, p. 212). A new law in Georgia in 1770 stipulated a fine of 20 pounds on anyone instructing a slave in reading and writing (Whiteaker 1990, p. 7). In 1819 Ohio excluded Negro children from the public schools (Erickson 1997, p. 208). Virginia followed suit with laws in 1819 and 1831 forbidding meetings to instruct Negros. In 1823 Mississippi outlawed "any gathering of six or more Negros for educational purposes" (Miller 1982, p. 212). An 1830 law in Louisiana prescribed imprisonment of from 1 to 12 months for anyone teaching a slave to read or write (Miller 1982, p. 212). A similar law was passed in Alabama in 1832 (Weinberg 1977, p. 3). In 1834, South Carolina "imposed a fine of $100 and imprisonment for 6 months on any white person involved in slave instruction" (Whiteaker 1990, pp. 7–8). In 1847 a law of the same kind was passed in Missouri (Gamerith 2002, p. 121). By the 1840s nearly every slave state had a compulsory ignorance law forbidding slave instruction (Whiteaker 1990, p. 8). "Slaves attempting to educate themselves received throughout the slave era beatings, loss of privileges, and even mutilation" (p. 9). As a consequence of these compulsory ignorance laws, literacy among African Americans was only about 5 % in 1865 according to an estimate by Franklin and Brooks Higginbotham (2011, p. 289).[17]

---

[15] However, a form of compulsory ignorance for girls has reemerged in areas controlled by the Taliban.

[16] In some regions reading was allowed because reading the Bible was considered to be a purely religious activity (Erickson 1997, p. 207). As time passed, reading, too, was forbidden for fear that slaves "would find literature that would incite them to rebel" (Whiteaker 1990, p. 7).

[17] For information on the education of African Americans in nineteenth century in general, see Albanese (1976), Franklin and Brooks Higginbotham (2011), and Gamerith (2002).

Isolation of Minority Children in Distant Boarding Schools

In another brutal form of forceful assimilation, which stemmed from a doctrine of social determinism, minority children were taken away from their parents and schooled in distant boarding schools where they were isolated from any influence of their native culture, family, language, and homeland. Such examples existed in the United States and Norway, among other countries. Frantz (1993, 1994, 1996, 1999) described how a U.S. government agency and groups within the Christian church treated the young American Indians from a viewpoint of social Darwinism (see also Blinderman 1978). According to Frantz (1996),

> [t]he boarding schools set up by the *Bureau of Indian Affairs (BIA)* and by various Christian denominations, which were often on or near the reservation but also often hundreds of miles away, were, in the parents' opinion, systematic reeducation institutions where the Indian children were robbed of their cultural identity. When they began such schools not only did the pupils have to wear white people's clothes, but they were given a military haircut, an Anglo-American name and they were not even allowed to speak their own language to name just some of the treatment these children had to undergo…
> …The extent to which this type of school as such was regarded as a foreign body is indicated by the fact that for a long time the only way to tear some of the children away from their parents was for the police to intervene and escort them to school. This boycott of "white" educational institutions is clearly reflected in the low percentage of Navajo children attending school; in 1912 it was only 12 percent. (p. 226)

Frantz (1999) continued by noting how

> [t]hese children have been wrenched away from their familiar sociocultural environment at a very early age, and forced to live separated from their families for years. What is more, they are subjected to strong pressures to assimilate by the school and the white majority. Many of these students are inevitably affected socially and psychologically as a result. (pp. 136–137)

According to Tippeconnic and Gipp (1982), this kind of boarding school education had very detrimental results: tribal and family disorganization; breakdown of tribal culture; lack of involvement of Indian parents in the education of their children; and students' emotional, psychological, and mental anguish. They described the outcome with phrases such as "mental genocide" or "national tragedy" (p. 126). The same type of method was used in Norway against Sami and Kven[18] children. "Through boarding schools established at a distance from home areas, [the Norwegian Director of Schools[19]] hoped to reduce the influence of Sami or Kven parents over their children" (Eriksen 1991, p. 70).

Language of Instruction

It is well known that the context of education—in multiethnic (multilingual) areas first of all the language of instruction and the qualification and cultural background and engagement of teachers—is decisive for the outcome of learning. For a minority

---

[18] Kven is a Norwegian expression for Finn.
[19] This function was established by the Norwegian government in 1902.

that wants to protect and develop its own cultural identity, language is one of its culture's most basic components and requires protection (Bilmes and Boggs 1979; Canadian Commission for UNESCO 1977, p. 78). For students belonging to an ethnic minority, it makes a difference whether they are educated in schools where teachers and peers have the same mother tongue and cultural background as their parents or whether they are confronted with a completely unfamiliar world in which the culture, knowledge, values, and narratives of their parents are denigrated or suppressed.

Language is the "heart of knowledge" (Strohmayer 2003, p. 525), a significant carrier of culture, and a central element of identity. For some advocates of bilingual education, usage of the first language is a civil right as well as the best guarantee for high educational attainment. For others, the mother tongue has only the temporary aim of cultural transition and adjustment to instruction in the mainstream language (see Trueba 1989, p. 88). The important question is whether this transition and adjustment to the language of the majority is voluntary, whether it happens abruptly or gradually, and which price the minority has to pay in the course of its transition. Bernstein (1964) discussed some of the conditions under which this language shift is possible and points out its consequences.

> Resistance, especially unconscious resistance, to such change in the case of a *public* language speaker is likely to be very high, for there is every probability that attempts to modify the linguistic orientation will be perceived by the speaker as attempts to change the means whereby he has been socialized. Such language change may involve for the speaker the experiences of isolation, bewilderment and defenselessness; whilst the structure of the 'teaching' situation may well be felt as persecutory. The individual's normal orientating system will no longer be appropriate and the level of anxiety will be raised considerably. This inhibits new learning. (p. 257)

One of the most severe discontinuities in the cognitive functioning, social adjustment, and motivation of minority children occurs when they have not learned the language of instruction before the age at which they enter school. When minority children are placed in a school system in which they cannot understand their teachers during the first year(s) of schooling, they suffer from abiding discrimination that nearly automatically leads to low motivation, resistance, high dropout rates, poor educational achievement, and long-term unemployment.

The language of instruction in schools emerged as "a focal point of conflict between the state building and the nation emancipation process" (Eriksen et al. 1991, p. 414). The nineteenth and twentieth centuries were replete with conflicts about the sanctioned use of languages in ethnically mixed areas. State authorities, usually composed of members of the dominating groups, declared the use of a single "national" language a functional necessity, a way to unite groups previously separated, and a means to create a homogeneous national culture. However, those minorities whose language was forbidden in school experienced these arguments as cultural imperialism and discrimination and as the most efficient instrument of forceful acculturation.

A number of countries went so far as to prohibit the use of minority languages in schools altogether. In late nineteenth-century Norway the children of the Sami and

Kven minority were forbidden to use their mother tongue even during breaks between classes (Eriksen 1991, p. 68). In the United States, the Bureau of Indian Affairs strictly forbade the use of Native American languages in schools under its control and enforced the rule until the end of World War II (Frantz 1993, p. 137). Yet, according to Feest (1976, pp. 198–199), even in the early 1970s, approximately 30 % of Navajo children did not know any English when they entered elementary school, and another 39 % had such poor English skills that they could not understand the teacher, 21 % were bilingual, and only 10 % spoke predominantly English. On the Fort Apache Indian reservation in 1986, Frantz (1999) found that "three quarters of the 5-year-old Apache children had either no knowledge of English or only a very poor command of the language" (p. 145).

Many challenges of bilingual education that today's immigration countries have to cope with (see e.g., Glazer 1981; Gunderson 1982) already existed in the multicultural and multilingual Habsburg Empire. Count Leo von Thun und Hohenstein (1811–1888), who became Minister of Education and Religion under the Habsburg monarchy in 1849, stated as early as 1838 that non-German-speaking children in the realm were to be taught in their mother tongue: "Children unable to speak any other language shall be instructed exclusively in Bohemian. If they are to learn German, then it shall be through the Bohemian language. Teachers are to be found who are completely fluent in [the children's] mother tongue" (Frankfurter 1894, p. 1).[20] In his capacity as minister, he also mandated that elementary school children were to receive their instruction in their mother tongue. The minister gave didactic tips on how multilingualism could be used to advantage (Frommelt 1963, p. 76). He realized that the language question could not be dealt with centrally, that regional particularities and the will of the affected populations had to be kept in mind. Intending his decrees on language use in the classroom to be understood not as rigid norms but rather only as guidelines to be adapted to the relevant local circumstances, the count's ministry avoided issuing any detailed directives on how to carry them out. The actual decisions lay with the subordinate educational authorities, the school inspectors, and the local school principals, or headmasters (Frommelt 1963, p. 84). Austria's Imperial Elementary School Law of 1869, which radically curtailed the church's influence on the education system and granted minorities considerable freedom in the approach to education, contained no clear provision regulating instruction in mixed-language regions (those in which two or more languages are spoken). Article 6 empowered the competent provincial educational authorities to decide on the language of instruction and on teaching in a second language after consultation with the entity responsible for maintaining the schools (usually the local communities). By today's standards, this educational policy was highly progressive and sound because it took the special regional conditions and the cultural context of the schools into consideration. However, most of the Slavic minorities were not prepared for this linguistic parity in the educational system (Frommelt

---

[20] "Kindern, die keine andere Sprache können, werde der Unterricht ausschließend böhmisch erteilt: sollen sie deutsch lernen, so geschehe es mittelst der böhmischen Sprache; man sorge für Lehrer, die ihrer Muttersprache vollkommen mächtig sind."

1963, p. 90). There were neither textbooks nor a sufficient number of teachers fluent in the standard form of the respective languages. Moreover, the nationalistic currents that intensified after Europe was swept by revolution in 1848 followed their own directions (for details see Havranek 1991).

Just how tolerant the educational policy of the Habsburg monarchy was is evident from a comparison between Austria and two other lands in Europe—nineteenth-century Poland as partitioned between Russia, Prussia and Austria; and South Tyrol after World War I. Additional examples are offered by Eriksen (1991), Garcia Garrido (1991), Miąso (1991), and others. After the failure of the 1863 January uprising in the Russian part of Poland, the educational system there came under the Ministry of Education in St. Petersburg. In 1869 Russian was decreed the language of instruction in schools and official business in the Russian sector of Poland (Kuikka 1991, p. 95; Miąso 1991, p. 172). In 1885 primary education became completely Russified.

> From then on the teaching of all subjects was performed in Russian.... The fact that children were taught in an alien language did not stimulate their educational interests. The teaching, restricted to the mere memorization of knowledge, did not prepare pupils for further studies. As a result of the russification of the primary school, Polish society as a whole was growing increasingly reluctant to learn Russian and even showing open hostility towards it. The dissatisfaction of the population with government primary schools led to the spread of private teaching, both legal and clandestine. (Miąso 1991, p. 173)

The German part of partitioned Poland suffered a similar fate, especially after the founding of the German Reich, when the German state's language policy toward the Polish population became more aggressive (Heinemann 1991, p. 142).

> So, from 1872, a programme of germanisation of the whole educational system was carried out [in the Grand Duchy of Poznań (Posen)[21]], and was done in a manner which was more drastic than that in the Russian sector of partitioned Poland. In the first place the teaching of the Polish language was forbidden, being replaced by German in both elementary and secondary schools. Polish still remained a school subject for a dozen or so years. Then, in 1887, it was completely removed from elementary schools. It remained an optional subject in secondary schools till 1900. (Miąso 1991, pp. 176–177)

In Galicia,[22] which belonged to the Habsburg Empire between 1772 and 1918, educational policy toward minorities was much more tolerant than in the Russian and Prussian (later German) parts. From 1772 until 1848, the language of instruction in Galicia was German. In 1869 the Polish language was restored in some parts of Galicia. After 1873 instruction in primary schools was given in the children's native language. The Provincial School Board, which was dominated by Polish administrators, controlled primary and secondary schools, with the central Austrian

---

[21] The Grand Duchy of Posen (Poznań) was part of the Kingdom of Prussia and was created from territories annexed by Prussia after the Partitions of Poland. It was formally established after Napoleon's final defeat in 1815. Per agreements derived at the Congress of Vienna, the duchy was to have a degree of autonomy. In reality, however, the duchy was subordinated to Prussia, and the proclaimed rights for Polish subjects were not fully implemented. Based on http://en.wikipedia.org/wiki/Grand_Duchy_of_Posen

[22] A historical region of central Europe in what is today southeast Poland and western Ukraine.

authorities looking after the universities and retaining the prerogative of appointing directors and teachers of secondary schools (Miąso 1991, p. 169).

In the Polish republic between 1918 and 1939, Polish authors and politicians turned the tables on the non-Polish minorities. The influential writer Jędrze Giertych (1903–1992) maintained that the Poles had a historic mission to bring the blessings of western civilization to the minorities living in Poland. He believed that Ukrainians, Byelorussians, Lithuanians, Germans, and other minorities living in Poland should be assimilated and should ultimately become part of the Polish nation (Tomiak 1991, pp. 187–188). Roman Dmowski (1864–1929), the leader of the Polish National Democratic Party, "viewed the national minority movements based upon linguistic distinctions as hostile, by their very nature, to the Polish nation state,… [H]e believed that they would gradually, but inevitably succumb to the Polish cultural influence" (Tomiak 1991, p. 187). Stanisław Grabski (1871–1949), another member of the Polish National Democratic Party, argued "that multi-ethnic states were not viable political structures and that the consciousness of national distinctiveness must necessarily evolve with time into the demand for political independence" (Tomiak 1991, p. 187).

Another example is South Tyrol, which Italy annexed together with parts of Slovenia and Croatia after World War I. In order to Italianize the northern and northeastern regions of Italy, where the vast majority of the inhabitants spoke German, Slovenian, or Croatian, the Gentile reform of Fascist Italy in 1924 forbade the use of non-Italian languages in schools and kindergartens (Ara 1991, p. 285). Tens of thousands of German-speaking children were instructed by teachers they did not understand at all. A decree issued by the prefect of Trent on November 27, 1925, prohibited even private tutoring in German within families if it was given to more than three children at the same time. Provincial and police authorities watched diligently for any form of private or clandestine instruction, and they intervened frequently (Ara 1991, p. 280).

In some multilingual areas without bilingual education, minority children may voluntarily attend additional, after-school education in their mother tongue. Attendance at such voluntary education is an informative indicator of the ethnopolitical awareness of the minority families. But it also depends on the skills and engagement of the teachers and on the degree of social control within the community. Table 2.2 summarizes this attendance in the bilingual areas of South Carinthia, which are predominantly rural. As it shows, the smaller the primary school, the higher the voluntary attendance in Slovenian language courses was and the greater the trend toward Slovenian during the given period. The percentage of Slovenian children voluntary attending classes taught in Slovenian was higher in small schools than in larger ones, as was the percentage of teachers who had absolved an examination credentialing them to teach in Slovenian. Table 2.3 shows that in one-classroom schools, all the teachers had absolved an examination in Slovenian, whereas in four-classroom schools only 51.2 % had passed such an examination and 44.4 % of the teachers had no knowledge of Slovenian at all.

There are six main questions about the language of instruction: Where and to what extent are minorities allowed to use their mother tongue as a language of

Table 2.2 Percentage of students in the Bilingual Region of South Carinthia who voluntarily attend classes taught in Slovenian, by size of Primary School

Organization of schools	Academic year and percentage		Number of schools offering classes taught in Slovenian	
	1960–1961	1976–1977	1960–1961	1976–1977
One class	35.9	44.5	15	11
Two classes	26.9	28.5	27	18
Three classes	24.5	12.4	16	8
Four classes	12.5	9.1	19	39

From Meusburger (1979, p. 248), Copyright 1979 by Department of Geography, Innsbruck University. Reprinted with permission
In a one-classroom school four grade levels of elementary school (i.e., 6- to 10-years old pupils) are taught in one classroom by one teacher

Table 2.3 Slovenian language skills of the elementary school teachers in the Bilingual Region of South Carinthia, by size of school (academic year, 1976–1977)

Organization of schools	No. of teachers	Percentage of teachers and their proficiency in Slovenien			
		Credentialed to teach in Slovenian	Passed secondary school graduation exams in Slovenian	Knowledge of Slovenian, no exam	No knowledge of Slovenian
One class	12	75.0	25.0	0.0	0.0
Two classes	36	86.2	8.3	2.8	2.8
Three classes	29	68.9	0.0	13.8	17.2
Four classes	261	47.5	4.2	3.8	44.4

From Meusburger (1979, p. 250). Reprinted with permission

instruction, at least in the initial years of schooling? What is the percentage of minority parents who favor bilingual education? What institution is responsible for establishing minority schools or deciding about bilingual education? What is the spatial distribution and size of schools in which the minorities can use their mother tongues? Which social, economic, and political factors influence the expansion or shrinking of areas in which minority languages are used in schools? What skills do minority children have in the language of instruction when entering primary schools?

The Impact of Teachers on the Development of Minority Cultures

With regard to the influence of elementary school teachers on the development of minority culture, one must distinguish between urban and peripheral rural areas. In villages teachers have a much more important extracurricular function and greater long-term impact on the community than their counterparts in large cities do. If sparsely inhabited minority areas are successful in attracting skilled teachers proficient in the minority's language and sympathetic to the minority's culture, then their schools will be indispensable elements of the community's cultural development. Such schools will be places that form and strengthen local and ethnic identity, social networks, and social cohesion and that nurture cultural activities and sensitivity to regional environmental problems. In social milieus of that kind, skilled teachers are still pivotal individuals in the community and have many functions to perform beyond the school. In urban areas the professionalization of these functions markedly reduces the need for the extracurricular part that teachers play (for details see Kramer 1993; Meusburger 1998, 2015a).

The development of ethnic self-esteem depends on the availability of adult role models. "Achieving group identity is difficult without the consistent guidance of important socializing agents" (Spencer and Markstrom-Adams 1990, p. 302; Spencer 1983). The ethnic identity, cultural roots, skills, and experience of teachers decide whether they will act as vital persons in the social networks of the ethnic community and as role models for their students. The quality of attachment and mutual understanding between teachers, students, and parents is also important for the vertical transmission of culture. Interpersonal relationships and interactions between teachers and minority students (including their parents) do not evolve automatically but rather as a matter of mutual respect and trust that the curriculum will provide the best development for students, and for society in general (see Pieke 1991, p. 163). Teachers have the authority to draw attention to certain events and to legitimate narratives. Insiders who belong to the same ethnic group and who share the same values, fate, and memories are more trusted than outsiders and thus are more reliable as sources of information. Emotional support from teachers and sound relationships between children and teachers enhance children's openness and tendency to comply with these agents of socialization.

Teachers are able to create, sustain, or destroy processes at work in the "cultural grounding of the self" (Miller 2001, p. 3139). The development of identity involves personal reflection and observation of oneself in relation to others.

> Identity is about belonging, about what we have in common with some people and what differentiates us from others. Identity gives one a sense of personal location, and provides a stable core of one's individuality; but it is also about one's social relationships, one's complex involvement with others, and in the modern world these have become even more complex and confusing. (Bulmer and Solomos 1998, p. 826)

If teachers belong to a "different world," it is difficult for them to perform their function as agents of socialization. They will not help students find their cultural identity, gain a sense of belonging, and feel pride in their ethnic group but rather will create ambiguity, build frustration, and bring about a lack of self-confidence. It is particularly important in the early grades of elementary school that the teachers belong to the ethnic group of their students, speak their language, and therefore act as a kind of role model. If teachers are ignorant of, indifferent to, or even opposed to the culture and experiences of their students and have low expectations of them, the learning process will be much slower.

Frantz (1999, p. 146) found during his field studies in 1986 that only 6 out of almost 200 teachers in the schools of Whiteriver, the capital of the Fort Apache reservation, were Apaches and that 5 were American Indians from other reservations. More than half of the white teachers commuted from outside the reservation, and the rest lived in a mobile home ghetto established by the Bureau of Indian affairs. In essence, 94.5 % of the teachers had almost no contact with the everyday life of the Apaches. Most of these white teachers had an extremely high turnover rate, which indicates that they were uncommitted to their school. The situation was much better in the one school of the Hualapai reservation in Peach Springs, where a quarter of the teachers were members of the tribe (Frantz 1999, p. 146). "While the Fort Apache students have only the standard textbooks used nationwide, the Hualapai Indians have developed their own books, which are based on their cultural values and tribal heritage" (Frantz 1999, p. 146).

The cultural and political importance of teachers has always made them the first to be dismissed by the new administration when a territory was occupied by another nation or when a state administration started a nationalization campaign against a minority. Although more than 95 % of the population in South Tyrol in the 1920s spoke German, most of the German-speaking teachers were replaced by Italian teachers after fascist Italy had annexed the region. Only 111 out of 745 German-speaking teachers in service in 1921 remained at the end of 1927–1928 school year, and most of the rest were forced to retire or move to other parts of Italy in the following years. Children with German as a mother tongue, many of whom knew no Italian at all, were thus not only confronted with being taught in Italian, at least when they entered school, but also with teachers who could not communicate with them (Ara 1991, p. 279). Similar circumstances prevailed in the Grand Duchy of Posen (Poznań) after 1872, when most teachers of Polish origin were transferred to Germany proper and replaced by German teachers in order to facilitate the Germanization of Polish schools (Miąso 1991, p. 177).

A host of key questions surround the role of teachers. Do teachers encourage or oppose the cultural interests of a minority group? What skills and teaching experience do they have? To what extent are the teachers embedded in the minority culture? How many teachers are residents of the community in which they teach and how many commute from elsewhere? For how many years have they taught in the ethnic environment they serve? How large is the turnover rate of teachers in schools with a large percentage of minority students? What is the age and gender distribution in the teaching staff in such schools? What are the academic achievements of the teachers' students? Which functions do teachers perform in the community apart from teaching?

Location and Size of Primary Schools

Some readers might argue that, in the age of globalization and transculturality (Welsch 1999), ethnic identities and practices can easily flow to other continents and that relationships between members of an ethnic group can be maintained over long distances. However, globalization does not mean that the relation between ethnicity and space is henceforth completely irrelevant. Common codes of communication, shared understanding, value systems, narratives, and cultural practices have to be learned, developed, and practiced in order to be internalized. Orientation knowledge and cultural knowledge are not only conditioned by power and rhetorical or textual construction but also by social solidarity and trust created by common rituals and face-to-face communication. Therefore, the size of ethnic communities and their rootedness to places or cultural environments do matter in certain periods of socialization and cultural learning. Cultural knowledge is created through practices and interaction in particular places and is embedded in specific social contexts. In order to secure the survival of their cultures, ethnic and religious minorities have to organize learning opportunities and practices for their children to socialize them in particular ways. Such learning practices operate informally at home, among peer groups, and in neighborhoods, but they also shape formalized interaction in cultural institutions such as schools and universities, and in other settings.

Minority schools are more than a place of instruction and formal education. They are sites and contexts where social relations evolve and where identities, social cohesion, self-awareness, attitudes, cultural preferences, discourses, stereotypes, and social inequalities emerge. They offer a place where parents, teachers, and other role models interact and where cultural performances are organized. The spatial pattern of minority schools that use the mother tongue of the students as a language of instruction resembles a kind of cultural marker, a flag defining political territories. Unlike ethnic awareness or ethnic identity, the language of instruction is easy to map and document. Local decision-makers from minorities therefore often insist on the use of the mother tongue in the initial years of elementary school, even when children grow up bilingually.

Even when the right to use one's mother tongue in the first grades of compulsory school is guaranteed by law, the group entitled to do so and a space where its members can exercise their minority rights still have to exist. The powers-that-be can manipulate the definition of an entitled group and of a place or territory where the group's rights can be applied. Even if blunt discrimination against ethnic minorities within the school system has recently become less frequent than it was in nineteenth or twentieth centuries, many subtle instruments for administering seemingly equal treatment are still used to discriminate against ethnic minorities. In sparsely inhabited and ethnically mixed areas it is to the advantage of minorities when small classes and small schools are allowed. Otherwise, they have little chance to maintain their autonomy in educational affairs, to employ their own teachers, and to use their mother tongue as a language of instruction. In many cases, the centralization of the elementary school system, the creation of larger school units and larger catchment areas, and the gerrymandering of catchment areas spelled the end of minority schools.

In rural areas of many states, two different concepts of regional educational planning compete (for details see Kramer 1993; Meusburger 1998, pp. 405–411). That of large centralized elementary schools to which children are transported over long distances by school buses or parents is claimed to increase financial efficiency, enhance the flexible assignment of teachers, augment the variety of courses, broaden social equality, improve the quality of education, and exemplify modernity.

The alternative concept of regional educational planning promotes small decentralized elementary schools located in or near the villages and towns where the 6–10-year-old students live. Designed for small classes and schools, this approach requires more teachers, and thus greater financial resources, per hundred students than large urban schools do. It is less flexible in the assignment of teachers. However, supporters of this concept assert that its pedagogical advantages and many positive extracurricular effects outweigh the higher financial costs. Its adherents have a different view of childhood, are more sensitive to the needs of minorities than advocates of centralized elementary schooling are, acknowledge the negative experience many children have with long-distance busing, and are primarily interested in the needs of communities (see Bunar 2010; Kramer 1993; Meusburger 2015a).

The arguments that financial resources are scarce and that national standards (e.g., teacher–student ratio) must be maintained are frequently used to close small schools of minorities and integrate minority children into large schools dominated by the mainstream culture. By setting a high minimum number of children required for establishing a school, minorities are often prevented in rural areas from opening a school of their own. However, standards for the minimum number of children required to open or close an elementary school are frequently changed according to the political interests of those in power. The 1924 educational reform in Poland by Stanisław Grabski (1871–1949)

> stipulated the conditions under which elementary public schools teaching in non-Polish languages could be established. These required that in a *gmina* (local administrative unit) where at least 25 per cent of [the] population was of a given minority nationality, parents of at least 40 children of school age in a school district could formally demand teaching in the mother tongue. If there were fewer than 40, children were to be taught in the official language, that is, in Polish. On the other hand, if there were, in a given school district, at least 20 children whose parents demanded teaching in Polish, irrespective of the number of children of any other nationality, the children were to be taught in two languages. If there were fewer than 20 [Polish children], the minority language was to be the language of instruction. (Tomiak 1991, p. 193)

In the early 1920s any community in South Tyrol that had at least 12 Italian children of school age received an Italian elementary school in addition to the already existing German one (Ara 1991, p. 273). During the 1970s and 1980s the Italian school authorities in South Tyrol maintained elementary schools teaching in Italian even when there were fewer than five Italian-speaking children in that commune or school district (Höfle and Meusburger 1983; Meusburger 1996).

When analyzing educational policies in bilingual areas, researchers find it revealing to compare the size of elementary schools shortly before they were closed for lack of students. In South Tyrol, the Italian school authorities in the 1970s and 1980s maintained elementary schools teaching in Italian as long as possible. Whereas 71.2 % of Italian elementary schools had but 1–3 students in the year before they were shut down, only 5.1 % of schools with German as the language of instruction were allowed to continue until they reached such a small size. Some

Table 2.4 The size of elementary schools in Tyrol before their closure with regard to the language of instruction

Number of students before the school was closed	Language of instruction			
	North Tyrol (Austria) (1965–1981)	South Tyrol (Italy) (1971–1981)		
	German	German	Italian	Ladin[a]
1	–	–	21.2	–
2	–	–	27.3	–
3	1.4	5.1	22.7	–
4	4.1	5.1	12.1	–
5–10	24.7	47.4	15.2	75.0
11–20	37.0	21.8	–	25.0
21–30	16.4	6.4	–	–
More than 30	16.4	14.1	–	–
Total	100.0	100.0	100.0	100.0
Absolute number of schools closed	73	78	66	4

Source: Höfle and Meusburger (1983). Reprinted with permission
[a]Ladin is a Rhaeto-Romance language mainly spoken in the Dolomite-Mountains in South Tyrol, the Trentino and the province of Belluno. It has connections with the Friulian and Swiss Romansh

42.3 % of the German schools and none of the Italian schools had more than 10 students in the year preceding closure (Table 2.4).

In South Tyrol numerous Italian one-class schools were kept open as long as possible. Of the 66 elementary schools with Italian as the language of instruction that were given up between 1971 and 1981, 71.2 % had only 1–3 students in the year preceding closure, and not a single one had more than 10 students. In comparison, only 5.1 % of the 78 schools with German as the language of instruction were allowed to continue until they had reached such a small size. Some 69.8 % of the schools in North Tyrol, 42.3 % of the German schools in South Tyrol, but not a single Italian school in South Tyrol had more than 10 students in the year preceding closure.

Responses of Ethnic Minorities to Measures of Forced Assimilation

Harsh educational policies have sometimes been quite successful at annihilating the culture and identity of ethnic minorities. In other cases they have had the opposite effect of what was intended (for details see Ara 1991; Eriksen 1991; Villgrater 1984). The responses of ethnic groups to assimilation forcibly effected through the school system range from compliance to lethargy, manifest or latent opposition, and antagonism (for details see Tomiak 1991). Such different reactions to pressure exerted by the educational system of the majority reveal much about the self-esteem, collective social status, and political self-organization of ethnic minorities; their internal coherence or fragmentation; their desire or disinterest to maintain their culture; and their vulnerability, lethargy, powerlessness, and abdication in cultural and political affairs. At least five categories of response have emerged in recent history:

- Acquiescing to the pressure of state institutions, renouncing schools in one's mother tongue, and more or less giving up one's cultural identity and practices after one or two generations
- Boycotting public educational institutions, keeping children at home, and—a practice of many American Indians tribes—hiding school-age children from authorities in charge of enforcing attendance (for details see Frantz 1996, 1999)
- Shifting the education of children in their mother-tongue and culture to religious institutions (e.g., Sunday schools), as done by the Jewish population in Czarist Russia[23] and the Habsburg Empire and, during the Russification of Poland in the late nineteenth century, by the Polish population
- Establishing secret schools or other means of clandestine education and thereby circumventing the majority's policy of forced assimilation or forced ignorance (a response that is discussed in the following paragraphs)

Cornelius (1991), Franklin and Moss (1994), Whiteaker (1990), Gamerith (see Chap. 4 in this volume), and others describe how in the United States daring slaves opened secret schools in order to pass on their knowledge to fellow slaves and thereby undermine the law of compulsory ignorance. Gatherings "under the trees" were intended to simulate harmless social events, but in effect they served to teach the attending people how to read and write. At night secret schools popped up to disseminate banned knowledge (Franklin and Moss 1994, p. 160; Whiteaker 1990, p. 8).

The spreading Russification in the Russian part of partitioned Poland made people reluctant to send their children to government schools. As a consequence, more than 3,100 underground schools were operating in the Kingdom of Poland between 1864 and 1914 (Miąso 1991, p. 174). "According to data gathered at the end of nineteenth century, 33 per cent of the Kingdom's inhabitants who could read and write owed these skills not to a government school but to clandestine teaching" (p. 174).

In South Tyrol the establishment of secret schools (so-called catacomb schools) began in late autumn 1924, after Italian had been introduced as the only acceptable language of instruction in schools and kindergartens and instruction in German had been largely banned even at home if it was given to more than three children. According to Villgrater (1984), the catacomb school network in South Tyrol ranged between 200 and 210 teachers, mainly women who were unmarried or still unemployed and men who had been dismissed by the Italian administration. Between 3,500 and 5,300 pupils attended these secret schools every year in the 1930s

---

[23] The basics of Judaism and the Hebrew language were taught in traditional Jewish elementary schools known as cheders, which were generally small and financed exclusively by the Jewish religious communities. "Any attempt to put them under control [of the Russian administration] failed as a rule. Only a few of them taught Russian, for instance" (Miąso 1991, p. 173).

(pp. 142–143).[24] Textbooks were smuggled into South Tyrol from Austria and Germany mainly over the Ötztal glaciers by men carrying up to 45 kg (nearly 100 lbs) of books (pp. 95–270). Those teachers whom police had seen to be teaching in catacomb schools were—depending on their activities—interrogated, imprisoned, put under police supervision, forced to leave their village, or banned to distant islands.[25]

Conclusion

This chapter has shown that learning processes in the educational system can be heavily influenced by cultural factors, teachers, and administrative measures. Repeatedly misappropriated to force the assimilation of minorities, the school in multiethnic nation-states has represented a contentious political and cultural arena. In some countries of Eastern and Southeastern Europe, the fall of the Iron Curtain showed the rekindling of ethnic conflicts that for decades were believed to have de-escalated. The political and cultural importance of ethnic identity is unlikely to decline in the globalized world. With the number of the world's refugees having soared to unprecedented levels in the second decade of the twenty-first century, additional tests still await multicultural societies. According to Joppke (1996) multiculturalism "is one of the most pervasive and controversial intellectual and political movements in contemporary Western democracies" (p. 449). This issue, however, is not only relevant to the policy-makers and educators, it confronts the scientific community with special challenged as well.

The schooling of nondominant ethnic minorities exemplifies the insight that normative rules, laws, and administrative measures are not universally applicable and that scientism[26] is not a good advisor when it comes to cultural diversity and power relations between ethnic minorities. An administrative measure that is beneficial for the schooling of minority A can be detrimental for minority B. Philosophers, legal scholars, and cultural scientists do not agree on collective cultural rights of minorities or on affirmative action. Not all nondominant minorities have the same cultural awareness and group solidarity; not all of them face the same degree of discrimination; and minorities also differ in their vision about the optimal schooling of their children. Not least, the geographical distribution and size of ethnic minorities influence the prospects of various schooling alternatives.

---

[24] The secret coordination and supervision of these catacomb schools, which were located mainly in remote farm houses, was organized by the Catholic priest Michael Gamper (1885–1956) and Josef Noldin (1888–1929). Gamper was succeeded by Maria Nicolussi (1882–1961).

[25] Rudolf Riedl (1876–1965) from Tramin was banned to the distant Italian island of Pantelleria, located between Sicily and Africa. Josef Noldin was banned to the island of Lipari (for details see Villgrater 1984).

[26] "We forget at our own peril that scientism is not a matter of specific paradigms, but about authority and the power of particular speaking positions, as well as about what counts as knowledge and evidence" (Grossberg 2010, p. 47).

The large variety of categories, histories, and geographies of minorities shifts the focus of research to the situatedness of ethnic groups (the *spatial context* in which they live), their cultural heritage, collective social status, internal solidarity, political awareness, and, primarily, their position in national and international power relations.

> [W]hile power operates in institutions and in the state, it also operates where people live their daily lives, and in the spaces where these fields intersect. Cultural studies is always interested in how power infiltrates, contaminates, limits, and empowers the possibilities that people have to live their lives in just, dignified, and secure ways. (Grossberg 2010, p. 29)

Conducting research about the schooling and the educational achievement of ethnic minorities, one finds it almost impossible to discover causalities in the sense of universal laws. More important is the causal reconstruction of interactions and the focus on milieus, social environments, or spatial contexts. What is sought in the empirical analysis of social macrophenomena is "not abstraction and maximal simplification but rather specification and adequate complexity of the explanation" (Mayntz 2002, p. 13). "Causal reconstruction is not about seeking statistical relations between variables but rather about explaining the macrophenomenon by identifying the processes and interdependencies involved in its having arisen" (p. 13).

Until recently, discourses in educational research and sociology of education about the schooling of minorities focused mainly on educational (pedagogical) and political arguments (see e.g., Pratte 1982; Glazer 1981). The interrelationships of minority schooling with regional or urban structures, the importance of spatial contexts and spatial disparities, the spatiality of communication and action, and the symbolic meaning of places for social relations—topics in the geography of education since the early 1970s—attracted little attention in these disciplines. In the early twenty-first century, however, scholars began "spatializing" the sociology of education (Ferrare and Apple 2010; Ford 2014; Gulson and Symes 2007; Robertson 2010). The importance of the spatial context was also underlined by representatives of cultural studies: "[A]ny discussion of issues of race and ethnicity cannot be separated from the particular context in which it is located and into which it is directed" (Grossberg 2010, p. 21).

Some of the key issues that Glazer (1981), Suárez-Orozco (1991), and others raised about schooling in multicultural societies are still unanswered. Hard questions to grapple with in future research are numerous: Can lessons be learned from the past and applied to future multicultural societies? What rights do autochtonous minorities have in the educational system? Is it legimiate for autochtonous ethnic minorities with strong territorial rooting to be treated differently than people who have migrated to those spaces? How can public education help integrate children into mainstream society without forcing them to shed their cultural identity when they are not fluent enough in the target language to follow classroom instruction? How many of the tasks facing a multicultural society can the public education system cope with? Are the teachers and the school administration prepared to take on these tasks? Who will cover the additional costs thereof? With which tasks can or should private initiatives (by parents or cultural associations, for example) or religious institutions help? As thorny as these matters are, responses that respect the geography, history, and culture of minorities—and will therefore vary from one country to the next—are potentially rewarding. Therein lies the inspiration for further inquiry on behalf of the multicultural society.

References

Achor, S., & Morales, A. (1990). Chicanas holding doctoral degrees: Social reproduction and cultural ecology. *Anthropology and Education Quarterly, 21,* 269–287. Retrieved from http://www.jstor.org/stable/3195877.

Albanese, A. G. (1976). *The plantation school.* New York: Vantage Press.

Ara, A. (1991). Italian educational policy towards national minorities, 1860–1940. In J. Tomiak, K. Eriksen, A. Kazamias, & R. Okey (Eds.), *Schooling, educational policy and ethnic identity: Vol. 1. Comparative studies on governments and non-dominant ethnic groups in Europe, 1850–1940* (pp. 263–290). New York: New York University Press.

Bernstein, B. (1964). Aspects of language and learning in the genesis of the social process. In D. Hymes (Ed.), *Language in culture and society: A reader in linguistics and anthropology* (pp. 251–263). New York: Harper & Row.

Bilmes, J., & Boggs, S. T. (1979). Language and communication: The foundations of culture. In A. J. Marsella, R. G. Tharp, & T. J. Ciborowski (Eds.), *Perspectives on cross-cultural psychology* (pp. 47–76). New York: Academic.

Blinderman, A. (1978). Congressional social Darwinism and the American Indian. *The Indian Historian, 11*(2), 15–17.

Boyer, P. (1997). *Native American colleges: Progress and prospects.* Princeton: Carnegie Foundation.

Boyer, P. (2001). Cultural assimilation. In N. J. Smelser & P. B. Baltes (Eds.), *International encyclopedia of the social & behavioral sciences* (Vol. 5, pp. 3032–3035). Amsterdam: Elsevier. doi:10.1016/B0-08-043076-7/00364-8.

Brett, N. (1991). Language laws and collective rights. *Canadian Journal of Law and Jurisprudence, 4,* 347–360.

Brown, R. H. (1993). Cultural representation and ideological domination. *Social Forces, 71,* 657–676. doi:10.1093/sf/71.3.657.

Brown v. Board of Education, 347 U.S. 483 (1954). http:www.ourdocuments.gov/doc.php?flash=false&doc=87

Bulmer, M., & Solomos, J. (1998). Introduction: Re-thinking ethnic and racial studies. *Ethnic and Racial Studies, 21,* 819–837. doi:10.1080/014198798329667.

Bunar, N. (2010). The geographies of education and relationships in a multicultural city. Enrolling in high-poverty, low-performing urban schools and choosing to stay there. *Acta Sociologica, 53,* 141–159. doi:10.1177/0001699310365732.

Canadian Commission for UNESCO. (1977). A working definition of 'culture'. *Cultures, 4*(4), 78–85.

Churchill, W., & Hill, N., Jr. (1979). Indian education at the university level: An historical survey. *Journal of Ethnic Studies, 7*(3), 43–58.

Cornelius, J. D. (1991). *"When I can read my title clear": Literacy, slavery, and religion in the antebellum South.* Columbia: University of South Carolina Press.

Craig, J. E. (1984). *Scholarship and nation building: The universities of Strasbourg and Alsatian society, 1870–1939.* Chicago: University of Chicago Press.

de Vroede, M. (1991). Language in education in Belgium up to 1940. In J. Tomiak, K. Eriksen, A. Kazamias, & R. Okey (Eds.), *Schooling, educational policy and ethnic identity: Vol. 1. Comparative studies on governments and non-dominant ethnic groups in Europe, 1850–1940* (pp. 111–131). New York: New York University Press.

Dierkes, J. B. (2010). *Postwar history education in Japan and the Germanys: Guilty lessons.* London: Routledge.

Driver, F. (2001). *Geography militant: Cultures of exploration and empire.* Oxford: Blackwell.

Erickson, R. (1997). The laws of ignorance designed to keep slaves (Blacks) illiterate and powerless. *Education, 118*, 206–209, 220.

Eriksen, K. (1991). Norwegian and Swedish educational policies vis-à-vis non-dominant ethnic groups, 1850–1940. In J. Tomiak, K. Eriksen, A. Kazamias, & R. Okey (Eds.), *Schooling, educational policy and ethnic identity: Vol. 1. Comparative studies on governments and non-dominant ethnic groups in Europe, 1850–1940* (pp. 63–85). New York: New York University Press.

Eriksen, K., Kazamias, A., Okey, R., & Tomiak, J. (1991). Governments and the education of non-dominant ethnic groups in comparative perspective. In J. Tomiak, K. Eriksen, A. Kazamias, & R. Okey (Eds.), *Schooling, educational policy and ethnic identity: Vol. 1. Comparative studies on governments and non-dominant ethnic groups in Europe, 1850–1940* (pp. 389–417). New York: New York University Press.

Feest, C. (1976). *Das Rote Amerika. Nordamerikas Indianer [Red America: North America's Indians].* Vienna: Europa.

Ferrare, J. J., & Apple, M. W. (2010). Spatializing critical education: Progress and cautions. *Critical Studies in Education, 51*, 209–221. doi:10.1080/17508481003731075.

Ford, D. R. (2014). Spatializing Marxist educational theory: School, the built environment, fixed capital and (relational) space. *Policy Futures in Education, 12*, 784–793.

Frankfurter, S. (1894). Thun-Hohenstein, Graf Leo. In *Allgemeine Deutsche Biographie. Thienemann–Tunicius* (Vol. 38, pp. 178–212). Leipzig: Duncker & Humblot. Retrieved from http://de.wikisource.org/w/index.php?title=ADB:Thun,_Leo_Graf_von&oldid=2101992

Franklin, J. H., & Brooks Higginbotham, E. (2011). *From slavery to freedom: A history of African Americans* (9th ed.). New York: McGraw-Hill.

Franklin, J. H., & Moss, A. A., Jr. (1994). *From slavery to freedom: A history of African Americans* (7th ed.). New York: McGraw-Hill.

Frantz, K. (1993). *Die Indianerreservationen in den USA. Aspekte der territorialen Entwicklung und des sozio-ökonomischen Wandels [The Indian reservations in the United States: Aspects of territorial development and socioeconomic change]* (Erdkundliches Wissen, Vol. 109). Stuttgart: Steiner.

Frantz, K. (1994). Washington Schools, Little White Man Schools und Indian Schools—bildungsgeographische Fragestellungen dargestellt am Beispiel der Navajo Indianerreservation [Washington Schools, Little White Man Schools, and Indian Schools—Issues of educational geographics as exemplified by the Navajo Indian Reservation]. *Die Erde, 125*, 299–314.

Frantz, K. (1996). Education on the Navajo Indian reservation: Aspects of the maintenance of cultural identity as seen from a geographical point of view. In K. Frantz & R. A. Sauder (Eds.), *Ethnic persistence and change in Europe and America* (Veröffentlichungen der Universität Innsbruck, Vol. 213, pp. 223–245). Innsbruck: Innsbruck University.

Frantz, K. (1999). *Indian reservations in the United States: Territory, sovereignty, and socioeconomic change* (University of Chicago geography research paper, Vol. 242). Chicago: The University of Chicago Press.

Freytag, T. (2003). *Bildungswesen, Bildungsverhalten und kulturelle Identität: Ursachen für das unterdurchschnittliche Ausbildungsniveau der hispanischen Bevölkerung in New Mexico [Education, educational behavior, and cultural identity: Causes of the below-average level of education in Hispanic population in New Mexico]* (Heidelberger Geographische Arbeiten, Vol. 118). Heidelberg: Selbstverlag des Geographischen Instituts der Universität Heidelberg.

Frommelt, K. (1963). *Die Sprachenfrage im österreichischen Unterrichtswesen, 1848–1859 [The language of instruction issue in the Austrian educational system, 1848–1859].* Graz: Hermann Böhlau.

Gamerith, W. (1998a). Education in the United States: How ethnic minorities are faring. In F. J. Kemper & P. Gans (Eds.), *Ethnische Minoritäten in Europa und Amerika. Geographische Perspektiven und empirische Fallstudien* (Berliner Geographische Arbeiten, Vol. 86, pp. 89–104). Berlin: Department of Geography, Humboldt University.

Gamerith, W. (1998b). Das US-amerikanische Bildungswesen. Räumlich-soziale Disparitäten im Spannungsfeld zwischen egalitären und elitären Prinzipien [The U.S. education system: Spatiosocial disparities between egalitarian and elitist principles]. *Mitteilungen der Österreichischen Geographischen Gesellschaft, 140*, 161–196.

Gamerith, W. (2002). *Ethnizität und Schule. Eine Analyse regionaler und sozialer Disparitäten der Ausbildungs- und Qualifikationsstrukturen US-amerikanischer Minderheiten* [*Ethnicity and school: An analysis of regional and social disparities in structures of education and qualification among minorities in the United States*], Postdoctoral dissertation, Heidelberg University, Heidelberg.

Garcia Garrido, J. L. (1991). Spanish educational policy towards non-dominant linguistic groups, 1850–1940. In J. Tomiak, K. Eriksen, A. Kazamias, & R. Okey (Eds.), *Schooling, educational policy and ethnic identity: Vol. 1. Comparative studies on governments and non-dominant ethnic groups in Europe, 1850–1940* (pp. 291–318). New York: New York University Press.

Gillborn, D. (2005). Education policy as an act of white supremacy: Whiteness, critical race theory and education reform. *Journal of Education Policy, 20*, 485–505.

Gillborn, D. (2008). *Racism and education: Coincidence or conspiracy?* London: Routledge.

Gillborn, D., & Ladson-Billings, G. (2010). Education and critical race theory. In M. W. Apple, S. J. Ball, & L. A. Gandin (Eds.), *The Routledge international handbook of the sociology of education* (pp. 37–47). London: Routledge.

Glazer, N. (1981). Ethnicity and education: Some hard questions. *Phi Delta Kappan, 62*, 386–389. Retrieved from http://www.jstor.org/stable/20385901.

Grossberg, L. (2010). *Cultural studies in the future tense*. Durham: Duke University Press.

Gulson, K. N., & Symes, C. (2007). *Spatial theories of education: Policy and geography matters*. New York: Routledge.

Gunderson, D. V. (1982). Bilingual education. In H. E. Mitzel (Ed.), *Encyclopedia of educational research: Vol. 1* (5th ed., pp. 202–211). New York: Free Press.

Gutiérrez, K. D., Ali, A., & Henríquez, C. (2010). Syncretism and hybridity: Schooling, language, and race and students from non-dominant communities. In M. W. Apple, S. J. Ball, & L. A. Gandin (Eds.), *The Routledge international handbook of the sociology of education* (pp. 358–369). London: Routledge.

Hall, S. (1996). The question of cultural identity. In S. Hall, D. Held, D. Hubert, & K. Thomson (Eds.), *Modernity. An introduction to modern societies* (pp. 595–634). Oxford: Blackwell.

Havranek, J. (1991). The education of Czechs and Slovaks under foreign domination, 1850–1918. In J. Tomiak, K. Eriksen, A. Kazamias, & R. Okey (Eds.), *Schooling, educational policy and ethnic identity: Vol. 1. Comparative studies on governments and non-dominant ethnic groups in Europe, 1850–1940* (pp. 235–261). New York: New York University Press.

Heffernan, M. J. (1994). The science of empire: The French geographical movement and the forms of French imperialism, 1870–1920. In A. Godlewska & N. Smith (Eds.), *Geography and empire* (pp. 92–114). Oxford: Blackwell.

Heinemann, M. (1991). State, school and ethnic minorities in Prussia, 1860–1914. In J. Tomiak, K. Eriksen, A. Kazamias, & R. Okey (Eds.), *Schooling, educational policy and ethnic identity: Vol. 1. Comparative studies on governments and non-dominant ethnic groups in Europe, 1850–1940* (pp. 133–161). New York: New York University Press.

Hobsbawm, E. (1989). *The age of empire*. New York: Vintage Books.

Höfle, K., & Meusburger, P. (1983). *Jüngere Strukturveränderungen des Pflichtschulwesens in Nord- und Südtirol* [*Recent structural changes of compulsory schooling in North and South Tyrol*] (Begleittexte zum Tirol Atlas, Vol. 8, pp. 25–32). Innsbruck: Wagner.

Hörning, K. H., & Winter, R. (1999). Widerspenstige Kulturen. Cultural Studies als Herausforderung [Intractable cultures: Cultural studies as a challenge]. In R. Winter & K. H. Hörning (Eds.), *Widerspenstige Kulturen: Cultural Studies als Herausforderung* (pp. 7–12). Frankfurt am Main: Suhrkamp.

Jacobs, L. A. (1991). Bridging the gap between individual and collective rights with the idea of integrity. *Canadian Journal of Law and Jurisprudence, 4*, 375–386.

Jarausch, K. H. (1983). Higher education and social change: Some comparative perspectives. In K. H. Jarausch (Ed.), *The transformation of higher learning, 1860–1930: Expansion, diversification, social opening, and professionalization in England, Germany, Russia, and the United States* (pp. 9–36). Chicago: University of Chicago Press.

Joppke, C. (1996). Multiculturalism and immigration: A comparison of the United States, Germany, and Great Britain. *Theory and Society, 25,* 449–500. Retrieved from http://www.jstor.org/stable/657908.

Jordan, C., & Tharp, R. G. (1979). Culture and education. In A. J. Marsella, R. G. Tharp, & T. J. Ciborowski (Eds.), *Perspectives on cross-cultural psychology* (pp. 265–285). New York: Academic.

Kramer, C. (1993). *Die Entwicklung des Standortnetzes von Grundschulen im ländlichen Raum. Vorarlberg und Baden-Württemberg im Vergleich [The development of the local network of primary schools in rural areas: Vorarlberg and Baden-Württemberg compared]* (Heidelberger Geographische Arbeiten, Vol. 93). Heidelberg: Selbstverlag des Geographischen Instituts der Universität Heidelberg.

Kuikka, M. T. (1991). Educational policy in Finland under Russian domination, 1850–1917. In J. Tomiak, K. Eriksen, A. Kazamias, & R. Okey (Eds.), *Schooling, educational policy and ethnic identity: Vol. 1. Comparative studies on governments and non-dominant ethnic groups in Europe, 1850–1940* (pp. 87–110). New York: New York University Press.

Kukathas, C. (1992). Are there any cultural rights? *Political Theory, 20,* 105–139. Retrieved from http://www.jstor.org/stable/191781.

Kymlicka, W. (1989). *Liberalism, community, and culture.* Oxford: Oxford University Press.

Kymlicka, W. (1991). Liberalism and the politicization of ethnicity. *Canadian Journal of Law and Jurisprudence, 4,* 239–256.

Kymlicka, W. (1992). The rights of minority cultures: Reply to Kukathas. *Political Theory, 20,* 140–146. doi:10.2307/191782.

Lau v. Nichols, 414 U.S. 563 (1974). http://usedulaw.com/362-lau-v-nichols.html

Lenihan, D. (1991). Liberalism and the problem of cultural membership: A critical study of Kymlicka. *Canadian Journal of Law and Jurisprudence, 4,* 401–422.

Lewis, J. (2002). From culturalism to transculturalism. *Iowa Journal of Cultural Studies, 1,* 14–32.

Mannhardt, J. W. (1965). *Bausteine zur Volkswissenschaft. Aufsätze und Reden* [Building stones to ethnology. Papers and lectures]. In: M. Straka (Ed.), Stuttgart: Gustav Fischer.

Marsden, W. E. (1982). Diffusion and regional variation in elementary education in England and Wales, 1800–1879. *History of Education, 11,* 173–194. doi:10.1080/0046760820110303.

Marsden, W. E. (1987). *Unequal educational provision in England and Wales: The nineteenth-century roots.* London: Woburn.

Mayntz, R. (2002). Zur Theoriefähigkeit makro-sozialer Analysen [On the ability of macrosocial analyses to contribute to theory]. In R. Mayntz (Ed.), *Akteure—Mechanismen—Modelle. Zur Theoriefähigkeit makro-sozialer Analysen* (Schriften des Max-Planck-Instituts für Gesellschaftsforschung Köln, Vol. 42, pp. 7–43). Frankfurt am Main: Campus.

Melville, M. B. (1988). Hispanics: Race, class, or ethnicity? *The Journal of Ethnic Studies, 16*(1), 67–83.

Melville, M. B. (1994). "Hispanic" ethnicity, race and class. In T. Weaver (Ed.), *Handbook of Hispanic culture in the United States* (Anthropology, Vol. 4, pp. 85–106). Houston: Arte Público Press.

Meusburger, P. (1979). Ausbildungsniveau und schulische Situation der Slowenen in Kärnten von 1848 bis 1978 [Educational achievement and schooling situation of the Slovenes in Carinthia, 1848–1978]. In P. Haimayer, P. Meusburger, & H. Penz (Eds.), *Fragen geographischer Forschung. Festschrift des Instituts für Geographie zum 60. Geburtstag von Adolf Leidlmair* (Innsbrucker Geographische Studien, Vol. 5, pp. 229–264). Innsbruck: Department of Geography, Innsbruck University.

Meusburger, P. (1980). *Beiträge zur Geographie des Bildungs- und Qualifikationswesens. Regionale und soziale Unterschiede des Ausbildungsniveaus der österreichischen Bevölkerung [Contributions to the geography of knowledge and education: Regional and social disparities of educational attainment of the Austrian population*] (Innsbrucker Geographische Studien, Vol. 7). Innsbruck: Department of Geography, Innsbruck University.

Meusburger, P. (1996). Educational achievement, language of instruction, and school system as key elements of minority research. In K. Frantz & R. A. Sauder (Eds.), *Ethnic persistence and change in Europe and America* (Veröffentlichungen der Universität Innsbruck, Vol. 213, pp. 187–222). Innsbruck: Innsbruck University.

Meusburger, P. (1998). *Bildungsgeographie. Wissen und Ausbildung in der räumlichen Dimension [Geography of education: Knowledge and education in the spatial dimension].* Heidelberg: Spektrum Akademischer Verlag.

Meusburger, P. (2001). The role of knowledge in the socio-economic transformation of Hungary in the 1990s. In P. Meusburger & H. Jöns (Eds.), *Transformations in Hungary: Essays in economy and society* (pp. 1–38). Heidelberg: Physica.

Meusburger, P. (2015a). Education, geography of. In J. D. Wright (Editor-in-chief) (Ed.), *International encyclopedia of the social & behavioral sciences* (2nd ed., Vol. 7, pp. 165–171). Oxford: Elsevier.

Meusburger, P. (2015b). Relations between knowledge and power: An overview of research questions and concepts. In P. Meusburger, D. Gregory, & L. Suarsana (Eds.), *Geographies of knowledge and power* (Knowledge and space, Vol. 7, pp. 19–74). Dordrecht: Springer. doi:10.1007/978-94-017-9960-7_2.

Miąso, J. (1991). Educational policy and educational development in the Polish territories under Austrian, Russian and German rule, 1850–1918. In J. Tomiak, K. Eriksen, A. Kazamias, & R. Okey (Eds.), *Schooling, educational policy and ethnic identity: Vol. 1. Comparative studies on governments and non-dominant ethnic groups in Europe, 1850–1940* (pp. 163–184). New York: New York University Press.

Miller, L. P. (1982). Black education. In H. E. Mitzel (Ed.), *Encyclopedia of educational research: Vol. 1* (5th ed., pp. 211–219). New York: Free Press.

Miller, J. G. (2001). Culture and the self (implications for psychological theory): Cultural concerns. In N. J. Smelser & P. B. Baltes (Eds.), *International encyclopedia of the social & behavioral sciences* (Vol. 5, pp. 3139–3143). Amsterdam: Elsevier. doi:10.1016/B0-08-043076-7/04644-1.

Mitter, W. (1991). German schools in Czechoslovakia, 1918–1938. In J. Tomiak, K. Eriksen, A. Kazamias, & R. Okey (Eds.), *Schooling, educational policy and ethnic identity: Vol. 1. Comparative studies on governments and non-dominant ethnic groups in Europe, 1850–1940* (pp. 211–233). New York: New York University Press.

Pieke, F. N. (1991). Chinese educational achievement and "folk theories of success.". *Anthropology and Education Quarterly, 22,* 162–180.

Pratte, R. (1982). Culture and education policy. In H. E. Mitzel (Ed.), *Encyclopedia of educational research: Vol. 1* (5th ed., pp. 394–400). New York: Free Press.

Robertson, S. L. (2010). "Spatializing" the sociology of education: Stand-points, entry-points, vantage points. In M. W. Apple, S. J. Ball, & L. A. Gandin (Eds.), *The Routledge international handbook of the sociology of education* (pp. 15–26). London: Routledge.

Shaw, C. R. (1929). *Delinquency areas. A study of the geographic distribution of school truants, juvenile delinquents, and adult offenders in Chicago.* Chicago: Chicago University Press.

Shweder, R. A. (2001). Culture: Contemporary views. In N. J. Smelser & P. B. Baltes (Eds.), *International encyclopedia of the social & behavioral sciences* (Vol. 5, pp. 3151–3158). Amsterdam: Elsevier. doi:10.1016/B0-08-043076-7/00840-8.

Sigler, J. (1983). *Minority rights: A comparative analysis.* Westport: Greenwood.

Smith, A. (1979). *Nationalism in the twentieth century.* Oxford: Martin Robertson.

Spencer, M. B. (1983). Children's cultural values and parental childrearing strategies. *Developmental Review, 3,* 351–370. doi:10.1016/0273-2297(83)90020-5.

Spencer, M. B., & Markstrom-Adams, C. (1990). Identity processes among racial and ethnic minority children in America. *Child Development, 61,* 290–310. doi:10.1111/1467-8624. ep5878983.

Stovall, D. (2006). Forging community in race and class: Critical race theory and the quest for social justice in education. *Race Ethnicity and Education, 9,* 243–259. doi:10.1080/13613320600807550.

Strohmayer, U. (2003). The culture of epistemology. In K. Anderson, M. Domosh, S. Pile, & N. Thrift (Eds.), *Handbook of cultural geography* (pp. 520–531). London: Sage.

Suárez-Orozco, M. M. (1991). Migration, minority status, and education: European dilemmas and responses in the 1990s. *Anthropology & Education Quarterly, 22,* 99–120. Retrieved from http://www.jstor.org/stable/3195915.

Tippeconnic, J. W., III, & Gipp, G. E. (1982). American Indian education. In H. E. Mitzel (Ed.), *Encyclopedia of educational research: Vol. 1* (5th ed., pp. 125–128). New York: Free Press.

Tomiak, J. (1991). Education of the non-dominant ethnic groups in the Polish Republic, 1918– 1939. In J. Tomiak, K. Eriksen, A. Kazamias, & R. Okey (Eds.), *Schooling, educational policy and ethnic identity: Vol. 1. Comparative studies on governments and non-dominant ethnic groups in Europe, 1850–1940* (pp. 185–209). New York: New York University Press.

Tomiak, J., & Kazamias, A. (1991). Introduction. In J. Tomiak, K. Eriksen, A. Kazamias, & R. Okey (Eds.), *Schooling, educational policy and ethnic identity: Vol. 1. Comparative studies on governments and non-dominant ethnic groups in Europe, 1850–1940* (pp. 1–11). New York: New York University Press.

Tomlinson, S. (1991). Ethnicity and educational attainment in England: An overview. *Anthropology & Education Quarterly, 22,* 121–139. Retrieved from http://www.jstor.org/stable/3195916.

Tomlinson, S. (2008). *Race and education: Policy and politics in Britain.* Maidenhead: Open University Press.

Trueba, H. T. (1989). *Raising silent voices: Educating the linguistic minorities for the 21st century.* New York: Newbury House.

UNESCO. (1953). *Progress of literacy in various countries: A preliminary statistical study of available census data since 1900* (Monographs on fundamental education, Vol. 6). Paris: UNESCO.

Välimaa, J. (2004). Nationalization, localization and globalization in Finnish higher education. *Higher Education, 48,* 27–54.

Villgrater, M. (1984). *Katakombenschule. Faschismus und Schule in Südtirol [Catacomb school: Fascism and school in South Tyrol]* (Schriftenreihe des Südtiroler Kulturinstitutes, Vol. 11). Bozen: Athesia.

Weaver, T. (1994). The culture of Latinos in the United States. In T. Weaver (Ed.), *Handbook of Hispanic culture in the United States* (Anthropology, Vol. 4, pp. 15–38). Houston: Arte Público Press.

Weinberg, M. (1977). *A chance to learn: The history of race and education in the United States.* Cambridge: Cambridge University Press.

Welsch, W. (1999). Transculturality: The puzzling form of cultures today. In M. Featherstone & S. Lash (Eds.), *Spaces of culture: City, nation, world* (pp. 194–213). London: Sage.

Whiteaker, L. H. (1990). Adult education within the slave community. In H. G. Neufeldt & L. McGee (Eds.), *Education of the African American adult: An historical overview* (Contributions in Afro-American and African studies, Vol. 134, pp. 3–10). New York: Greenwood Press.

Chapter 3
Race, Politics, and Geography in the Development of Public Schools in the Southern United States

Adam Fairclough

The Development of Public Schools in the United States

Public schools in the United States have been largely shaped by four influences: local control, separation of church and state, the comprehensive principle, and racial segregation. In contrast to the centralized, national systems of Europe, American public schools are financed and administered by 13,500 independent districts—in 1949 they numbered 83,000—under the political control of locally elected school boards (Reese 2005). Local control rests upon local funding: It was not until the 1960s that money from Washington reached K–12 education (kindergarten through high school) in any appreciable quantity, and even today federal funds account for only 8 % of public school finances (U.S. Department of Education 2011). Federal oversight is weak. Congress did not create a separate Department of Education until 1979, and it remains the smallest cabinet-level department, with only 5000 employees. There is no national curriculum, no federal standards, and no federal inspection. Continuing commitment to local control produces enormous variations in per capita spending between different school districts.

The separation of church and state in the United States, rigidly policed by the Supreme Court, has produced a clear division between public schools, which are secular in character, and religious schools, all of which are private. The state cannot subsidize religious schools, and the public schools cannot teach religion or permit religious worship. However, in the nineteenth century public schools had a distinctly Protestant character, encouraging the formation of a Roman Catholic school system. Catholic schools, which receive no public money, teach almost half of the 5.3 million children enrolled in private schools. Protestant schools, which account for 28 % of private school enrollment, are especially numerous in the South, the

A. Fairclough (✉)
Institute for History, Leiden University, Doelensteeg 16, 2311 VL Leiden, The Netherlands
e-mail: a.fairclough@hum.leidenuniv.nl

© Springer International Publishing Switzerland 2016   55
P. Meusburger et al. (eds.), *Ethnic and Cultural Dimensions of Knowledge*,
Knowledge and Space 8, DOI 10.1007/978-3-319-21900-4_3

so-called Bible Belt, where many parents deplore the secular character of the public schools (Tyack 2003).

The public schools themselves have been structured around the comprehensive principle: the idea that children of all abilities should attend the same school. American schools have lacked the hierarchical, differentiated character of many European school systems, with their distinction between vocational and academic institutions. Paradoxically, however, the egalitarianism of the comprehensive high school has, like the ban on public support for religious schools, fostered a large private school sector that accentuates class inequalities. All told, private institutions enroll about 10 % of all school-age children in the United States (Tyack 2003).

Finally, American public schools have been characterized by a large degree of racial segregation. Before 1954 separate schools for Blacks and Whites were required by law in 17 states (mostly in the South) as well as in Washington, D.C. In the North, too, Blacks attended separate schools wherever they constituted a substantial portion of the population. Although this type of de facto segregation was never as rigid as the South's *de jure* segregation, and depended upon custom rather than law, it was nevertheless deliberately fostered by White-controlled school boards and city governments. In some northern cities Blacks accepted segregated schools in return for the appointment of Black teachers in those schools. Nevertheless, racial segregation has nearly always had the effect of accentuating inequalities between Black and White public schools. To a large extent, moreover, such inequalities lay beyond the reach of government action due to Supreme Court decisions that affirm the principle of local control. School districts are independent entities, the Court has ruled, and the wealthier districts have no obligation to subsidize the poorer ones. Similarly, the Court has decided that no school district can be compelled to facilitate racial integration in a neighboring district, a ruling that makes it impossible to treat city and suburbs as a single entity (Reese 2005).

The South as a Special Case

Public schools in the South have always been worse than in the rest of the nation. Because of large Hispanic and other non-English speaking immigration, California and New York now have the lowest rates of literacy. Historically and consistently, however, the southern states have displayed the lowest levels of educational achievement. This record is partly a reflection of the South's relative poverty, but it also testifies to a weaker commitment to public education compared to other regions of the nation.

The reformers who built the public school systems of the North were imbued with the democratic, egalitarian ethos of the early nineteenth century, the age of the common man. Public education was considered the great equalizer, common schools providing opportunity for all. In addition, the commercially minded and industrializing North regarded an educated population as a vital source of economic growth and prosperity as well as a means of mitigating ethnic and class antagonisms (Cremin 1980; Kaestle 1983; Katz 1968).

The South, by contrast, showed little interest in developing public schools before the Civil War (1861–1865). For one thing, the region was overwhelmingly rural, and its population lived on scattered farms rather than in towns and villages. Schools were difficult to organize, and many farmers questioned the relevance of book learning to agricultural pursuits, which required the labor of the entire family, especially sons, for much of the year. Moreover, the wealthy slave owners who dominated southern society saw little benefit in expanding government services; wedded to small government and low taxes, they did not wish to spend money on public schools for non-slaveholding Whites, especially when they sent their own children to private schools. Although New Orleans, Baltimore, and some smaller cities had well-developed school systems for Whites before the Civil War, in the vast rural hinterland public schools were few and far between (Reese 2005).

Whites attempted to keep the South's four million slaves—more than a third of the region's population in 1860—illiterate. Their reasoning was simple: Literate slaves would be able to think for themselves, absorb political ideas, and become difficult to control. By denying slaves knowledge of and access to the outside world, masters discouraged slave initiative and reinforced their isolation within the plantation. The Nat Turner slave rebellion of 1832, which coincided with the start of an all-out campaign led by William Lloyd Garrison to abolish slavery, encouraged southern states to enact laws making it a crime to teach slaves to read and write (Williams 2005). Whites also banned abolitionist literature from circulating in the South.

The laws against slave literacy were not always rigidly enforced, and it was not uncommon for masters to educate their favorite slaves (who were often the mulatto children they had fathered through illicit relationships with slave women). Masters also recognized that key slaves with management skills could assist them in running the plantation. In some cases slaves surreptitiously taught themselves to read, often with the inadvertent assistance of the master's young children. Nevertheless, 90–95 % of the former slaves were completely illiterate upon their emancipation in 1865 (Cornelius 1991; Williams 2005).

Literacy among the South's approximately 261,000 free people of color was far higher. Although subject to extensive discrimination, the region's free Blacks enjoyed a good deal of independence, and some acquired considerable wealth (including the ownership of slaves). In New Orleans, Mobile, and elsewhere free Blacks attended private schools. Most slave states banned schools for free Blacks, but underground or secret schools flourished, sometimes with the silent acquiescence of the White authorities. In addition, free Blacks could buy books and newspapers, form clubs and societies, and educate their children within the home (Fairclough 2007). After the Civil War, these free-born Blacks furnished many of the first teachers for the ex-slaves, as well as the first Black ministers and politicians.

Consequences of the Civil War

After its victory in the Civil War, the North attempted to remake southern society along more democratic and egalitarian lines. In 1867 Congress enfranchised the freedmen (former slaves), a move that enabled the northern-based Republican party to gain temporary political control of the ex-Confederate states. Writing new state constitutions based upon equal rights, the Republicans created public school systems that, although segregated by race, included Blacks. Indeed, the former slaves provided the principal demand and constituency for public schools. Having seen Whites deny them education for so long—a fact that in their eyes demonstrated the importance of literacy—Blacks of all ages crowded the primitive classrooms. For many ex-slaves, a desire to read the Bible drove their enthusiasm (Anderson 1988; Butchart 2010).

However, the legacy of slavery and the political consequences of the Civil War gravely handicapped the South's public schools. Devastated by invading armies and afflicted by a succession of postwar crop failures, the South could ill afford the level of taxation that a fully developed public school system would have required. Moreover, because ex-slaves supplied the main demand for public schools, Whites resented being taxed at all to fund these schools. In the eyes of defeated ex-Confederates the public schools were the progeny of Yankee conquerors, their principal function to train an army of Black Republican voters. They hated the northern teachers who encouraged Black aspirations to equality. Dependent upon Black labor to cultivate their cotton, sugar, rice, and tobacco fields, White landlords feared that schooling would render Blacks unfit for manual labor. Education made Blacks unhappy with their lot and less susceptible to White authority. "Educate a nigger," the saying went, "and you spoil a good field hand." In the years after the Civil War, White terrorists, loosely organized into groups like the Ku Klux Klan, burned down Black schools and threatened, whipped, and even murdered their teachers (Butchart 2010). "I slept with a pistol under my head, an Enfield rifle at my side, and a knife at the door" (Marrs 1885), recalled one Black teacher.

In time, White landlords and employers decided that suppressing Black schools was both impractical and undesirable. Blacks were more likely to quit a plantation if they lacked access to a school; the presence of a school made for a more stable and contented workforce. However, most Whites wished to restrict Black education to basic literacy, and many continued to believe that schooling for Blacks damaged the interests of Whites. When the Democratic Party regained power in the South after the collapse of Reconstruction (1865–1877), it slashed public spending on public education and made racial segregation mandatory.

The disfranchisement of Black voters between 1890 and 1915 opened up vast disparities in per capita spending between Blacks and Whites (Harlan 1958). Moreover, a new breed of White politician, sometimes labeled *the demagogues*, exploited Blacks' vulnerability by stoking the racist sentiments of the all-White electorate. They branded Black education a failure and sought to degrade the quality of Black schools. Augusta, Georgia, for example, abolished its Black public high school. New Orleans eliminated grades 6–8 in the Black schools. States stripped Black colleges of degree-granting powers and forbade them to use the title *university*. Some Black colleges received so little public funding that they had no books in their libraries. Contending that schooling should train Blacks to be farmers, laborers, and domestic servants, Whites tried to expunge Latin (which symbolized the uselessness of classical or literary education for Blacks) from the curriculum. The superintendent of schools in Birmingham, Alabama, J. H. Phillips—a man widely regarded as an enlightened administrator—contended that the "Negro brain" stopped developing in adolescence, making it pointless for Blacks to remain in school after the age of 14 (Anderson 1988; Fairclough 2007).

The worst Black schools could be found in the majority-Black rural counties of the Deep South, where Black families worked as sharecroppers for White landlords. White political control meant that per capita spending on White children exceeded spending on Black children by a factor of ten-to-one or more (Harlan 1958). Local school boards contributed little to the education of Blacks beyond the payment of a teacher's salary for 3 or 4 months in the year. The teachers themselves, however, were often the products of the same kind of primitive one-room, single-teacher elementary school. Many had never attended a secondary school, let alone a college or university. Rural schools suffered the additional handicap of a constantly shifting population. In the cotton South, Black sharecroppers often moved every year in their search for a better landlord, an annual migration that continually disrupted the already meager and intermittent education of their children. Even when sharecroppers stayed put, the demands of agriculture took priority over education, making school attendance erratic (Johnson 1941).

In the South's cities the schools were invariably much better; here Whites had no need for the labor of Black children, and they recognized that living and working in the city required literacy. Nevertheless, White schools always received the lion's share of public funding; and whereas Whites had access to high schools, Blacks, outside Texas and Virginia, did not (Anderson 1988). Hence, Blacks had to look outside the public sector for quality education. Of the 24,000 African Americans who attended secondary schools in 1916—a mere 2 % of the Black school-age population—half were enrolled in private schools. The private colleges and universities funded by northern churches provided the only real higher education for Blacks. The small minority of well-trained teachers came from institutions like Atlanta University, Yankee implants that White southerners tolerated but disliked (Anderson 1988; Fairclough 2007).

White hostility to the education of Blacks damaged public education in general. Whites were so reluctant to see taxes spent on Blacks schools that they often rejected local taxation—school taxes had to be voted upon in local referenda—that would have improved their own schools. Public schools for White children were far worse than their northern counterparts, and in many rural areas were scarcely better than the Black schools. In 1900 only half of the South's children attended school regularly, and the South had the highest illiteracy rate in the nation.

Philanthropic Foundations and Southern Schools

The dire condition of the South's public schools elicited little help from the federal government. Although some Blacks advocated a Prussian model of education—attributing Prussia's military success to the excellence of its public school system—control of public education remained with the states and localities. The Republican Party, which first came to power in 1861, strengthened the national government but did not create a centralized state along European lines. The Freedmen's Bureau, created by Congress in 1865 to assist Blacks in the transition from slavery to freedom, was instrumental in the creation of freedmen's schools in the South, but the Bureau was a temporary expedient that ceased operations in 1872 (Cimbala 1997; McFeely 1968). The federal Office of Education, created in 1868, did little more than collect statistics.

The only substantial federal initiative in the field of education before the 1950s consisted of grants of land to the states in order to subsidize the creation of agricultural colleges. The Second Morrill Act of 1890 required that these subsidies be "equitably divided" between Whites and Blacks, leading southern states to create separate Black colleges (Wright 1981). These land-grant colleges (which continue today as state universities) often provided the only publicly funded secondary and higher education available to Blacks in the South. Despite their emphasis on agriculture, they came to function mainly as teacher-training institutions and as such slowly raised educational standards within the Black teaching force. When it came to public schools, however, the federal government contributed virtually nothing, leaving Blacks in the South at the mercy of White politicians and administrators who continued to systematically neglect Black schools.

This vacuum at the center provided an opening for philanthropic foundations like the General Education Board (GEB), founded in 1902. The foundations were the products of northern industrial capitalism. The GEB, for example, was funded by Standard Oil magnate John D. Rockefeller; the Carnegie Foundation, by steel magnate Andrew Carnegie. Consistent with their origins, these new foundations approached philanthropy from a businessman's perspective, viewing the South's ignorant and politically volatile population as a drag on economic development. They saw that the South was lagging behind the North in terms of industrialization and that the agricultural sector, the traditional mainstay of the South's economy, was in decline. In order to adapt and catch up, the foundations believed, the South needed a better-educated and more productive workforce. Hence, the GEB set out to improve the South's public schools. It gave money directly to state and local governments and, by making its grants contingent upon matching public funds, stimulated higher taxation for public education (Anderson and Moss 1999; Harlan 1958). The GEB, along with the Carnegie Foundation, also developed national rating scales to bring some kind of uniformity to the business of comparing schools across so many different systems.

Booker T. Washington (1856–1915), who headed a Black school in Alabama, Tuskegee Institute, acted as the principal intermediary between the foundations and

Black southerners. A skilled diplomat and prodigious fundraiser as well as an inspirational speaker, Washington attempted to increase White support for Black education during a time of intense racism, North and South. By urging Blacks to focus on self-improvement rather than politics and by accepting racial segregation, he tried to soften the hostility of southern Whites to Black education. Championing basic vocational training for Blacks (industrial education) and preaching the virtues of agriculture, he presented the Negro as a reliable worker rather than a threatening competitor. Critics charged that Washington was capitulating to racism and accepting second-class citizenship. Defenders argued that he was simply making the best of a bad situation. Historians later discovered that, behind the scenes, Washington fought the disfranchisement of Black voters and was up to his neck in politics. Controversial to this day, he secured philanthropic support for Black schools at a time when leading politicians in the South were declaring Black education a "failure" (Harlan 1972, 1983; Norrell 2009).

Improvements to Black schools, nevertheless, were slow in coming. In order to sell its philanthropy to the southern states, the GEB not only accepted White supremacy but also made the improvement of *White* public schools its top priority. The philanthropic foundations had no intention of working against racial segregation or challenging Blacks' exclusion from the franchise, believing that these things, however morally unjust, diminished conflict between the races and promoted political stability (Stanfield 1985). After World War I philanthropic giving to Black schools increased: The foundations improved teacher training, gave money to Black colleges, and subsidized the construction of 5357 schoolhouses between 1917 and 1932, one-fifth of all Black schools in the South (Anderson 1988; Anderson and Moss 1999; Hoffschwelle 2006). Nevertheless, the inequalities between Black and White public education remained as great as ever. Improvements to Black schools always lagged two or three decades behind improvements to White schools (Bond 1934/1966).

Segregated Schools as Anchors of the Black Community

Blacks in the South became intensely attached to their schools, whatever their inadequacies. The term *public school*, in fact, obscures the enormous investment that ordinary Blacks made in them. School boards typically provided little more than a teacher's salary: Everything else, including the schoolhouse, Blacks furnished themselves. Many Black schools, especially the high schools, had originated as private initiatives by individual teachers—men and women who devoted selfless lives to their development—and depended upon the surrounding community to sustain their work. Blacks donated labor and building materials, provided board and lodging for teachers, and engaged in continual fundraising for school improvements (Hoffschwelle 2006). Along with churches, schools were social anchors of the Black community; indeed, many schools were held in church buildings, and in the early decades of emancipation the teacher was often the preacher.

Segregation gave teachers prestige within the Black community. Black teachers infused their work with a missionary spirit to uplift the masses; they viewed themselves as the "talented tenth" that could provide inspiration to the race (Fairclough 2000; Fultz 1995). Like preachers, they commanded respect as leaders and role models. With so many other career paths closed to them, Blacks who might otherwise become doctors, lawyers, businessmen, or politicians went into teaching. Although teachers were the backbone of the Black middle class, they were nevertheless closely integrated into their communities. Through the practice of home visiting, they maintained close contact with their pupils' families. Many Black parents, perhaps most, could not imagine White teachers functioning in the same way and doubted that even the best-intentioned Whites could completely overcome their racial prejudice; such parents tended to prefer Black teachers (Rabinowitz 1974). Inside the classroom, teachers maintained rigid discipline, making frequent resort to corporal punishment (Fairclough 2007).

Hence, segregated schools made a substantial contribution to the development of Black culture in the South. Although segregated at the insistence of Whites, Blacks had no desire for integration per se. Their all-Black schools provided a safe haven within which they were shielded from White condescension and White bullying. It was an environment that fostered a clear sense of racial identity and encouraged, to some extent, racial solidarity. This is not to say that Black racial identity did not contain ambiguities and tensions. Dark-skinned children often complained that light-skinned teachers placed them at a disadvantage; light-skinned children complained of being bullied by darker-skinned classmates (Fairclough 2007). Nevertheless, through the teaching of Negro history and the daily singing of *Lift Every Voice and Sing*—the Negro national anthem, written in 1900—Black teachers encouraged racial pride. Although rarely in a position to engage in political action or open protest, teachers helped Black children resist the dehumanizing ideology of White supremacy by equipping them with knowledge and inculcating self-respect (Fairclough 2000; Fultz 1995).

In the 1940s the gap between Black and White public education began slowly to narrow. The change came partly as a consequence of developments in the southern economy. The Great Depression and the collapse of world trade dealt a near-mortal blow to the South's plantation system. President Franklin Roosevelt's New Deal paid farmers to grow less, increasing the surplus of Black labor. The introduction of the mechanical cotton picker in the 1940s further diminished White landlords' labor needs. Black sharecroppers and agricultural laborers found themselves without land and without work. During the 1930s they had had nowhere to go, but World War II sparked an economic boom that not only ended mass unemployment but also created employment shortages that could not be met by White labor alone. Black sharecroppers quit farming and moved to the cities; huge numbers migrated to the North. For the first time, Blacks entered the industrial workforce in substantial numbers. The proportion of Blacks living in rural areas declined from 51 % in 1940 to 27 % in 1960; the proportion living on farms diminished from 35 % to 2 % (Mandle 1978). With no pressing need to keep Blacks tied to the land, White opposition to secondary and higher education for Blacks diminished.

By 1945 the South's political leaders recognized that unless they took steps to reduce the most glaring disparities between Black and White schools, the Supreme Court might declare segregation to be unconstitutional. In 1896 the Court had upheld racial segregation by declaring that separate but equal facilities did not violate the Fourteenth Amendment, even though that measure, enacted after the Civil War, provided for equal citizenship and banned state-sponsored racial discrimination. In the late 1930s, however, Roosevelt began to mold a liberal Supreme Court that was increasingly sensitive to issues of civil liberties and racial discrimination. The Court would no longer turn a blind eye to the way that southern Whites practiced segregation as a form of systematic inequality (McMahon 2004). In 1938 the Court ruled that if a state did not provide genuinely separate but equal facilities for graduate education (in this case a law school), then it must admit Black students to the White university. Although this decision did not strike down segregation per se, it did serve notice to segregating states that the federal courts would no longer permit blatant discrimination (Tushnet 1987).

The Supreme Court also handed a powerful weapon to the nation's leading civil rights organization, the National Association for the Advancement of Colored People (NAACP). Documenting the inequalities between Black and White public education, the NAACP sued southern officials in federal court for violating the Fourteenth Amendment. Unsure that the Supreme Court was prepared to rule against the principle of racial segregation, the NAACP did not ask for integration but instead demanded the equalization of teachers' salaries and educational facilities. Nevertheless, the threat of integration proved persuasive. Desperate to preserve racial segregation, and fearful that the Supreme Court would repudiate the "separate but equal" doctrine if directly presented with the question, White politicians approved crash programs to upgrade Black schools and colleges (Bartley 1995). But the spending was too little, and it came too late. In 1950, for example, Mississippi spent $122 a year on each White child but only $22 on each Black child (Ashmore 1954; Bolton 2000).

The Integration Era, 1954–1997

The NAACP had hesitated to attack segregation head-on only because it could not be certain that the Supreme Court would decide in its favor. But after 1941 the political climate changed with such rapidity that a Supreme Court ruling against segregated schools became, if not a certainty, a probability. World War II discredited racism, and in 1948 the United Nations adopted the Universal Declaration of Human Rights, which denounced distinctions based upon race and color. As the Cold War unfolded and America posed as the champion of democracy and leader of the Free World, the South's system of racial segregation became politically embarrassing. In addition, the Democratic Party was having to accommodate a substantial bloc of northern Black voters, its size daily swelled by new migrants from the South. In his successful election campaign of 1948, President Truman endorsed the NAACP's civil rights agenda: Winning Black support was more important than losing five southern states. In the same year he initiated the integration of America's armed forces (Dudziak 2000). In 1950, confident that the President and the Supreme Court were sympathetic allies, the NAACP decided to attack the principle of racially seg-regated education (Tushnet 1987, 1994). Four years later, in the case of *Brown v. Board of Education*, the Supreme Court ruled that the South's system of segregated public schools violated the Constitution. The decision foreshadowed the abolition of all state-supported segregation.

Whites in the South did not accept the *Brown* decision with good grace. On the contrary, they staged a long delaying action that produced memorable confronta-tions between state and federal power in Little Rock, Arkansas; New Orleans, Louisiana; Oxford, Mississippi; and elsewhere. But the strength of the civil rights movement led by Martin Luther King and others, and the consequent buildup of federal pressure from the courts, the President, and the Congress, proved irresist-ible. By 1970 the South's public schools were substantially integrated—far more so than in the North—and Black children were more likely to have a White teacher than a Black one (Patterson 2001).

Many Blacks, however, came to regard integration as a dubious gain. Integration brought about the abolition of a wastefully expensive dual school system and allowed for efficiency savings. In implementing integration, however, White politi-cians and administrators usually made savings by closing Black schools, demoting Black principals, and dismissing Black teachers (Fultz 2004). In addition, the notion that integration meant, ideally, minority status within a majority-White school bred a sense of alienation. Black children found it difficult to feel at home in integrated schools where the principal and most of the teachers were White. The traditions of the Blacks schools disappeared; those of the integrated (formerly White) schools had little appeal. It was clear, moreover, that Whites bowed to integration out of necessity, not because they accepted integration as an ideal. Classroom integration was much easier to achieve than social integration (Fairclough 2004). "Physical desegregation occurred," complained one Black teacher, "without any spiritual or emotional desegregation" (Robinson 1978). Many Whites sought to reproduce all-White classrooms by moving to White suburbs or by sending their children to over-whelmingly White private schools (Kruse 2005). In some rural counties, *all* the White children enrolled in private schools—dubbed *segregation academies*—after the federal courts forced the public schools to integrate (Nevin and Bills 1976).

By the 1990s "White flight" had made the South's big-city public school systems overwhelmingly Black (Bayor 1996; Biles 1986; Kellar 1999; Pratt 1991). In New Orleans, for example, decades of litigation led to a Pyrrhic victory: By 1993 Whites accounted for a mere 8 % of total public school enrollment (Fairclough 2008). Legal segregation had given way to integration only to evolve into northern-style de facto segregation. With the Supreme Court limiting the scope of integration—no busing between cities and suburbs, no equalization of spending between rich and poor school districts—the onward march of resegregation appeared unstoppable. By 2007, when the Court foreclosed the possibility of even voluntary action to promote integration, litigation under *Brown* had virtually ended (Wolters 2008).

The current nostalgia for the Black schools of the pre-*Brown* era reflects profound disappointment that the predicted academic and social benefits of integration were not forthcoming. Some even argue that *Brown* was a mistake and that Blacks would have been better off pressing for equal schools, which Whites might have accepted, rather than integrated schools, which Whites fought tooth and nail. The Supreme Court's contention in *Brown* that Black schools bred a sense of inferiority that hampered children's ability to learn was based upon dubious psychology (Wolters 2008). It was also contradicted by facts. The existence of high-performing segregated schools—and in some numbers—showed that it was indeed possible to motivate children in an all-Black environment (Siddle Walker 1996; Sowell 1986). Moreover, Black academic achievement greatly improved between 1950 and 1970, the period *before* wholesale integration (Fairclough 2007).

The nostalgia is nevertheless misplaced. In the same way that some citizens of Russia and the former eastern bloc countries yearn for the certitudes of communism, some Blacks in the South hark back to the era of segregation as a time of community cohesion and racial solidarity. But whatever those Black schools achieved, they had done so in spite of segregation, not because of it. They had maintained discipline, moreover, through the kind of brutal corporal punishment that is not tolerated today. And if teachers had acted as community leaders, they retreated into silence after the *Brown* decision. Dependent upon Whites for their employment, public school teachers, especially high school principals, sometimes acted as the eyes and ears of the White community inside the Black community. It was true, also, that many Black teachers, especially the worst-educated, feared being thrown into competition with White teachers. As a professional group, Black teachers had a direct stake in the outcome of integration. Some felt threatened by it and resisted it; others saw it as an opportunity and embraced it (Fairclough 2007; Reese 2005).

All social change involves profit and loss, and when a bad system breaks down the new one rarely lives up to expectations. Many mistakes were made during the drawn-out, conflict-ridden process of school integration. Some were the inevitable by-products of America's cumbersome and fragmented political system, which makes all major political change a protracted and untidy business. Nevertheless, the NAACP was correct in calculating that separate but equal was a contradiction in terms. The Supreme Court may have erred in its use of psychological evidence, but the basic argument in *Brown* was correct: Segregated schools *did* imply Black inferiority. Whites had designed the system on the assumption that association with Blacks on terms of equality was degrading. It buttressed a legal definition of race that pivoted upon a criminalization of sexual relations between Blacks and Whites, a ban that was not lifted until the Supreme Court ruled it unconstitutional in 1967 (Wallenstein 2002). The entire system facilitated White privilege and White supremacy. With their monopoly on political power, as well as their monopoly on all jobs from bank president down to bus driver, Whites had no incentive to allow Blacks equal educational advantages. Even the postwar equalization programs were intended to perpetuate these inequalities. Moreover, these programs never came close to closing the gap: White schools and colleges always received more funding

(Bolton 2000). Suffering no political penalty for neglecting Black schools, White administrators took less interest in them and held them to a lower standard.

Sadly, the assumption that political control by Blacks would lead to increased devotion to raising standards in Black schools has not always been justified. In New Orleans, for example, federal prosecutors convicted two dozen school board officials, both elected and appointed, for taking bribes and stealing money. In 2003 the state obtained the authority to take over the city's failing schools. The mushrooming of charter schools since Hurricane Katrina reflects utter frustration with the traditional pattern of local political control of public schools (Fairclough 2008). Hailed by some as a way of liberating schools from the dead hand of school board bureaucracies and teachers' unions, criticized by others as a devious form of privatization, charter schools have yet to prove their worth.

Critics of integration emphasize a persisting achievement gap: The average 17-year-old Black child is still about 4 years behind the average White child in reading and math. Yet a victory can be incomplete and remain a victory. Integration actually *reduced* the achievement gap by ushering in a period of spectacular academic progress by Black students. By 1990, when the first cohort of Black children had graduated from integrated schools, two thirds of Blacks aged 25 and over had completed high school or college. That was double the 1970 figure. White achievement had also risen, but not as fast. Similarly, the gap between White and Black test scores narrowed by 20 points, the equivalent of two age grades. Finally, despite the advance of resegregation, public schools in the South remain substantially more integrated than those in the North. The most segregated school systems today are to be found in California, New York, Michigan, and Illinois (Fairclough 2007).

References

Anderson, J. D. (1988). *The education of blacks in the South, 1860–1935*. Chapel Hill: University of North Carolina Press.

Anderson, E., & Moss, A. A., Jr. (1999). *Dangerous donations: Northern philanthropy and southern black education, 1902–1930*. Columbia: University of Missouri Press.

Ashmore, H. S. (1954). *The Negro and the schools*. Chapel Hill: University of North Carolina Press.

Bartley, N. V. (1995). *The new South, 1945–1980*. Baton Rouge: Louisiana State University Press.

Bayor, R. H. (1996). *Race and the shaping of twentieth-century Atlanta*. Chapel Hill: University of North Carolina Press.

Biles, R. (1986). A bittersweet victory: Public school desegregation in Memphis. *Journal of Negro Education, 55*, 470–483.

Bolton, C. C. (2000). Mississippi's school equalization program, 1945–1954: A last gasp to try to maintain a segregated educational system. *Journal of Southern History, 66*, 781–814.

Bond, H. M. (1966). *The education of the Negro in the American social order*. New York: Octagon Books. (Original work published 1934).

Butchart, R. E. (2010). *Schooling the freed people: Teaching, learning, and the struggle for black freedom, 1861–1876*. Chapel Hill: University of North Carolina Press.

Cimbala, P. (1997). *Under the guardianship of the nation: The Freedmen's Bureau and the reconstruction of Georgia, 1865–1870*. Athens: University of Georgia Press.

Cornelius, J. D. (1991). *"When I can read my title clear": Literacy, slavery, and religion in the antebellum South*. Columbia: University of South Carolina Press.

Cremin, L. A. (1980). *American education: The national experience, 1783–1876*. New York: Harper & Row.

Dudziak, M. C. (2000). *Cold War civil rights: Race and the image of American democracy*. Princeton: Princeton University Press.

Fairclough, A. (2000). "Being in the field of education and also being a Negro … seems … tragic": Black teachers in the Jim Crow South. *Journal of American History, 87*, 65–91.

Fairclough, A. (2004). The costs of *Brown*: Black teachers and school integration. *Journal of American History, 91*, 43–55.

Fairclough, A. (2007). *A class of their own: Black teachers in the degregated South*. Cambridge, MA: Belknap Press.

Fairclough, A. (2008). *Race & democracy: The civil rights struggle in Louisiana, 1915–1972* (2nd ed.). Athens: University of Georgia Press.

Fultz, M. (1995). African American teachers in the South, 1890–1940: Powerlessness and the ironies of expectations and protest. *History of Education Quarterly, 35*, 401–422.

Fultz, M. (2004). The displacement of black educators post-*Brown*: An overview and analysis. *History of Education Quarterly, 44*, 11–45.

Harlan, L. R. (1958). *Separate and unequal: Public school campaigns and racism in the southern seaboard states, 1901–1915*. Chapel Hill: University of North Carolina Press.

Harlan, L. R. (1972). *Booker T. Washington: The making of a black leader*. Oxford: Oxford University Press.

Harlan, L. R. (1983). *Booker T. Washington: The wizard of Tuskegee, 1901–1915.*. Oxford: Oxford University Press.

Hoffschwelle, M. S. (2006). *The Rosenwald schools of the American South*. Gainesville: University Press of Florida.

Johnson, C. S. (1941). *Growing up in the black belt: Negro youth in the rural South*. Washington, DC: American Council on Education.

Kaestle, C. F. (1983). *Pillars of the republic: Common schools and American society, 1780–1860*. New York: Hill & Wang.

Katz, M. B. (1968). *The irony of early school reform: Educational innovation in mid-nineteenth century Massachusetts*. Cambridge, MA: Harvard University Press.

Kellar, W. H. (1999). *Make haste slowly: Moderates, conservatives, and school desegregation in Houston*. College Station: Texas A&M University Press.

Kruse, K. M. (2005). *White flight: Atlanta and the making of modern conservatism*. Princeton: Princeton University Press.

Mandle, J. R. (1978). *The roots of black poverty: The southern plantation economy after the Civil War*. Durham: Duke University Press.

Marrs, E. P. (1885). *Life and history of the Rev. Elijah P. Marrs: First Pastor of Beargrass Baptist*. Louisville: Bradley and Gilbert. Retrieved from http://docsouth.unc.edu/neh/marrs/menu.html

McFeely, W. S. (1968). *Yankee stepfather: General O. O. Howard and the freedmen*. New Haven: Yale University Press.

McMahon, K. J. (2004). *Reconsidering Roosevelt on race: How the presidency paved the road to Brown*. Chicago: University of Chicago Press.

Nevin, D., & Bills, R. E. (1976). *The schools that fear built: Segregationist academies in the South*. Washington, DC: Acropolis Books.

Norrell, R. J. (2009). *Up from history: The life of Booker T. Washington*. Cambridge, MA: Belknap Press of Harvard University Press.

Patterson, J. T. (2001). *Brown v. Board of Education: A civil rights milestone and its troubled legacy*. Oxford: Oxford University Press.

Pratt, R. A. (1991). A promise unfulfilled: School desegregation in Richmond, Virginia, 1956–1986. *Virginia Magazine of History and Biography, 99*, 415–448.

Rabinowitz, H. N. (1974). Half a loaf: The shift from white to black teachers in the Negro schools of the urban South, 1865–1890. *Journal of Southern History, 40*, 565–594.

Reese, W. J. (2005). *America's public schools: From the common school to "No Child Left Behind."*. Baltimore: Johns Hopkins University Press.

Robinson, D. R. (1978). *The bell rings at four: A black teacher's chronicle of change*. Austin: Madrona Press.

Second Morrill Act of 1890, ch. 841, 26 Stat. 417, 7 U.S.C. 322 et seq. Retrieved from http://www. csrees.usda.gov/about/offices/legis/secondmorrill.html

Siddle Walker, V. (1996). *Their highest potential: An African American school community in the segregated South*. Chapel Hill: University of North Carolina Press.

Sowell, T. (1986). *Education: Assumptions versus history: Collected papers*. Stanford: Hoover Institution Press.

Stanfield, J. H. (1985). *Philanthropy and Jim Crow in American social science*. Westport: Greenwood Press.

Tushnet, M. V. (1987). *The NAACP's legal strategy against segregated education, 1925–1950*. Chapel Hill: University of North Carolina Press.

Tushnet, M. V. (1994). *Making civil rights law: Thurgood Marshall and the Supreme Court, 1936–1961*. New York: Oxford University Press.

Tyack, D. (2003). *Seeking common ground: Public schools in a diverse society*. Cambridge, MA: Harvard University Press.

U.S. Department of Education. (2011). 10 facts about K–12 education funding. Retrieved 24 Mar 2011, from http://www2.ed.gov/about/overview/fed/10facts/index.html

Wallenstein, P. (2002). *Tell the court I love my wife: Race, marriage, and law: An American history*. New York: Palgrave Macmillan.

Williams, H. A. (2005). *Self-taught: African American education in slavery and freedom*. Chapel Hill: University of North Carolina Press.

Wolters, R. (2008). *Race and education, 1954–2007*. Columbia: University of Missouri Press.

Wright, C. W. (1981). *A history of the black land-grant colleges, 1890–1916*. Doctoral dissertation. Retrieved from ProQuest, UMI Dissertations Publishing (Order No. 8125432).

Chapter 4
Spatial Traditions of Knowledge and Education: Ethnic Groups in the United States Reconsidered

Werner Gamerith

The Ethnic Quandary

America has constantly struggled to develop a schooling system intended to undergird the political message of egalitarianism framed from the very beginning of the United States as an independent nation. The claim of abstaining from feudalism and monarchic traditions as they had been common in many parts of Europe called for corresponding schools. Public education in the United States thus came to be seen as an instrument for implanting political attitudes and ideological concepts into American society. Yet the plan has always remained different from reality, which soon became riddled with racial and ethnic tensions transforming public (and private) education into a battlefield of ethnic strife and conflict. Knowledge and access to it are organized and severely constricted by ethnic factors. There is practically no social category that better helps explain and understand differences in educational achievement than racial and ethnic ascription and identity (Fairclough 2011; Gamerith 2005b).

There seems to be hardly any aspect in American daily routines that does not have something to do with racial and ethnic relations. America's deeply stratified society is reflected by a broad range of attributes, not the least of which is anchored in race and ethnicity. The long-standing tradition of egalitarianism going back to the founding fathers has been accompanied by flaws from the outset. These shortcomings, too, have engendered specific traditions of social inequality and tremendous gaps between several racial and ethnic groups. America's vision for well-being has come true only through the subordination of some social groups (Pincus and Ehrlich 1994).

W. Gamerith (✉)
Department of Geography, University of Passau, Innstr. 40, 94032 Passau, Germany
e-mail: werner.gamerith@uni-passau.de

© Springer International Publishing Switzerland 2016
P. Meusburger et al. (eds.), *Ethnic and Cultural Dimensions of Knowledge*,
Knowledge and Space 8, DOI 10.1007/978-3-319-21900-4_4

Easy earmarking to produce and sustain different social strata has followed several lines, the first being color. Human labor, particularly in the American South, strongly became entwined with concepts of skin color and racial lineage (Cravins 1997; Moore 2010). Initially, indentured servants who had just arrived from Europe provided cheap labor for a colonial society that relied heavily on such poorly paid work. Instead of paying the transatlantic passage, the indentured laborer would succumb to the obligation of menial work for a certain time. This relationship had not yet been defined in matters of race and ethnic origin, but it took that direction just when the first "cargoes" of Black Caribbean labor had been discharged at the seaports along the East Coast in the early decades of the seventeenth century (Finzsch et al. 1999). Labor relations gradually developed into the institution of slavery, which not only widened the Black–White gap but also entrenched the further disfranchisement of African Americans. One need not know of the reports of social unrest and urban rioting from August through November 2014 even in medium-sized cities like Ferguson, Missouri,[1] to realize that color lines in the United States have not been completely swept away yet. Neither has the Civil Rights Movement been able to make the color lines entirely invisible (Romano and Raiford 2006). Even if they were to disappear completely, the memorial landscapes created in the wake of the Civil Rights Movement would still linger (Dwyer 2002; Dwyer and Alderman 2008).

Another strategy to maintain and legitimate social cleavages adopted ethnic criteria. Different ethnic elements within the same "racial" background defined by color (e.g., "White") began to be viewed as constituents of various identities separated from each other by a subtle web of cultural, religious, linguistic, or political lines. European immigrants to America not only identified themselves as Whites (the group with which the American-born population undoubtedly associated them) but often developed strong feelings for the national and ethnic heritage of their country of origin (Alba 1990). Sicilians or Neapolitans, when stuck in the ethnic ghettoes of New York City or Boston, realized the Italian flavor to which they were beginning to be linked by the American-born elites (Alba 1985, 1996; Cavaioli 1992). Settlers from Prussia, Hesse, and the Palatinate felt the repercussions of World War I on the European scene and were quickly confronted with a new kind of German identity (Schlemper 2007). Even in America, they could not evade the negative stereotypes that came with Germany's entry into the war. They witnessed the emergence of new ethnic and cultural lines ensuing from nationalist tensions in Europe and elsewhere. Nation-building in the immigrant enclave was thus a process as sudden and unexpected as it was in the emerging nation-states back home (Bayor 1988). Given the cultural lines established within White-European America, it took but a small step to imbue ethnic traits with distinct values and causal explanations. Italians (or Italian-Americans if they were granted the privilege of trying to climb up the ladders of social mobility and integrate with the American mainstream) were

---

[1] Most of the "classic" urban rioting in the 1960s, but also in the 1990s, took place in the ghettoes of large cities such as Detroit, New York, and Los Angeles. Ferguson has destroyed the illusion that suburban communities will never encounter ethnic unrest.

not thought of as excelling in agriculture but rather as performing the low-paying jobs in the urban economy. The same opinion was held of the Irish, except that their proficiency in the English language helped them land jobs in the bureaucracy much more easily than their Italian counterparts could. Both groups' rootedness in Catholicism, especially that of the Irish, made them suspect to the general public. Ethnic labeling quickly keyed into these kinds of cultural or religious traits. Never in American history has ethnicity among Whites figured more prominently than during this heyday of immigration from the 1880s to the 1910s. With it came a host of cultural stereotypes and prejudice. Although ethnicity did not become an objective social category because of these clichés, it acquired social importance and power—to such an extent that official and semiofficial rhetoric tried to eliminate these identities and let them evaporate in the famous (and infamous) "melting pot." The cultural lines drawn between different European ancestry ethnicities were gradually eroded by assimilation. In its most refined form, that is to say acculturation, it became the standard notion of American politics up until the interwar years.

Aside from a color line that had served as a rather reliable means of demarcation between slaves and nonslaves, and from culture lines that had engendered a variety of European ancestral ethnicities, there was arguably also a political line along which European and African Americans were distinguished from Native Americans. Until World War II Native American people were kept strictly segregated from all other Americans (Frantz 1993, 1994, 1999). Stringent political measures assured the establishment of reservations, the most obvious of any politically and ethnically based territorial assignment. This policy was turned completely upside down at the end of World War II, when assimilation of Native America into the U.S. mainstream was declared the explicit target. In the early 1970s this policy was changed again with a declaration of self-determination and the establishment of autonomous administrative bodies inside the recreated reservations. Along with the cultural revival that Native Americans embarked on during the 1980s, political demarcation has been one, if not *the*, pillar of Native American ethnicity to this day.

There is still an ethnic story to tell in the United States, although critics have argued that the ethnic category has given way to a more subjective one: self-identity among Americans. Ethnicity, they claim, has acquired something of a private feeling embedded in nostalgia, particularly for the descendants of immigrants who came from Europe a century or so ago. Some commentators today think that globalization and transnationalism have undermined the value of ethnic labeling (e.g., Glassman 2010; Rex 1997). They stress the flexibility and the various mixed categories of ethnic identification in a society that has undergone rapid change. Even without the ideological backgrounds of the claim that differences and inequality are inconsistent with the American principle of equality of opportunity, these considerations may be true, but ethnic terminology is still alive in the United States, and in some aspects it figures more prominently than ever before (Gamerith 1998).

In this chapter I aim to address one specific field that has long been affected by America's ethnic quandary: schooling and education (Gamerith 2005a). In the United States there is possibly no experience as fraught with ethnic tensions as the access to knowledge and education.

Obama and a Black Awakening?

In U.S. politics President Obama, who did not play the ethnic card himself when he ran for the presidency in 2008 and 2012, nonetheless aroused concerns and hopes for an African American renaissance. Official statistics are still broken down by various ethnic groups despite the frequent change of their official definition in the census over the last four or five decades, which has restricted the comparability of the data. Discrimination against ethnic groups is a poignant issue that attracts much public attention. Fighting against ethnic bias has spawned a broad range of initiatives, programs, and legislation. It was the Clinton administration that put schooling issues on the federal agenda. Authorities were charged with reducing discrimination against distinct ethnic groups, and public agencies such as the National Center for Education Statistics (NCES) received substantial funding to provide detailed statistical data on public and private education. With Barack Obama being elected the first non-White President of the United States, advocates of ethnic equality hoped for an improvement in educational outcomes among ethnic minorities, particularly after the rather impassive response to ethnic matters during George W. Bush's presidency (2001–2009).

Studies on how the ethnic gap in achievement has lessened seem to be inconclusive and somewhat contradictory. There is also no clear indication of whether Obama's presidency has contributed to reducing ethnic differences in educational outcomes. Although some commentators claim to see an "Obama effect" in schooling (e.g., Wiarda 2006), with a gradual reduction in the disparity between reading and math test scores among African Americans and Whites, equality still lies far off. To whatever extent an Obama effect may exist empirically, the early enthusiasm for the first non-White president in American history has yielded to disillusionment. Whether Barack Obama's ascent may serve as a role model and source of inspiration for members of the African American community depends on the general assessment of his presidency and has yet to be seen. There is no doubt, however, that many African Americans are disappointed by what the Obama administration did not manage to do: reduce overall poverty levels, bridge the ethnic gap, relieve segregation, and, above all, improve educational achievement. Reading skills tested among 12th graders showed a significant ethnic bias in 2013, and there has not been much change since 2002, except for the worse when it comes to African Americans (Nation's Report Card 2014; Fig. 4.1).

On a reading scale from *basic* to *proficient* to *advanced*, 40 % of White 12th graders performed at the proficient level; 7 % at the advanced. Among African American students of the same age group, only 15 % and 1 % made it into the proficient and advanced categories, respectively. There is absolutely no sign of increasing educational outcomes among African Americans in terms of reading skills: The percentage of performers with reading skills above the basic level has remained largely static since at least 2002 (Fig. 4.1) and has sometimes even slightly declined in some years since 1992 (Nation's Report Card 2014). Only among fourth graders

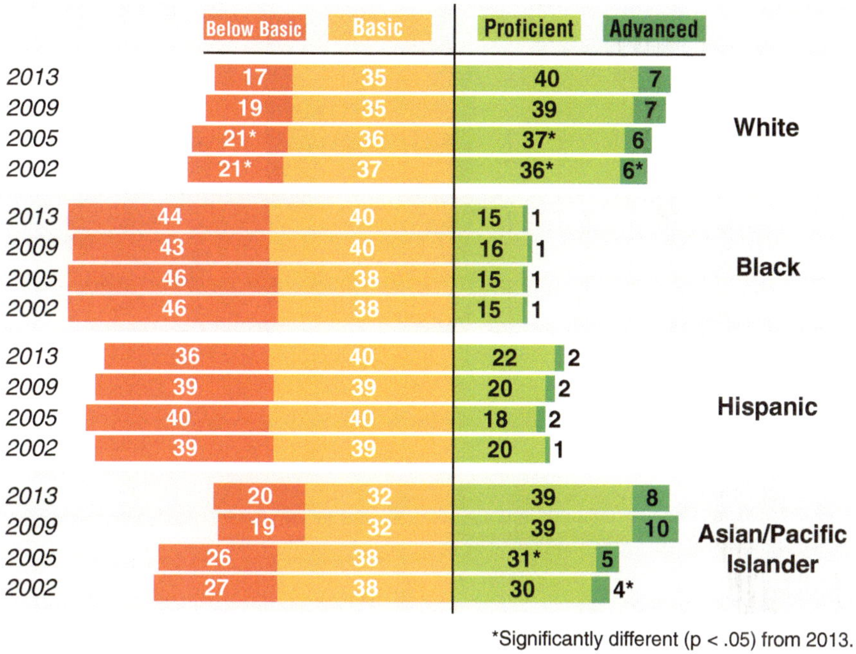

	Below Basic	Basic		Proficient	Advanced	
2013	17	35		40	7	
2009	19	35		39	7	White
2005	21*	36		37*	6	
2002	21*	37		36*	6*	
2013	44	40		15	1	
2009	43	40		16	1	Black
2005	46	38		15	1	
2002	46	38		15	1	
2013	36	40		22	2	
2009	39	39		20	2	Hispanic
2005	40	40		18	2	
2002	39	39		20	1	
2013	20	32		39	8	
2009	19	32		39	10	Asian/Pacific
2005	26	38		31*	5	Islander
2002	27	38		30	4*	

*Significantly different (p < .05) from 2013.

Fig. 4.1 Trend in U.S. 12th-grade NAEP reading achievement-level results at or above proficient, by race/ethnicity. 2002–2013 (Adapted from The Nation's Report Card 2014, Subject: Reading, second table, 2014, by the National Center for Education Statistics (NCES). Cartographer: Volker Schniepp)

has the gap between the average reading scores of Whites and African Americans somewhat narrowed within the last couple of years (National Center for Education Statistics 2012, p. 11).[2]

Whether or not an Obama effect has set in, and no matter how it can eventually be measured, the call for new, inspiring, and feasible role models for African American students is overdue (Stewart et al. 1989). Role models strongly influence occupational aspirations, which, in turn, are not equally represented among different ethnic groups. In a sample of nearly four thousand 12th graders representing several locations in the United States and various ethnic backgrounds (Gamerith 2002), African Americans were far more inclined to cling to dreamlike expectations of occupational success than were students from any other ethnic group. On the one hand, they frequently view athletes as their personal heroes and emulate their

---

[2] Even if African Americans were to fare better in reading achievement, this improvement would not immediately translate into success for them on the job market. Reading proficiency is only the most basic of qualifications any job applicant has to come up with. Literacy alone will not open higher positions to ethnic minority people. Ethnic differences will just be transferred to the next highest level of educational achievement.

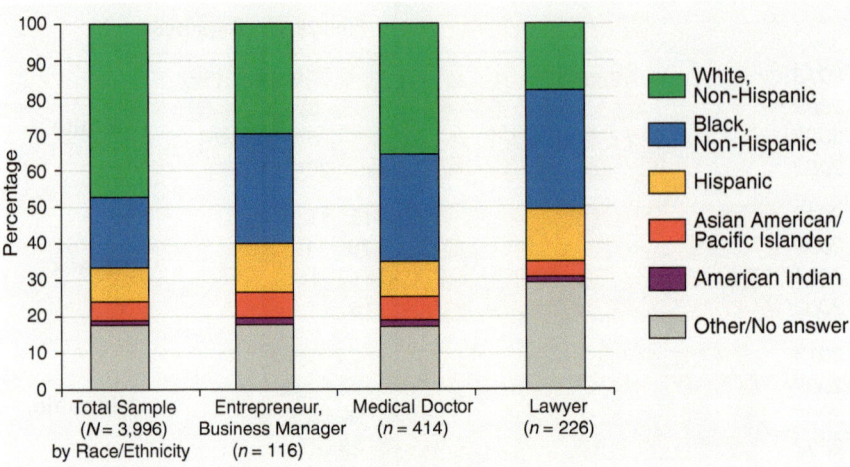

Fig. 4.2 Occupational aspirations among U.S. 12th graders, by race/ethnicity (Data from
W. Gamerith (2002). Copyright 2002 by W. Gamerith)

behavior as often as possible. Of course, these aspirations are just as often shattered,
and they keep students from adopting realistic perspectives for their future lives. On
the other hand, African American students do not strongly identify themselves with
role models that are anchored in the local community and might do it some good,
such as teachers. The willingness to follow that path is less pronounced among them
than among students from other ethnic backgrounds. Aspiring to become a teacher
obviously is not what counts for African American high school students (Cole 1986;
Dupré 1986; Garibaldi 1991; King 1993; for higher education see Fechter 1990;
Jackson 1991). One could surmise that they suppress their school experiences and
will never allow them to become the focus of their occupational aspirations. The
teaching profession lacks great attraction for African American students. They
often do not even think of becoming a teacher. But they are aware of the attractive
jobs anyway. Lawyers, medical doctors, and business managers are highly esteemed
by young African Americans despite their probable suspicion that such dreams are
hopeless (Fig. 4.2). Cultural values such as close family ties, which can help children succeed at school, or occupational aspirations conditioned by religious traditions that condemn job-related failure may be responsible for active occupational
ambitions—or the lack thereof.

Data on the ethnic and racial background of American lawyers suggest it is a
dream for African American students to aspire for this opportunity. A mere 4.8 % of
licensed lawyers in the United States were African Americans in 2010[3]; Of the law-

---

[3] http://www.americanbar.org/content/dam/aba/migrated/marketresearch/PublicDocuments/law-
yer_demographics_2012_revised.authcheckdam.pdf

yers working in large law firms, African Americans accounted for only 3 % in 2014.[4] And whereas the percentage of lawyers accounted for by all other ethnic groups increased in the big law firms in 2013, that of African Americans dropped even further.[5]

Geographical Reflections of Educational Attainment

The uneven distribution of reading and mathematics skills among fourth and eighth graders in the United States is not surprising. One can anticipate that the ethnic factor translates into a specific spatiality that results from regional disparities between poor inner cities and affluent suburbs. Results on learning and educational attainment in economically weak regions differ from those in growing and dynamic areas. However, two things *are* surprising from a geographical perspective—the striking bipolar regionalization of the United States when it comes to schooling and education (particularly between the North and the South) and the persistence of this spatiality over the centuries. Mapping the combined features of ethnicity and education uncovers a social geography deeply implanted in stories of excellent and ambitious schooling but also of traditional undereducation, forced ignorance, and poor endowment (Gamerith and Messow 2003). Cultural values and social hierarchies are reflected in a distinct ethnic geography of learning, schooling, and education.

Reading skills among public school fourth- and eighth-graders from all over the United States provide an initial outstanding example. Represented on a map (Fig. 4.3), the result could raise the question of whether the United States is still united—at least when reading skills are depicted at the aggregate state level. A sharp East–West line separates states with average or above-average reading skills in the North from a band of states in the South from Alabama to California (in addition to South Carolina, West Virginia, Washington, D.C., Hawaii, and Alaska), where the percentage of students performing at least at the proficient level ranks below the national average for both grades 4 and 8. Not only the "Old South" and its noteworthy percentages of minority people, as in the "Black Belt," but also parts of the Southwest and the West, with their high percentages of immigrant students, fail to achieve national standards in reading skills among fourth and eighth graders. Problems of inner city schools obstruct educational achievement in the capital, and poor rural West Virginia faces similar shortages.

Spatial representations of reading skills as highlighted in Fig. 4.3 correspond with the geography of a phenomenon that frames the deplorable culmination of school failure—leaving school before graduation. Dropping out does not occur evenly across the United States; it is connected to a variety of geographical and

---

[4] http://www.abajournal.com/news/article/only_3_percent_of_lawyers_in_biglaw_are_black_which_firms_were_most_diverse/?utm_source=feeds&utm_medium=rss&utm_campaign=site_rss_feeds

[5] http://www.americanlawyer.com/id=1202657037862?slreturn=20150025093610

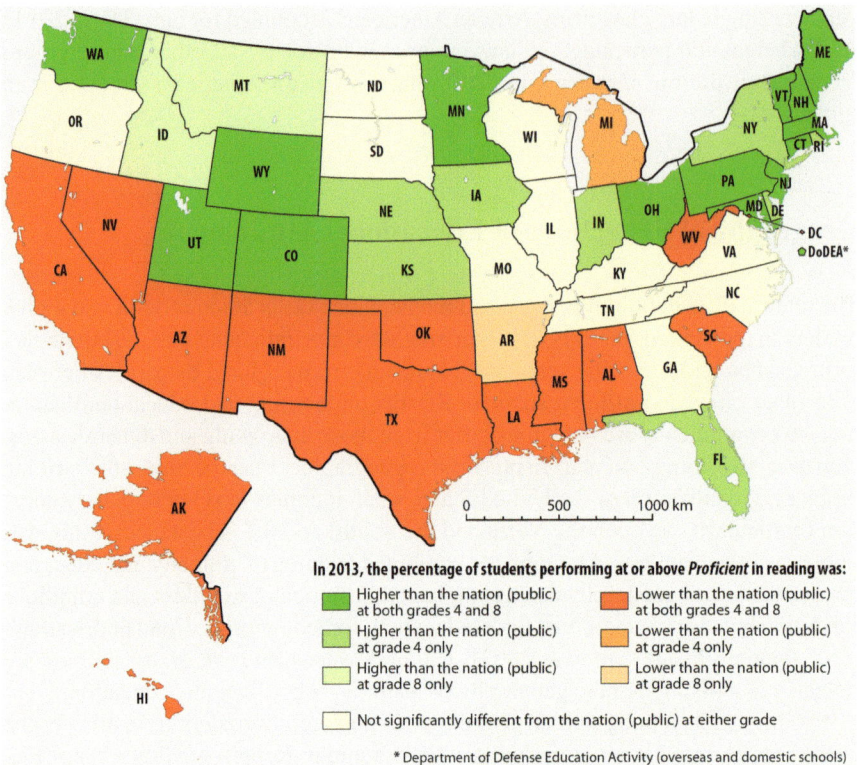

Fig. 4.3 Percentage comparison between states and the nation for public school students at or above proficient in fourth- and eighth-grade NAEP reading (2013) (From National Center for Education Statistics (NCES) (2014). Cartographer: Volker Schniepp. Reprinted with permission)

contextual factors (Altenbaugh et al. 1995). Some authors have argued that it is the neighborhood as one very decisive parameter that puts students in jeopardy of dropping out of school (Apple 1989; Kozol 1991). Negative influences from the family, the peer group, and the school itself coalesce into a spatial metaphor such as the inner city neighborhood, which some experts look upon as a trap for endangered students. Social geography becomes a pernicious predictor of an individual's achievement. Such arguments might seem deterministic, but they have a strong spatial correlation that is superimposed by an intricate web of social, political, and economic indicators. The geography of education is not deterministic; it validates the contextual factors. Mapping the dropout phenomenon reveals a pattern similar to that of other schooling attributes: The South has the highest dropout rate, whereas the Midwest and parts of the Northeast do much better. In the West however, the dropout phenomenon appears in a more irregular way with notable regional differences dependent also upon political or religious influences: Immigrant communities with higher portions of *Hispanics*, particularly in the inner city *barrios*, are inclined to generate more dropouts than, say, rural communities affected by Mormon belief systems.[6] These disparities at a national level are certainly modified on a local scale because dropping out of school happens primarily within the jurisdiction of the local school district. Once again, neighborhood and the dropout phenomenon closely correspond and interfere with each other.

A stigma is attached to students who leave high school without diploma. Supposedly, they are at risk of being caught in the lower ranks of society without any chance to climb the social ladder. In a globalized service economy dropouts have no major assets and thus are often associated with the low-paying job sector. They can be deemed the losers in a polarized and uneven school system, and they are challenged by economic restructuring and rising minimum standards of educational attainment, which are required even by the most elementary service jobs. The dropout phenomenon lies at the very center of ethnic inequality and constitutes both cause and effect. Poverty escalates attrition from school, which, in turn, increases the risk of staying poor.

Lamenting dropouts and seeing them at permanent risk of failure with restricted or no participation in the labor market constitutes a rather new approach to dealing with the problem. Yet dropping out of school is not at all a new phenomenon; it happened quite frequently in the twentieth century. Dropout rates have constantly fallen, yet social scientists still focus on this rather small, but increasingly stigmatized segment of educational attainment. This interest parallels the preoccupation with the so-called underclass and rests on a discourse that was introduced, or reintroduced, into American sociology in the early 1960s and that periodically centered more on academic than practical considerations (Mingione 1996; Schneider-Sliwa 1996; Theodore 2010). In some respects the dropout "problem" has been construed by both public and scholarly debates, with the side effect of stigmatizing and sometimes even criminalizing the school dropout (Wacquant 2008). This explanation, of course, does not mean that there is no problem at all and that dropping out of school does not warrant serious public attention.

Yet there is a clear positive trend in the general dropout rates (Fig. 4.4), although data-collection procedures have been changed over the years in ways that reduce the ability to compare the figures prior to 1992 with those of successive years. The late 1970s saw dropout rates of more than 14 %, and general figures in the 1980s still hovered between 12 % and 14 %. Some remarkable decline has occurred since 2005, with 6.6 % (2012) being the lowest dropout rate ever recorded in the United States.[7] The risk of leaving school before graduation varies widely with the ethnic background of the students, even if one controls for other variables such as parents' socioeconomic status, income, or occupation. Educational achievement thus has a clear ethnic undertone. With a 7.5 % dropout rate in 2012, African Americans fare much better than they did in the 1970s, when figures tended to exceed 20 %. The dropout rate among Hispanics has also improved significantly in recent years, but remains at a staggering 12.7 %. For decades it ranged in the 30s and higher 20s.

---

[6] On the sociocultural aspects of educational achievement among Hispanics (or the Hispanos as the direct descendants of Spanish settlers from the sixteenth century onward), see Freytag's contribution in this volume.

[7] Retrieved from http://nces.ed.gov/programs/digest/d13/tables/dt13_219.70.asp

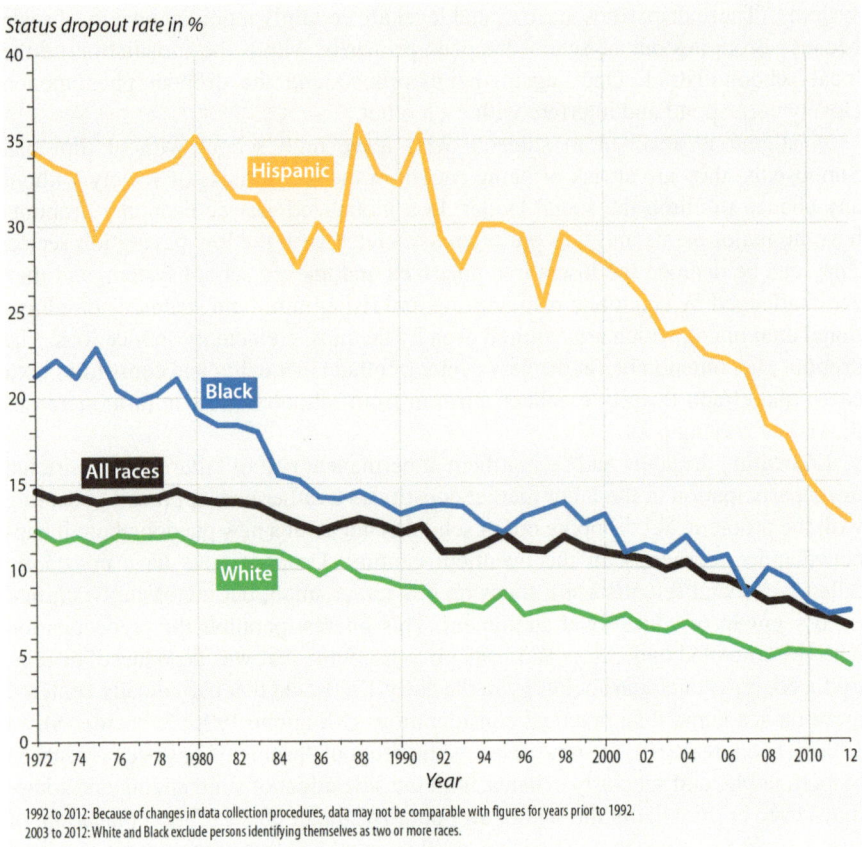

Status dropout rate in %

1992 to 2012: Because of changes in data collection procedures, data may not be comparable with figures for years prior to 1992.
2003 to 2012: White and Black exclude persons identifying themselves as two or more races.

Fig. 4.4 Percentage of U.S. high school dropouts among persons 16–24 years old (status dropout rate), by sex and race/ethnicity: selected years, 1972–2012 (Cartographer: Volker Schniepp. Based on National Center for Education Statistics (2013))

Certain ethnic groups succeed in America's public schools; others are more susceptible to failure.

Public Schools: The "Great Equalizer"?

Clear ethnic and regional divergence between the results of educational achievement by students in the United States is inconsistent with the original intention behind the public school system when it was established in the nineteenth century.[8]

---

[8] Unlike authorities in many European states, those in the United States do not require that children attend a public school. Children are supposed to be educated and trained but not necessarily at a public school. Home schooling and private schools can be an option, and public school attendance is not compulsory anywhere in the United States. Officially, public school enrollment is voluntary. The public school system has never materialized as a mandatory tool to get students to class.

Public schools appeared only after the founding of private institutions, which had been criticized as instruments of social inequality (Cremin 1980). The first private schools were either denominational or nondenominational institutions, and they dated from the colonial period (Rippa 1988). What they had in common was their advocators' spirit to keep state or church influences (or both) out of the schools in order to guarantee the form of instruction that was closest to the groups' belief and value systems. Hundreds of renegades, freethinkers, and eccentrics among the early North American colonists in what was still part of the British domain would never have allowed the state (or the church) to have a say in what was going to be taught in schools. Accordingly, these institutions developed into the basic system of early American education—fully on a private basis, that is, without regulation by the colonial or parochial administration in this field. Other people resented the state indeed, but were willing to open schools to provide for the needs of the church and whichever denomination was at stake. Puritans in New England and Anglicans in Virginia established colleges and thus began to nourish the dualism between religiously affiliated and state institutions of higher education.

Yet it was the elementary level that became the principal domain of the state in the early nineteenth century, whereas private efforts focused more on higher education and resulted in the founding of very elitist colleges and universities (Story 1988). Freedom of choice, an American principle upheld by the public during the nineteenth century, warranted the existence of a private school sector. Even though public funding of Catholic schools provoked fierce fighting, as in New York (Ravitch 1974), commitment to maintaining a system of private education was beyond question until the 1920s. With rousing patriotic sentiment during the interwar period, some states passed laws to declare the attendance of public schools obligatory (Ravitch 1985, p. 187). However, President Johnson's Great Society program in the 1960s put the responsibility for equal opportunity back to the public schools. Private schools with limited access and raised demands would not correspond to the idea of a society based on freedom and equality. Only in the wake of the conservative turn in the 1980s did private education receive a boost again. Protestant schools in particular, sometimes of fundamentalist orientation, thrived, though Catholic institutions still dominated the private sector, at least with regard to student enrollment. Of the conservative Protestant schools operating in 1990–1991, only 68.4 % were older than 10 years; just 18 % of those Protestant institutions had been established prior to 1970 (McLaughlin et al. 1995, p. 13).

American democratic values helped establish a public school system during the nineteenth century. Although school authorities never enforced school attendance, they did stipulate that education and learning (and hence the process of acquiring knowledge) were compulsory. Equal opportunity for any citizen and across all social and ethnic lines was the leading principle for the formation of the public school sector. The public school eventually began to be seen as an instrument for reinforcing identification with America, the respective state, and the local community

---

However, it did have formal goals, such as to streamline the student body according to American cultural values or to provide the means for equal opportunity.

as well. At the same time, different administrative agencies connected with public schooling at various levels resulted, technically, in the emergence of hundreds or thousands of school systems, not just one. Thus the essential body of the school district came into existence; America has more than 13,000 of them.

The American public school has often been thought of as a cradle of democracy. Educating the people (not just children) and instilling them with democratic values is based based on the writings of Thomas Jefferson. The Louisiana Purchase in 1803 more than doubled the United States in area, and the Jefferson administration sought new ways to organize and manage the territorial gains (Kallgren 1999, p. 168). One product envisioned for public schooling was the reliable citizen who would help consolidate the vast new lands and make them an integral part of the nation. The Land Ordinance of 1785 stipulated that the revenues from selling 136th of a township (one square mile), would go toward the development of a public school system. America's westward expansion literally built the public schools. Republican values and bourgeois attitudes were also translated architecturally, with schools as grand, neo-Gothic structures resembling palaces or church buildings. "School architecture became one of the clearest expressions of bourgeois social values throughout the nineteenth century" (Reese 1995, p. 82). As architecture created a narrative of democratic ideals, so did history in continually referring to the supposedly nonprivileged access to public education. Yet such education *was* privileged, and hundreds of thousands of American people were denied the right to attend public schools—or forced to content themselves with second- or third-rate, underfunded schools, as the next section shows.

Public schooling is supposed to work as an incubator for democratic behavior, but the functionality attributed to schooling goes beyond the nurturing of democracy. The Progressive Movement in the late nineteenth century prompted proponents of public education to call for social and ethnic equality achieved by a uniform and standardized school system open to the general public (Yeakey 1988, p. 287). Coordinated curricula were thought to ensure social justice. Horace Mann, one of the leading nineteenth-century American pedagogues, is credited with the famous phrase extolling public education as "the great equalizer in the condition of man" (see Hurn 1978, p. 87). Yet the notion of schooling as a mechanism of social change and equality ignores the contextual factors that encompass and modify the school system. Incessant conflicts around the public school, its financial endowments, and its local value testify to the hypothetical character of the claim that it is an equalizer. Equal opportunity provided by public schooling sparked conflict in the past just as it does today. Only the location of these conflicts has changed:

> Equality of opportunity in the nineteenth century was to be the joint result of the receipt of an equal (common) education for all and the smooth functioning of competition in the marketplace. In the twentieth century, schools themselves would remain equally available to all, but the educations provided within them would be unequal. Competition would be shifted from the marketplace to the school itself. (Reynolds and Shelley 1990, p. 120)

However, the general belief in the social blessings of the public school system is widely accepted and endorsed by the civil rights movement (Kuran 1995, p. 138).

The public school was crucial in the process of assimilation, when millions of European immigrants flooded into the United States. Innumerable children, notably in the metropolitan ghettoes in the Northeast, were acquainted with the basic values of American society. Instruction in the language, civic education, and the commitment to both the Union and the individual state, symbolized in school rituals such as reciting the Pledge of Allegiance and saluting the flag (see Fig. 4.5), formed part of the daily routine in the public school (and some private institutions as well). Generations of immigrants could thus be Americanized very efficiently (Ravitch 1985, pp. 188–195). In the attempts to integrate the newcomers, the public school often did not limit itself to children but included adults as well. Attending a night school was something of a regular activity for many immigrant parents. In Chicago in 1909, 13 % of the adults *who* knew no English *were* enrolled in night school. Often, the first English words taught at these schools were, "I'm a good American" (Molesky 1988, p. 52). English language and civics constituted the main subjects. Sometimes, company-sponsored courses (dubbed "factory classes") were held on organization's own premises and conducted by "teachers" who had never completed any formal teacher training. The generous financial sponsorship of entrepreneurs helped sustain these schools. Standardized textbooks and other graphic teaching material glorifying the territorial expansion of the young nation were also responsible for successful connection with the American mainstream.

At the local level public schools stabilize the community and foster social identity and a feeling of solidarity, which have gained particular importance in ethnically diverse urban fields and in rural areas as well. The public school basically strengthens

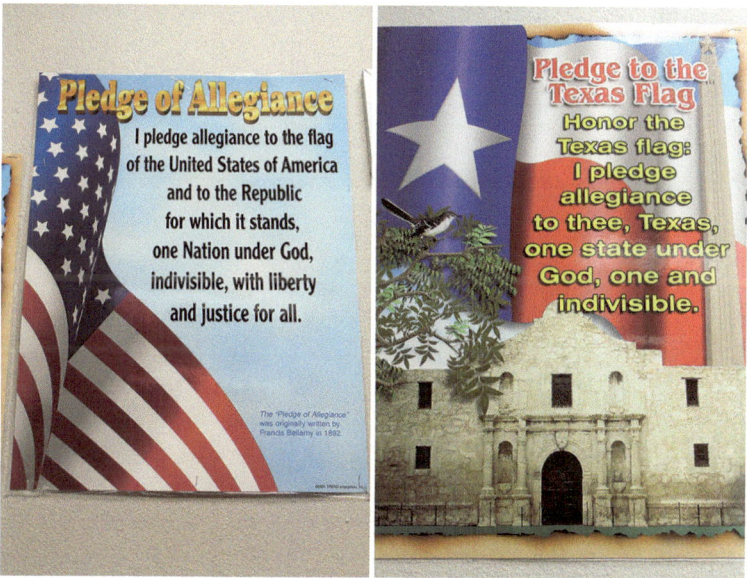

Fig. 4.5 Texas classroom posters for paying homage to the Union (the Pledge of Allegiance) and the State of Texas (the Pledge to the Texas Flag) (Photos by W. Gamerith 2012)

social cohesion in these communities. Local identity centers upon this institution, transforming it into a powerful focus of ethnic conflict. Small rural schools, mainly in sparsely populated areas of the Midwest and West, not only serve as sites of education and transfer of knowledge but also as local venues of political rallies, social gatherings, leisure, and adult training. Sometimes, the school building was used even as a temporary hospital. Identification with the school does not build without a respected teacher. Social control helps keep teachers on track. In the past, "people were interested in everything the teacher did or said or wore.... Citizens kept their country school teachers under close surveillance to ensure that they followed a stern moral code" (Gulliford 1996, p. 73). It is the same today. Local elites, who are involved in important decisions about the school authorities and the teaching personnel, may also improve identification with rural schools. The local community often encourages its small schools and wards to oppose any attempts to close them down. These institutions sometimes receive astonishing appreciation in isolated rural settlements. Local people agree to subsidize "their" one-room schoolhouses to prevent them from being boarded up. And when these efforts fail, the community turns these structures into local museums, steeping them in distinct feelings of identity, nostalgia, and home.

Junction School (Fig. 4.6) in Stonewall, Texas, for example, is linked with the biography of President Lyndon B. Johnson, who was raised in that part of the Texas Hill Country. Even though the young Johnson had attended this tiny one-room shack for only a few months and the school was closed shortly after World War II and fell into disrepair thereafter, it received national attention when it became the

Fig. 4.6 Junction School, Stonewall, Texas: a rural one-room schoolhouse, now part of the National Park Service (Photo by W. Gamerith 2013)

Fig. 4.7 President Lyndon B. Johnson signing the Elementary and Secondary Education Act, in front of Junction School, Stonewall, Texas, April 11, 1965. Public domain (Retrieved from https://commons.wikimedia.org/wiki/File:LBJ-education-act.jpg?uselang=de)

stage for the Great Society president when he signed the Elementary and Secondary Education Act on April 11, 1965. Sitting on a rustic wooden bench in front of the former school building, with an elderly lady, his first teacher, at his side (Fig. 4.7), the old one-room schoolhouse became a well-orchestrated setting for an all-encompassing national political endeavor to fight poverty, racial inequality, and educational disadvantages. This little wooden structure somewhere in the back country readopted the myth of the "good old days" and made this former world feel authentic to a national audience tied together by mass media. However, restaging the Great Equalizer was more symbolic and would never meet the sad reality that confronted American public education at that time.

The ideology of public education as an incubator of American civic values and as a ladder to social advancement has never fully come true. However, it has brought people of different ethnic and cultural backgrounds together. Native-born persons and immigrants alike have clustered around the institution of the public school out of reverence for their country. Public education has thus served as a symbol of allegiance to the state—with a message that has never really materialized within a society whose internal cleavages have been constantly deepening instead of being countered. In addition, public education has had to cope with an increasingly important private sector, which has further diminished the mission of the public school as the Great Equalizer.

Traditions of Ethnic Undereducation

Of course, many inadequacies in American public education are linked with lack of funding or disparate financial endowment. School funding comes mainly from local property taxes, which are characterized by high regional variability. Poor school districts with low public revenues thus exist right next to affluent school districts with high tax revenues. Equalization of financing the public school system is a political phrase at most but has not been achieved yet. Self-autonomy and local control dominate public opinion on how to manage the public school system.

But money or the lack thereof is not the only reason for the varying quality in public education and the low achievement of many ethnic groups. Additional institutional constraints and structural forces affect educational outcome. Social attitudes toward ethnic minorities also play their part in schooling. Discrimination, stereotypes, prejudice, and assimilation belong to that complex set of factors. Educational attainment is influenced by personal characteristics and individual dispositions—such as students' motivations, family and parental factors, and peer group pressure. Yet one can provide a kind of overarching explanation for the constant troubles of public education, particularly in regard to ethnic minority people, by looking at the historical development of public schooling and at the resulting traditions of undereducation.

A broad range of indicators testifies to the disadvantages of which certain minority groups (especially African Americans and Hispanics, but also Native Americans) have to be aware. The expression "separate but equal" has accompanied the public school system for a long time until the postwar period, however, ethnic barriers within the system were the norm, not the exception, and led to forms of compulsory ignorance among many ethnic minority students (Gamerith and Messow 2003). Free public schooling has never encompassed all of America. It has always been provided differently (and unequally) along different ethnic and racial lines. In the South, for instance, African Americans had to comply with only a very rudimentary system of public education. Their history of education resembles a sequence of bans and prohibitive orders (Anderson 1988). As slaves, African Americans were allowed only tightly restricted access to schooling and education, if any. Far into the nineteenth century, educational achievement among African Americans was remarkably low. Slaveholders often undermined any initiative for slaves to learn writing and reading. During the antebellum years, many states in the South had issued bans[9]

---

[9] However, some African Americans did receive a considerable amount of formal education. James W. C. Pennington, for instance, a former fugitive slave and the first African American student at Yale University, was awarded an honorary doctorate from Heidelberg University, Germany, in 1849 (Honeck and Meusburger 2012, p. 296).

against teaching African Americans (Cornelius 1991). Regular school attendance did not exist among members of that group. Some church or philanthropic initiatives tried to stave off the devastating effects such regulations imposed on the African American community (Anderson and Moss 1999). In many hidden places, clandestine meetings were the only opportunity to gain a little insight into how to read and write. Gatherings "under the trees" were intended to simulate harmless social events, but in effect they served to teach the attending crowd. At night, secret "schools" popped up to disseminate banned knowledge (Franklin and Moss 1994, p. 160). Even after slavery had been declared illegal, schooling for African Americans remained bleak (Butchart 2010). While a public system of secondary education rapidly developed between 1880 and World War I, African Americans were further marginalized. In 1916 the United States had fewer than 70 high schools for African Americans (Low and Clift 1981, p. 335). The state of Maryland, with an African American population of 200,000 in 1911, ran just a single public high school for African Americans (in Baltimore). New Orleans offered four high schools for an urban population of 25,000 Whites, while 9,000 African Americans in the city had no such institution at all.

When a high school for African Americans existed, it could by no means be seen as equal to the institutions for Whites. In the rural South poor and paltry cabins without heating or plumbing sheltered schools and students, often in only one room. Personal descriptions provide vivid impressions on how everyday life in an African American school must have looked. Reporting on the school buildings and their facilities in the mid-1930s, the superintendent of schools for African Americans in Mississippi stated,

> Of the 3,753 Negro schoolhouses in Mississippi, 2,313 are owned by public school authorities. The other 1,440 schools are conducted in churches, lodges, old stores, tenant houses, or whatever building is available.... The Negroes themselves, in some cases, are building and repairing their schoolhouses out of their own meager savings and with their own labor. School buildings need to be erected to displace the many little shanties and churches now being used....
>
> There is also dire need for school furniture and teaching materials—comfortable seating facilities, stoves, blackboards, erasers, crayon, supplementary reading materials, maps, flash cards, and charts. In many of the 3,753 colored schools of the State there is not a decent specimen of any one of the above mentioned items. In hundreds of rural schools there are just four blank, unpainted walls, a few old rickety benches, an old stove propped up on brickbats, and two or three boards nailed together and painted black for a blackboard. In many cases, this constitutes the sum total of the furniture and teaching equipment. (Quoted in Wilkerson 1939/1970, pp. 28–29)

W. E. B. DuBois, one of the leading first-generation African American civil rights activists, remembered his schooldays in a rural school in Tennessee in 1896 in his book *The Souls of Black Folk* (1903):

> The schoolhouse was a log hut, where Colonel Wheeler used to shelter his corn. It sat in a lot behind a rail fence and thorn bushes, near the sweetest of springs. There was an entrance where a door once was, and within, a massive rickety fireplace; great chinks between the logs served as windows. Furniture was scarce. A pale blackboard crouched in the corner.

My desk was made of three boards, reinforced at critical points, and my chair, borrowed from the landlady, had to be returned every night. Seats for the children—these puzzled me much. I was haunted by a New England vision of neat little desks and chairs, but, alas! the reality was rough plank benches without backs, at times without legs. They had the one virtue of making naps dangerous—possibly fatal, for the floor was not to be trusted. (quoted in Gulliford 1996, p. 103)

Additionally, scarce financial means reduced the quality of learning still further, for instance, by a shortened school year. Differences in the duration of the school year between White and African Americans schools could vary substantially. In the early twentieth century, the average length of a school year in African American public schools in Wilcox County, Alabama, was 80 days. White students in the same county attended school for 140 days a year. According to reports on school "years" in some rural areas in the South, schools were open for not more than 2 months a year (Grossman 1995, p. 108).

Hispanics, Native Americans, and Asian Americans have had educational experiences similar to those of African Americans. In the Southwest both old Spanish settlers and later immigrants from Mexico were denied full access to the public school system (see Freytag 2003, for New Mexico). Some of the structural processes already shown for the history of public schooling among African Americans were also at work in the Spanish–American borderlands. A public school sector had not been established before the late nineteenth century, and it was mainly reserved for Anglos. Even though segregation at school was not enforced by law, it apparently happened arbitrarily. In a process implicitly aimed at ostracizing, the Spanish-speaking children were sent to isolated and second-rate school buildings. Lack of money was a chronic problem. In 1896 New Mexico introduced the mandatory and exclusive use of English at public schools (Trueba 1976, p. 12). With the complete elimination of Spanish from the classroom, the conflict entered a new stage. In Texas, Hispanics were relegated to inferior social positions as well, and the school performed a useful task in keeping the minority at that lower status. Around 1900 only one fifth of the native Mexican children in Texas attended school, and even in 1944 almost 50 % of Hispanic children in Texas received no teaching or training at all (Cockroft 1995, pp. 12–13). Illiteracy rates of 50 % or more remained all too common for this group until the 1930s.

Asian Americans and particularly immigrants from China were victimized by nativist tendencies in the nineteenth century, and establishing their own schools proved to be difficult and controversial. But open resistance by Whites was not strong enough to completely thwart all Chinese efforts to create for a school system of their own. White schools in California, for example, would not tolerate the admission of Chinese students, but they did not bother so much about the Chinese initiatives to start some small Chinese schools, the first of which opened in San Francisco in 1859 (Chan 1991, pp. 57–58). Compared with Hispanics and African Americans, the Chinese had the advantage of settling in an urban context, nestled in compact ethnic neighborhoods, which facilitated mutual aid and the organization and administration of schools. And because many of the Chinese immigrants were urban merchants, businessmen, and artisans, they had more money available than, say, Hispanics with their agricultural way of life.

Restrictive codes of schooling pertained to African Americans in particular but also affected Hispanics and Asian Americans—whereas Native Americans were suppressed not by the denial, but by the establishment, of a school system that stood in contrast to their system of values and beliefs. To be specific, White-Anglo America was convinced that the best way to assimilate Native American children was to put them into boarding schools hundreds of miles away from the reservations. Enforced attendance at boarding schools thereby became a viable tool for discriminating against this ethnic group. This policy tore children out of their families and local cultures and was perhaps even more brutal than refusing students' access to solid education, the method used against African Americans, Hispanics, and Asian Americans.

These deep-seated practices of keeping distinct ethnic minorities from participating in the public school sector or of assigning them to specific boarding schools took their toll. Much smaller enrollment figures and restricted chances on the job market were the immediate consequences. The collateral message was that these minority people were not seen as an integral part of American society. They were regarded as barred from the promises of the American dream. Cultural barriers and purported individual deficiencies were held responsible for an alleged unwillingness and inability among ethnic minority groups to share American values. Their status as "visible" (and thus "unassimilable") minorities apparently legitimated these proscriptive attitudes with respect to public schooling. The effects were devastating, with school enrollment figures at extremely low levels in many states and territories in the South and West in the late nineteenth century (Fig. 4.8), and many minority people acquired a status that, some scholars argue, results in a form of "internal colonialism"—another form of dependency and deprivation with limited chances to escape from it.

Public schooling thus became the arena where existing social power relations were maintained and reinforced, but also contested. It is anything but a coincidence that efforts by minority groups to emancipate themselves as reflected by the civil rights movement originated in schooling issues. Probably no one has better illustrated the close relationship between suppression, opposition, and schooling in terms of ethnic struggle than Norman Rockwell in his shattering 1964 painting *The Problem We All Live In* (Fig. 4.9).

An all-White public school is opened up to a young African American girl, but even though she is escorted by four officers on her way to school, she still has to endure denigration, violence, and mockery. The "N-word" is still scrawled on the wall, like the three Ks, signaling that the past has not yet passed—at least during the 1960s. As Rockwell shows, the light–dark contrast figured prominently even in such a trivial instance as that of a girl on her way to school. Still today, America's ethnic and schooling dilemma is not an easy task to contend with.

Shedding light on the provision of public schooling for ethnic minorities over the years reveals long-standing traditions of undereducation among members of specific ethnic groups. Bans and coercion allowed the dominant group to control

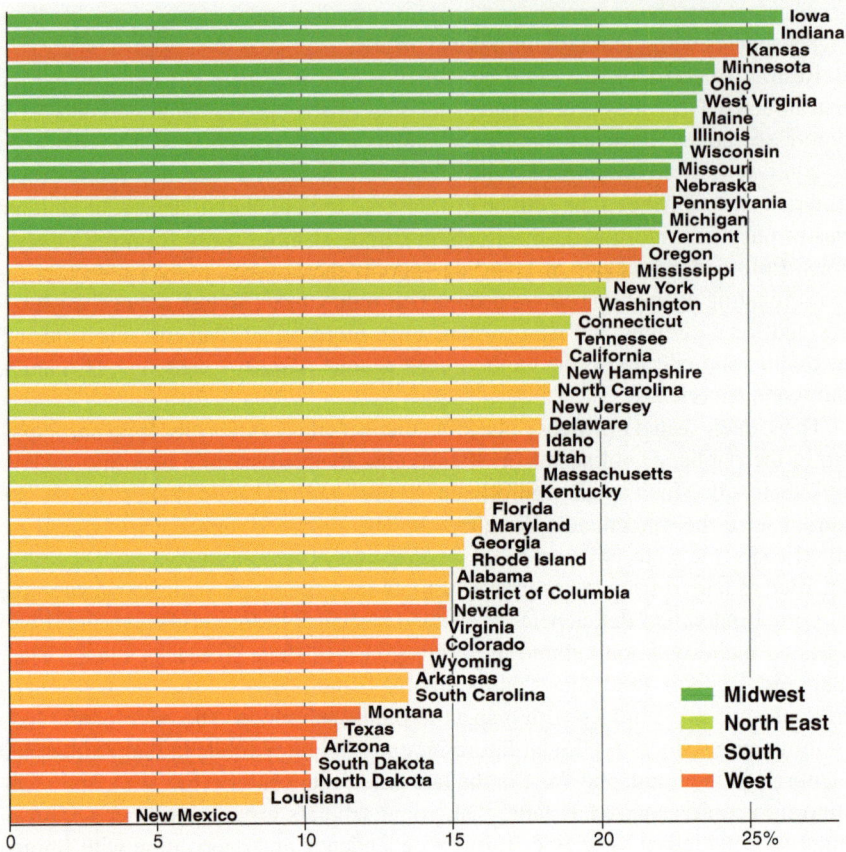

Fig. 4.8 Enrolled public-school students as a percentage of state population, by state (1880). (From by W. Gamerith (2002). Copyright 2002 by W. Gamerith. Cartographer: Volker Schniepp)

minorities' access to public education and to create and maintain whatever conditions seemed appropriate to keep these subordinate ethnic people ignorant. When the establishment of a public minority school became inevitable, it was kept structurally and financially dependent on the demands and needs of the White majority population. Although desegregation and integration have been introduced thanks to the impact that the Civil Rights Act of 1965 has had on U.S. legislation, patterns of suppression and the instruments for effecting them have not completely vanished. The past still shimmers in the present, or—to put it in the words of Nobel prize laureate William Faulkner (1950)—"The past is never dead. It's not even past."

Fig. 4.9 *The Problem We All Live In*. Painting by Norman Rockwell (1964). Oil on canvas. 58″×36″ (Reprinted with permission. Retrieved from https://abagond.files.wordpress.com/2011/07/norman-rockwell-the-problem-we-all-live-with1.jpg)

References

Alba, R. D. (1985). The twilight of ethnicity among Americans of European ancestry: The case of Italians. *Ethnic and Racial Studies, 8*, 134–158.

Alba, R. D. (1990). *Ethnic identity: The transformation of White America*. New Haven: Yale University Press.

Alba, R. D. (1996). Italian Americans: A century of ethnic change. In S. Pedraza & R. G. Rumbaut (Eds.), *Origins and destinies. Immigration, race, and ethnicity in America* (pp. 172–181). Belmont: Wadsworth Publishing Company.

Altenbaugh, R. J., Engel, D. E., & Martin, D. T. (1995). *Caring for kids: A critical study of urban school leavers*. London: Taylor & Francis.

Anderson, J. (1988). *The education of Blacks in the South, 1860–1935*. Chapel Hill: The University of North Carolina Press.

Anderson, E., & Moss, A. A., Jr. (1999). *Dangerous donations: Northern philanthropy and southern Black education, 1902-1930*. Columbia: University of Missouri Press.

Apple, M. W. (1989). American realities: Poverty, economy, and education. In L. Weis, E. Farrar, & H. G. Petrie (Eds.), *Dropouts from school: Issues, dilemmas, and solutions* (SUNY-series, frontiers in education, pp. 205–223). Albany: State University of New York Press.

Bayor, R. H. (1988). *Neighbors in conflict: The Irish, Germans, Jews, and Italians of New York City, 1929–1941*. Urbana: University of Illinois Press.

Butchart, R. E. (2010). *Schooling the freed people. Teaching, learning, and the struggle for Black freedom, 1861–1976*. Chapel Hill: The University of North Carolina Press.

Cavaioli, F. J. (1992). Group politics, ethnicity, and Italian-Americans. In M. D'Innocenzo & J. P. Sirefman (Eds.), *Immigration and ethnicity. American society—"Melting pot" or "salad bowl"?* (Contributions in sociology, Vol. 97, pp. 61–74). Westport: Praeger.

Chan, S. (1991). *Asian Americans: An interpretive history* (Twayne's immigrant heritage of America series). Boston: Twayne Publishers.

Cockroft, J. D. (1995). *Latinos in the struggle for equal education* (The Hispanic experience in the Americas). New York: Franklin Watts.

Cole, B. P. (1986). The Black educator: An endangered species. *Journal of Negro Education, 55*, 326–334. doi:10.2307/2295103.

Cornelius, J. D. (1991). *"When I can read my title clear": Literacy, slavery, and religion in the antebellum South*. Columbia: University of South Carolina Press.

Cravins, G. G. (1997). 'Race' as a scientific and organizational construct: A critique. *GeoJournal, 41*, 233–243.

Cremin, L. A. (1980). *American education: The national experience, 1783–1876*. New York: Harper Collins.

Dupré, B. B. (1986). Problems regarding the survival of future Black teachers in education. *Journal of Negro Education, 55*, 56–66. doi:10.2307/2294633.

Dwyer, O. J. (2002). Location, politics, and the production of civil rights memorial landscapes. *Urban Geography, 23*, 31–56. doi:10.2747/0272-3638.23.1.31.

Dwyer, O. J., & Alderman, D. H. (2008). *Civil rights memorials and the geography of memory* (Center books on the American South). Chicago: The Center for American Places at Columbia College Chicago.

Fairclough, A. (2011). Entwicklung des öffentlichen Bildungswesens im Süden der USA. Ethnizität, Politik und Geographie [Development of the public school system in the U.S. South. Ethnicity, politics, and geography]. *Geographische Rundschau, 63*(10), 52–59.

Faulkner, W. (1950). *Requiem for a Nun*. New York: Random House.

Fechter, A. (1990). The Black scholar: An endangered species. *The Review of Black Political Economy, 19*(2), 49–59.

Finzsch, N., Horton, J. O., & Horton, L. E. (1999). *Von Benin nach Baltimore. Die Geschichte der African Americans* [From Benin to Baltimore: The history of African Americans]. Hamburg: Hamburger Edition.

Franklin, J. H., & Moss, A. A., Jr. (1994). *From slavery to freedom: A history of African Americans* (7th ed.). New York: Knopf.

Frantz, K. (1993). *Die Indianerreservationen in den USA. Aspekte der territorialen Entwicklung und des sozio-ökonomischen Wandels* [Indian reservations in the United States: Aspects of territorial development and socioeconomic change] (Erdkundliches Wissen, Vol. 109). Stuttgart: Steiner.

Frantz, K. (1994). Washington Schools, Little White Man Schools und Indian Schools—bildungs-geographische Fragestellungen dargestellt am Beispiel der Navajo Indianerreservation [Washington Schools, Little White Man Schools, and Indian Schools—Issues of educational geographics as exemplified by the Navajo Indian Reservation]. *Die Erde, 125*, 299–314.

Frantz, K. (1999). *Indian reservations in the United States: Territory, sovereignty, and socioeconomic change* (University of Chicago geography research paper, Vol. 242). Chicago: The University of Chicago Press.

Freytag, T. (2003). *Bildungswesen, Bildungsverhalten und kulturelle Identität. Ursachen für das unterdurchschnittliche Ausbildungsniveau der hispanischen Bevölkerung in New Mexico* [*Schooling, educational achievement, and cultural identity: Causes of the below-average level of education in the Hispanic population in New Mexico*] (Heidelberger Geographische Arbeiten, Vol. 118). Heidelberg: Selbstverlag des Geographischen Instituts der Universität Heidelberg.

Gamerith, W. (1998). Das US-amerikanische Bildungswesen—Räumlich-soziale Disparitäten im Spannungsfeld zwischen egalitären und elitären Prinzipien [The U.S. educational system— Spatiosocial disparities between egalitarian and elitist principles]. *Mitteilungen der Österreichischen Geographischen Gesellschaft, 140*, 161–196.

Gamerith, W. (2002). *Ethnizität und Schule. Eine Analyse regionaler und sozialer Disparitäten der Ausbildungs- und Qualifikationsstrukturen US-amerikanischer Minderheiten* [Ethnicity and school: An analysis of regional and social disparities in structures of education and qualification among minorities in the United States]. Postdoctoral dissertation, Heidelberg University, Heidelberg.

Gamerith, W. (2005a). Das Bildungssystem in den USA. Gute Schulen für alle? [The educational system in the United States: Good schools for everyone?]. *Geographische Rundschau, 57*(1), 38–45.

Gamerith, W. (2005b). Ethnizität und Bildungsverhalten. Ein kritisches Plädoyer für eine "Neue" Kulturgeographie [Ethnicity and educational attainment: A critical plea for a "new" cultural geography]. In K. Kempter & P. Meusburger (Eds.), *Bildung und Wissensgesellschaft* (Heidelberger Jahrbücher, Vol. 49, pp. 309–332). Berlin: Springer.

Gamerith, W., & Messow, E. (2003). "Geography of ignorance". Ethnische Minoritäten und das öffentliche Schulwesen in den USA ["Geography of ignorance": Ethnic minorities and public education in the United States]. *Petermanns Geographische Mitteilungen, 147*, 88–90.

Garibaldi, A. M. (1991). Abating the shortage of Black teachers. In C. V. Willie, A. M. Garibaldi, & W. L. Reed (Eds.), *The education of African-Americans* (pp. 148–158). New York: Praeger.

Glassman, J. (2010). Critical geography III: Articulating race and radical politics. *Progress in Human Geography, 34*, 506–512. doi:10.1177/0309132509351766.

Grossman, J. R. (1995). A certain kind of soul. In J. O. Horton & L. E. Horton (Eds.), *A history of the African American people: The history, traditions & culture of African Americans* (pp. 94–117). New York: Smithmark.

Gulliford, A. (1996). *America's country schools* (3rd ed.). Niwot: University Press of Colorado.

Honeck, M., & Meusburger, P. (2012). American students up to 1914. In P. Meusburger & T. Schuch (Eds.), *Wissenschaftsatlas of Heidelberg University: Spatio-temporal relations of academic knowledge production* (pp. 296–299). Knittlingen: Bibliotheca Palatina.

Hurn, C. J. (1978). *The limits and possibilities of schooling: An introduction to the sociology of education*. Boston: Allyn & Bacon.

Jackson, K. W. (1991). *The racial crisis in American higher education* (SUNY series, frontiers in education, pp. 135–148). Albany: State University of New York Press.

Kallgren, D. (1999). Race, place, and poverty in the pattern of Southern school attendance, 1850–1950. *Historical Geography, 27*, 167–192.

King, S. H. (1993). The limited presence of African-American teachers. *Review of Educational Research, 63*, 115–149. doi:10.3102/00346543063002115.

Kozol, J. (1991). *Savage inequalities: Children in America's schools*. New York: Crown.

Kuran, T. (1995). *Private truths, public lies: The social consequences of preference falsification*. Cambridge, MA: Harvard University Press.

Low, W. A., & Clift, V. A. (Eds.). (1981). *Encyclopedia of Black America*. New York: McGraw-Hill.

McLaughlin, D., O'Donnell, C., & Ries, L. (1995). *Private schools in the United States: A statistical profile, 1990–91*. Washington, DC: U.S. Department of Education. NCES 95-330.

Mingione, E. (Ed.). (1996). *Urban poverty and the underclass: A reader* (Studies in urban and social change). Oxford: Blackwell Publishers.

Molesky, J. (1988). Understanding the American linguistic mosaic: A historical overview of language maintenance and language shift. In S. L. McKay & S. C. Wong (Eds.), *Language diversity: Problem or resource? A social and educational perspective on language minorities in the United States* (pp. 29–68). Cambridge: Newbury House Publishers.

Moore, J. (2010). Darwin's progress and the problem of slavery. *Progress in Human Geography, 34*, 555–582. doi:10.1177/0309132510362932.

National Center for Education Statistics (NCES). (2012). Reading 2011: National assessment of educational progress at grades 4 and 8 (NCES 2012–457). Retrieved from http://nces.ed.gov/nationsreportcard/pdf/main2011/2012457.pdf

National Center for Education Statistics (NCES). (2013). *Digest of Education Statistics: 2013 Tables and Figures (Table 219.70)*. Washington, DC: U.S. Department of Education. Retrieved from http://nces.ed.gov/programs/digest/d13/tables/dt13_219.70.asp

National Center for Education Statistics (NCES). (2014). *A first look at mathematics and reading: National Assessment of Educational Progress [NAEP] at Grades 4 and 8* (p. 9). Washington,

DC: U.S. Department of Education. Retrieved from http://nces.ed.gov/nationsreportcard/subject/publications/main2013/pdf/2014451.pdf

Pincus, F. L., & Ehrlich, H. J. (Eds.). (1994). *Race and ethnic conflict: Contending views on prejudice, discrimination, and ethnoviolence.* Boulder: Westview Press.

Ravitch, D. (1974). *The great school wars: New York City, 1805–1973: A history of the public schools as battlefield of social change.* New York: Basic Books.

Ravitch, D. (1985). *The schools we deserve: Reflections on the educational crises of our times.* New York: Basic Books.

Reese, W. J. (1995). *The origins of the American high school.* New Haven: Yale University Press.

Rex, J. (1997). The nature of ethnicity in the project of migration. In M. Guibernau & J. Rex (Eds.), *The ethnicity reader: Nationalism, multiculturalism and migration* (pp. 269–283). Cambridge: Polity Press.

Reynolds, D. R., & Shelley, F. M. (1990). Local control in American public education: Myth and reality. In J. E. Kodras & J. P. Jones III (Eds.), *Geographic dimensions of United States social policy* (pp. 107–133). London: Edward Arnold.

Rippa, S. A. (1988). *Education in a free society: An American history.* New York: Longman.

Romano, R. C., & Raiford, L. (Eds.). (2006). *The civil rights movement in American memory.* Athens: The University of Georgia Press.

Schlemper, M. B. (2007). From the Rhenish Prussian Eifel to the Wisconsin Holyland: Immigration, identity and acculturation at the regional scale. *Journal of Historical Geography, 33,* 377–402. doi:10.1016/j.jhg.2006.05.001.

Schneider-Sliwa, R. (1996). 'Hyper-Ghettos' in amerikanischen Großstädten: Lebensräume und Konstruktionsprinzip der urban underclass ["Hyper-Ghettoes" in large American cities: Living environments and construction principle of the urban underclass]. *Geographische Zeitschrift, 84,* 27–42.

Stewart, J., Jr., Meier, K. J., & England, R. E. (1989). In quest of role models: Change in Black teacher representation in urban school districts, 1968–1986. *Journal of Negro Education, 58,* 140–152. doi:10.2307/2295588.

Story, R. (1988). Harvard students, the Boston elite, and the New England preparatory system, 1800–1876. In B. E. McClellan & W. J. Reese (Eds.), *The social history of American education* (pp. 73–90). Urbana: University of Illinois Press.

The Nation's Report Card. (2014). 2013 mathematics and reading: Grade 12 assessments. Retrieved from http://www.nationsreportcard.gov/reading_math_g12_2013/#/reaching-proficient

Theodore, N. (2010). Urban underclass: The wayward travels of a chaotic concept. *Urban Geography, 31,* 169–174. doi:10.2747/0272-3638.31.2.169.

Trueba, E. T. (1976). Issues and problems in bilingual bicultural education today. *NABE, 1*(2), 11–19. doi:10.1080/08881685.1976.10668275.

Wacquant, L. (2008). *Urban outcasts. A comparative sociology of advanced marginality.* Cambridge: Polity Press.

Wiarda, J. -M. (2006, July 27). Einfach aufgegeben. In den USA versagen die Highschools: 30 Prozent der Schüler eines Jahrgangs verlassen ohne Abschluss die Schule [Simply given up: High schools failing in the United States: Thirty percent of one student's age group drop out]. *Die ZEIT,* p. 67. Retrieved from http://www.zeit.de/2006/31/C-Dropouts

Wilkerson, D. A. (1970). *Special problems of Negro education.* Westport: Negro Universities Press. (Original work published 1939).

Yeakey, C. C. (1988). The public school monopoly: Confronting major national policy issues. In D. T. Slaughter-Defoe & D. J. Johnson (Eds.), *Visible now: Blacks in private schools* (Contributions in Afro-American and African studies, Vol. 116, pp. 284–307). New York: Praeger.

Chapter 5
Educational Inequalities Reflecting Sociocultural and Geographical Embeddedness? Exploring the Place of Hispanics and Hispanic Cultures in Higher Education and Research Institutions in New Mexico

Tim Freytag

First and foremost, education is committed to the double aim of building human capital and fostering social cohesion and integration. Although educational institutions are instrumental in qualifying students and producing human capital that can serve as an economic resource, the extent to which educational institutions can promote social and cultural integration remains a question. As the French sociologists Pierre Bourdieu and Jean-Claude Passeron (1970, 1985) argued, one can understand education as a mode of social reproduction that perpetuates socioeconomic and cultural inequalities. From that perspective the family background of a child bears on his or her prospects for educational success. However, there are many children and students who, though having a less favorable background, have earned the highest academic degrees and have assumed leading professional positions. Obviously, educational institutions can play an ambiguous role in either fostering or hindering social mobility.

Exploration of the vast set of factors that determine the chances for educational success poses a particularly great challenge. First, these factors reside in different dimensions, such as a person's capacities and dispositions for learning, the individual's family background or social milieu, the availability and accessibility of educational infrastructure, the quality of the teachers and the schooling provided, and the given educational aspirations and related opportunities on the labor market. None of these factors stands alone; they all work together in a complex interplay that researchers can study through the conceptual lens of intersectionality (Collins 2000; Crenshaw 1989; McCall 2005).

T. Freytag (✉)
Faculty of Environment and Natural Resources, Freiburg University,
79085 Freiburg, Germany
e-mail: tim.freytag@geographie.uni-freiburg.de

© Springer International Publishing Switzerland 2016   93
P. Meusburger et al. (eds.), *Ethnic and Cultural Dimensions of Knowledge*,
Knowledge and Space 8, DOI 10.1007/978-3-319-21900-4_5

The aim of this chapter is to improve the understanding of the part that sociocultural and geographical embeddedness has in the production of educational inequalities. Drawn from the work of the economists Karl Polanyi (1944) and Mark Granovetter (1985), the concept of embeddedness refers to the economic development occurring within a specific framework of institutions and actors. The guiding idea is that one can explain spatial inequalities in an economy by identifying and analyzing specific frameworks within which economic activities take place. Similarly, one can explore the social and cultural embeddedness of educational institutions and actors (Freytag and Jahnke 2015). Revisiting one of my previous studies (Freytag 2003a, b), this chapter focuses on the place of Hispanics and Hispanic cultures in higher education and research institutions in the southwestern U.S. state of New Mexico. In the following pages the term *sociocultural and geographical embeddedness* means that actors and institutions are embedded in a context or milieu of sociocultural relations, including shared habits and values. It also means that this context or milieu has a geographical dimension—not a clearly delimited territory but rather a shared space of day-to-day practices. One may view this approach as a contribution to the wider field of the geographies of education (Holloway et al. 2010; Meusburger 1998).

I begin by trying to measure the share of the Hispanic population and their educational attainment and by comparing it with that of other ethnic and cultural population groups represented in U.S. census data. The results show the extent to which Hispanics are underrepresented in higher education in the United States, particularly at the University of New Mexico (UNM) in Albuquerque. The second part of this study presents an exploration of three different aspects of sociocultural and geographical embeddedness that can deepen the understanding of the Hispanic population's educational attainment. I focus on UNM to examine the embeddedness of the students, the teaching staff, and the educational institution itself. I conclude the chapter by outlining how all these three aspects of embeddedness can be seen as structural barriers instrumental in affecting the educational attainment of many, but by no means all, Hispanic students.

Educational Attainment of the Hispanic Population

The terms *Hispanic* and *Latino* are often used interchangeably, although they do not have exactly the same meaning. Hispanic was used in the U.S. census in 1970, 1980, and 1990 to designate people identifying with Spanish language or culture referring to Spain and to the countries and regions with a predominantly Spanish-speaking population in the Americas. In the 2000 and 2010 U.S. census *Latino* encompassed non-Spanish languages and cultures located in Central and South America but not in Spain and Portugal. In this chapter I use the term *Hispanic*, which designates a type of ethnicity, not a specific race (Hispanics are mostly associated with the racial category *White*). Combining the identifiers of ethnicity and race, one can speak of *Hispanic White*, for example (see Table 5.1).

Table 5.1 Population in New Mexico and the United States by Race and Ethnicity, 2010 (in Percentages)

	New Mexico			U.S.		
	Hispanic or Latino	Not Hispanic or Latino	Total	Hispanic or Latino	Not Hispanic or Latino	Total
One race	44.0	52.3	96.3	15.3	81.8	97.1
White	27.9	40.5	68.4	8.6	63.8	72.4
Black or African American	0.3	1.7	2.0	0.4	12.2	12.6
American Indian and Alaska Native	0.9	8.5	9.4	0.2	0.7	0.9
Asian	0.1	1.3	1.4	0.1	4.7	4.8
Native Hawaiian and other Pacific Islander	0.0	0.1	0.1	0.0	0.2	0.2
Some other race	14.8	0.2	15.0	6.0	0.2	6.2
Two or more races	2.3	1.4	3.7	1.0	1.9	2.9
Total	46.3	53.7	100.0	16.3	83.7	100.0

Data from U.S. Census Bureau (2010)

Table 5.1 shows that a notable share of Hispanics exists in the U.S. population, particularly in the state of New Mexico. The Hispanic population in New Mexico includes two particularly large subgroups: (a) immigrants from Mexico ("Mexican Americans") and (b) "Hispanos," who represent the descendants of Spanish immigrants and settlers from the sixteenth century until the end of the Spanish colonial era and the subsequent independence of Mexico in 1821. Because the U.S. census data widely derives from self-assessment, it is important to remember that the categories of race and ethnicity are socially constructed and appear closely connected with questions of identity. The respondents are free to express their belonging to one or more racial and ethnic categories.

With regard to educational attainment, Hispanics are not well represented in the U.S. population of persons 25 years old or more who hold a college or university degree (see Fig. 5.1). According to the American Community Survey (cited in Ryan and Siebens 2012, p. 6), more than one third of the Hispanic population in the United States in 2009 had no high school diploma, General Educational Development test (GED), or alternative credential. Only one third of the Hispanics in the United States had some college or university experience, including an associate's or bachelor's degree in some cases, with higher university degrees remaining an exception. Figure 5.1 illustrates that the non-Hispanic White population and the Asian population experience considerably higher educational success than Hispanics and Blacks do in the United States.

A closer look at the academic degrees awarded by UNM between 1992 and 2011 reveals an almost constant increase in the share of Hispanics receiving the bachelor's degree and the master's degree (see Fig. 5.2). However, the share of Hispanics

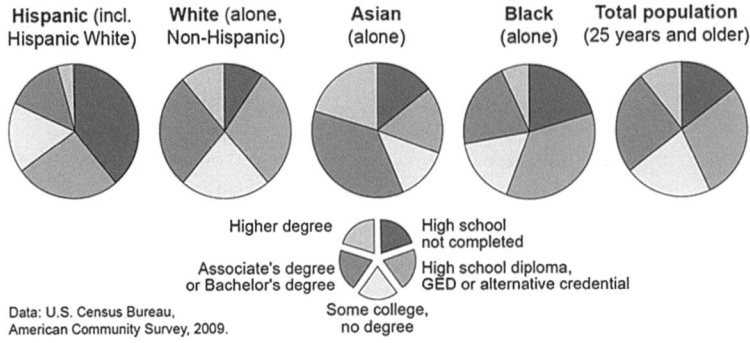

Fig. 5.1 Academic degrees earned by the U.S. population 25 years old or more, by race and ethnicity, 2009 (in percentages) (From U.S. Census Bureau and the American Community Survey, 2009, as cited in Ryan and Siebens (2012, p. 6))

completing a doctorate did not grow appreciably over the same period. Moreover, the share of higher-education enrollment among Hispanic students continues to lag behind their share of the overall Hispanic population in New Mexico, particularly as a percentage of all students acquiring higher degrees.

It is striking that Hispanics show comparatively low educational performance. Supposedly, the categories of ethnicity and race make a larger difference to educational success on average than do income or other socioeconomic characteristics. Other researchers, have explored this phenomenon (e.g., Herrnstein and Murray 1994), claiming that racial inequalities are a result of cognitive or intellectual abilities that vary according to race. In the following section, I take a different stance, arguing and illustrating that sociocultural and geographical embeddedness is a critical factor that can play a decisive role within the complex circumstances of educational attainment.

Educational Inequalities and the Sociocultural and Geographical Embeddedness of Students, Teachers, and Educational Institutions

The following sections are based on my fieldwork conducted in New Mexico from 1999 to 2002. Much of this work was published in a doctoral dissertation in German (Freytag 2003a) and in a contribution to a book in English (Freytag 2003b). The fieldwork in New Mexico consisted mainly of a series of 19 semistructured interviews with professors and decision-makers at UNM and was supplemented by observations and archival documents. The analysis also drew on U.S. census data and other statistical compilations. Writing now from the perspective of sociocultural and geographical embeddedness, I reflect on selected findings of that earlier research in order to explore how students, teachers, and teaching institutions can be instrumental in producing educational inequalities across the categories of a race and ethnicity.

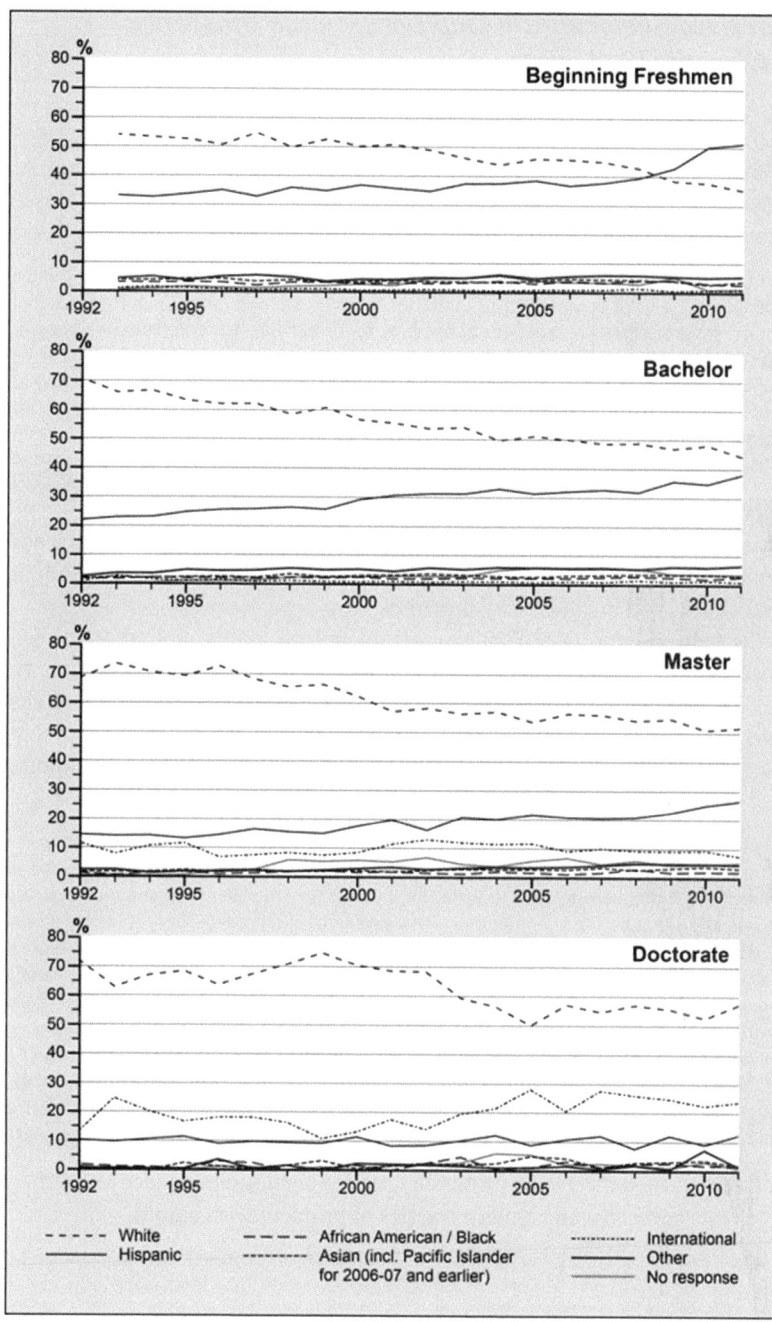

Fig. 5.2 Academic degrees earned at the University of New Mexico (UNM), main campus, by race and ethnicity, fall 1992–2011 (in percentages) (Data from UNM Factbooks and Official Enrollment Reports)

A Perceived Sociocultural Gap Between the University and Hispanic Students

With the label *Hispanic* denoting a category closely related to identity and self-assessment, the construction of difference between non-Hispanic White (or Anglo-Americans) and Hispanics rests primarily on the idea of opposed identities. This matter of identity was illustrated in the following statement by one of the interviewees:

> I grew up speaking both languages and the family home was definitely bicultural. I think we probably could have done a sort[ing] in terms of Mexican food / American food, Mexican music / American music, English / Spanish. A lot of our life was compartmentalized that way; I mean, I never thought that pork chops, for example, were Mexican food—well, Mexicans think that pork chops are Mexican, but in my classification system of life I would have said that's American food. Sliced bread, I would have said American food. And I think that we were very aware of what parts of our lives fit where. (Interview no. 14, Freytag 2003a, p. 156)

In conceptual terms, I would like to avoid thinking of ethnicity and identity as essentialisms and would rather take these notions as being part of an enduring dynamic process. They are in a constant process of reproduction (i.e., stabilization) and transformation in day-to-day practices.

In the field of education, Hispanic identities form partly around the difficulties supposedly occurring in educational attainment. This identity formation results from how Hispanics see themselves and how others see them. In some cases it appears to be rather difficult to combine university studies with the expectations and duties of the family. In addition, the family may not be able to give its children the kind of support they need during their university studies.

> When a student comes home and he just got a bad grade and he feels that he is in the process of losing a scholarship and his family doesn't understand what's going on, they don't get what all those numbers are and so forth. So, how do the students try to communicate with their families that they are running into some problems and not feel[ing]—because it is not so much that the parents don't care—they care very much, but they don't understand a system that is very complicated, that is very, very complicated. I know that when I came to college, it was up to me to get myself registered, it was up to me to pick my classes, it was up to me—and for as much as my parents wanted to help, they don't understand the system…So, our students struggle with how do I get my father to understand that I can't go to a family outing because I have to study. And yet for the family you come from it is extremely important. You have to go to baptisms and you have got to go to all the rosaries and funerals—and, so, your family can't understand. (Interview no. 17, Freytag 2003a, p. 166)

In the eyes of some UNM professors, the sociocultural distance and difference that students perceive can result in a series of problems on campus.

> Students can end up being very confused—I think, more our younger students. Our older students, our juniors and seniors are a little bit more steady, they have learnt how to even cope or deal with it and lose it. But, I think, our young ones are caught between—we were raised with obviously everybody older than we, we respected [them], and we gave [them] that respect. And so, teachers were always right, professors are always right. Having talked back to a professor here you wouldn't dare even tell your folks "*I don't agree with this professor*" because then your folks would say "*Don't ever speak like that. That's a PhD you*

are talking to, that's a doctor you are talking to. They know it, you don't." And it was just built on that respect that you never question that authority. I think, we get a little bit more steady. However, we still have that—we respect tremendously the positions of power and that just takes us back to where we were raised and what culture is here. And so, for a professor in the classroom to demand a certain conversation or to demand a certain debate or to demand "*You look me into the eyes*", and so on, and so on—he has no idea what this culture is about. If a professor who chooses to teach at this university took a look at the intense diversity or lived with a Hispanic, American Indian, Black family in this area, he might react differently while teaching classes. (Interview no. 17, Freytag 2003a, p. 165)

First-generation Hispanic students appear to be particularly vulnerable, especially if they come from rural regions and economically rather low family background.

[E]conomically the people in Kansas, not all of them but the majority come from middle class, middle middle class and very stable environments whereas here in New Mexico you don't find that as much. I think, especially Hispanic students, not all of them but a lot of them are first-time college students—not all of them but a lot of them are. And a lot of them come from small communities in New Mexico and, I suppose, coming to a place like UNM is a very frightening experience. [Albuquerque] is the only big city in New Mexico; there is one third of the state's population here. So, it is a very frightening experience and, I think, a lot of people don't come because of that and for financial reasons as well, and they go to New Mexico State University where it is all a bit smaller. (Interview no. 13, Freytag 2003a, pp. 162–163)

Several of the interviewees pointed out that Hispanics who grew up in New Mexico tend to feel particularly attached to the cultural and physical environment of this U.S. state.

[I]f you grew up in a Hispanic home it is that one feels with the warmth, the caring, the nurturing, I think, the nurturing that comes from not only a lot of good food and music and conversation and lessons from the grandparents but it is such a connected feeling—and a lot of our New Mexican, our native New Mexican students receiving their teaching degrees want to teach here. They are not interested in teaching on the East or West Coast, they want to teach in this state and so, "*once a New Mexican, always a New Mexican*" is that connection here. (Interview no. 17, Freytag 2003a, p. 157)

I knew I belonged to some place. I knew I belonged in New Mexico,...I belonged to some place in the world. There was never any doubt. And there were people who didn't, who didn't feel that. I have always felt that all this Spanish here, this is my *tierra*, this is my land. No, I don't mean in ownership sense, this is what I belong to. And that is the result of my family. That's not, and that's what I give to my girls, I think they have a place where they belong to. (Interview no. 14, Freytag 2003a, p. 159)

Conflict situations with the family tend to occur particularly when the prospect of further qualification or professional career opens up, bringing with it the challenge of geographical mobility. There may arise the question of whether qualification and professional career can be compatible with the family. Students can therefore find themselves caught between deciding whether to remain near the family or to opt for geographical and cultural separation.

[Our students] are primarily from New Mexico. When I was an undergraduate we were from all over the place. But I think that's a part of the University of New Mexico and the state of New Mexico more than it is a changing dynamic across the country. New Mexico students tend to stay, I think, relatively close to the Southwest. And many of them want to go to either UNM or New Mexico State because when they get their degrees they attempt to stay in New Mexico[.]…

Many of them have extended families in New Mexico, it is a very small-town mentality here. Money?—No, I think it really does have to do with the family and extended family, especially as most of the towns are so small. I have one student, for example, who is getting multiple honors. She is going to go to law school and has been accepted to Michigan, Washington, University of California in Berkeley, and UNM. And she still thinks that UNM may be the place that she goes, where UNM's reputation as a law school is not even close to the other places. I'm trying to encourage her very much to go out of state but she keeps telling me "*all my family is here, and the job I want to get is here*". I think that may be the case but I'm really not sure. (Interview no. 9, Freytag 2003a, p. 190)

The predicament that education may lead to geographical and social mobility surfaces in a traditional Hispanic saying: *Educar un muchacho es perder buen pastor* (Educating a boy means losing a good shepherd). As one of the interviewees confirmed, the period of studying at a university may be lived as an experience of cultural transformation and alienation:

Well, for example, I have a student who is from rural New Mexico from a rancher, his family has a ranch and he continues to come in his cowboy boots and with a big hat. But he is an exception because most of the kids who come in, whatever their racial background is, undergo barriers and change their clothing. Somebody is not going to arrive in cowboy boots and cowboy hats. I think that moving in and out of identities is—in fact, for some students it can be psychologically painful. I think for some students it can be psychologically testing, and for others it is a skill. They can go home and then their family is their way and they come here and they are in another way, it is not—for some of the students it is not an easy divide to cross. (Interview no. 14, Freytag 2003a, p. 182)

The general observation that Hispanic students tend to feel a sociocultural gap between the educational institution and themselves does not hold uniformly for all Hispanics. Considerable difference exists, particularly at the level of an individual's experiences. However, the collected data suggest that the experience of difference is common among many Hispanic students and the university teaching staff in how they see themselves and how others see them. In this context one can thus speak of ethnicity and cultural identity being functional as a structural barrier in the attainment of higher education.

The Role of Teaching Staff from Outside New Mexico

As shown by the career paths of the interviewees, a sizeable majority of the UNM professors did not grow up or study in New Mexico (see Fig. 5.3). A large percentage of them, especially among the non-Hispanic teaching staff, moved to UNM from other parts of the United States. Many of these academics pursued their own studies and early career at universities located, for example, on the East or West coast of the country. One of the interviewees confirmed and critically commented on this pattern:

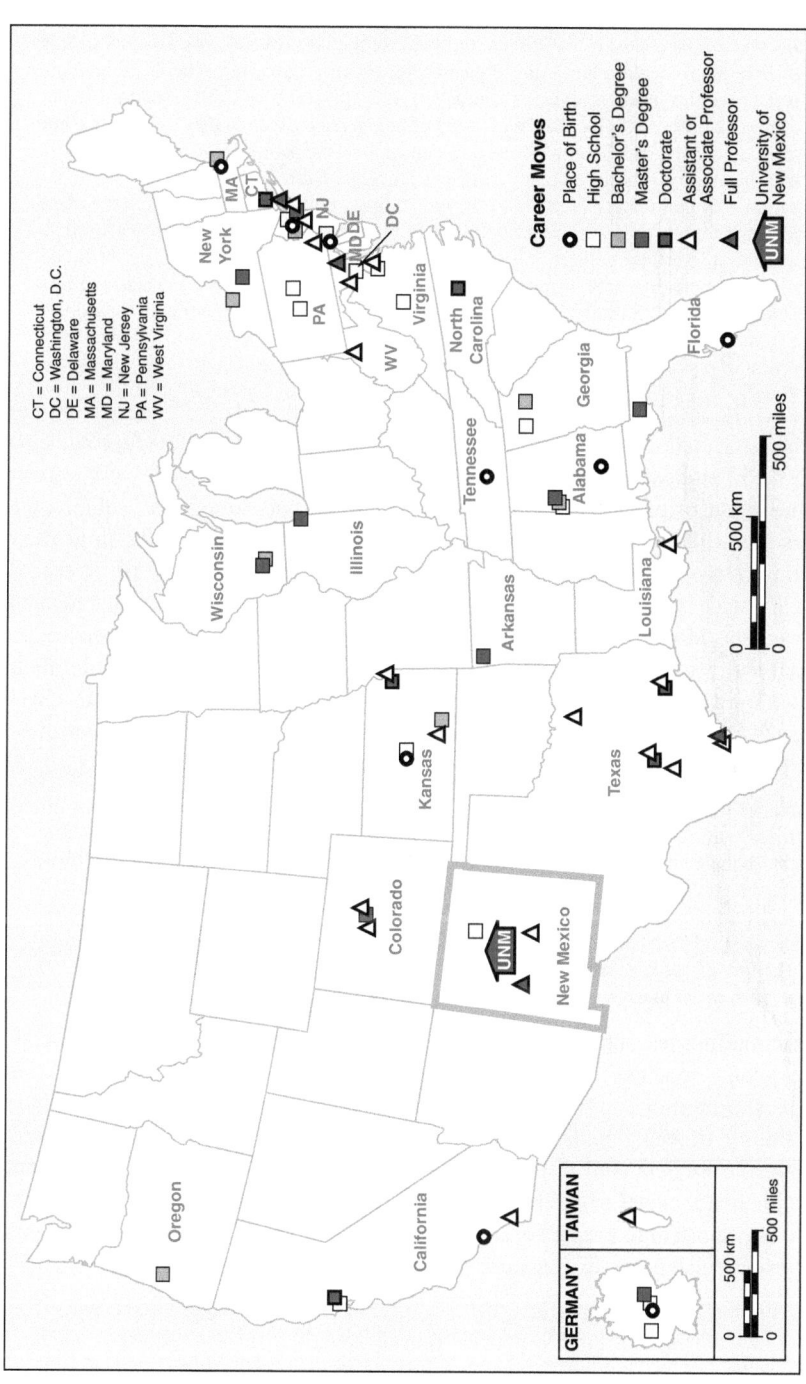

Fig. 5.3 Career moves of selected University of New Mexico professors. CT (Connecticut), DE (Delaware), DC (Washington, DC), MA (Massachusetts), MD (Maryland), NJ (New Jersey), PA (Pennsylvania), WV (West Virginia) (Adapted from Freytag (2003a, p. 175))

About 90 percent of our faculty come from other institutions in the United States—much, much more selective institutions. They come from these institutions because those are the schools that produce most of the faculty and they are the best schools, you know. And it is everywhere from Harvard and Yale, Princeton to the very good, big schools in Indiana, Pennsylvania, Wisconsin, Michigan to very, very good private schools like Stanford or big powerful state universities like Berkeley, you know, and the problem is that those people not only don't understand the Southwest and don't understand the nature of the Hispanic culture, but they have very little in common with our students. Most likely they were not first-generation college students themselves, they most likely come from an upper middle-class Anglo White family, they were very intelligent from the very beginning, they usually didn't have trouble to get into college and got into graduate school, got into the best graduate schools. They actually come from another world, but because they have a job teaching in a state university they think, well, it should be the same. But it is not. Actually they don't come from the same world because so many of our students are minority students. And only 13 percent of our faculty is minority. (Interview no. 19, Freytag 2003a, p. 177)

Apparently, this faculty member's colleagues from outside New Mexico do not share the same student experience (e.g., living on campus, being a full-time student) as their UNM students. Many of the students are older than their professors were at the same stage of education, and they already have part-time jobs or their own families and children to take care of, responsibilities that sometimes limit their opportunities to engage completely in a full-time program of study. The diverging experience with studying is often perceived as a difference separating students and their teachers. Many of the university teachers take their personal experience as universal and regret it when their standards and expectations are not met by their students. The gap, or divide, that then emerges between students and their teachers may lead to misunderstanding, frustration, and suboptimal student grades. As confirmed by one of the interviewees,

the motivation of students to work hard and to be successful has severely decreased over the years. And one seldom sees the same types of work ethic in undergraduate students and graduate students in the United States that one saw back in the 1960s and 1970s. I think this is particularly true in graduate education where American students come to places like the University of Maryland, the University of New Mexico and are not motivated at all. My own personal preference nowadays is to take students from foreign countries into my research group, whoever they are. The foreign students share the same work ethics that I had when I was younger and so I find it much easier to work with them. (Interview no. 1, Freytag 2003a, p. 177)

These findings are supported by the observation that the Hispanic teaching staff is often keenly sensitive to the particular situation of Hispanic students. Hispanic professors sometimes regard themselves as a role model. They exemplify the compatibility of being both Hispanic and successful in higher education. Such Hispanic role models tend to project their own career paths onto the educational paths of their students. For example, a Hispanic professor explained in the interview that it was important to prepare and expose the sometimes fairly parochial students from New Mexico to the experiences of a larger world:

[Our students] don't quite know, particularly the ones from first-generation, what's expected in a college education and in a broader perspective, what they will face in a larger world. They tend to be very parochial, you know, they tend to be from New Mexico—and in New Mexico they simply haven't been exposed to a lot of other experiences. They don't know what else is out there in the rest of the world. So, part of the challenge of teaching is trying to expose them to that. There are places in which the world doesn't operate like it does in New Mexico…And this is a state where education has not always been highly valued. (Interview no. 8, Freytag 2003a, p. 179)

So what is the underlying way in which Hispanic professors at UNM see and position themselves? How do they combine their role as a university teacher with their cultural identity as Hispanics? When conducting the interviews at UNM, I found it striking that many Hispanic professors furnished their offices with objects of Hispanic culture and tradition (e.g., figurines and all kinds of knick-knacks and art) that signify cultural belonging. Bilingualism, too, plays an important part in day-to-day working life at UNM. Hispanic instructors on the staff occasionally use Spanish expressions or phrases in their conversations, which they hold chiefly in English within the professional university context.

During my fieldwork, I observed that the Hispanic teaching staff seemed perfectly at ease in a principally Anglo-American setting at the university. From time to time, and in particular situations and particular places (especially in one's own office), many colleagues did have the wish to manifest their Hispanic belonging. It almost seemed as though they might be afraid of losing their cultural belonging. In some cases, it even seemed as though they were lamenting such a loss and were trying to compensate through an act of reconstruction—by artificially staging Hispanic culture in their working environment.

Observing the Hispanic professors and their Hispanic students makes it clear that educational institutions and the related educational system cannot be looked upon as "neutral" toward cultural norms and identities. In fact, a difference opposed to Hispanic culture and traditions comes across at UNM—and probably even more so at other institutions of higher education in the United States. This opposition does not stem exclusively from a host of university professors who were themselves socialized as students and academics at other universities outside New Mexico and who are now bringing their experiences into UNM. It has also to do with the history of how UNM and, more generally, the framework of higher education and research institutions in the United States have been shaped over time.

Tracing Back the History of Institutionalized Education in New Mexico

To advance the understanding of why and how the University of New Mexico and other educational institutions were established and how they have been culturally shaped over time, this section summarizes the history of education in New Mexico and the rise of the University of New Mexico.

Having been part of the Imperial Spanish Viceroyalty of New Spain (1598–1821) and the independent state of Mexico (1821–1848), New Mexico became a U.S. territory in 1850, 2 years after the Mexican–American war, and finally gained U.S. statehood in 1912. During the Spanish and Mexican eras, institutionalized education was limited to a few Catholic mission schools (Gallegos 1992). Under U.S. rule a slightly increased number of private and public schools were established in the 1870s and 1880s as the territory modernized and as more and more White Anglo-American and European settlers arrived (Moyers 1941).

The immigrants and settlers had their own cultural values and ideas about modern society. Subsequently, the process of modernization affected not just the economy but society as a whole. In the educational sector a public school system replaced most of the Catholic mission schools. Although this modern education system was open to Hispanic children, Hispanics felt increasingly marginalized and started to call themselves foreigners in their own land (Weber 1973). This trend intensified with the introduction of compulsory schooling in 1891. From 1870 to 1910, the literacy rate of the adult population rose from 20 % to 80 % (Seyfried 1934). The U.S. census indicates that the number of pupils increased from 4,755 in 1880 to 18,215 in 1890 and to 60,000 by the time New Mexico gained statehood in 1912. Comparing the literacy rate of the population in New Mexico with that in other parts of the United States, George Sánchez (1940) concluded that it was relatively low because New Mexico had not received sufficient support from the U.S. government since becoming a U.S. territory in 1850 (Freytag 2003b).

The establishment of a Jesuit college in Las Vegas, New Mexico, in 1877 was followed by the founding of New Mexico State University in Las Cruces (1888) and UNM in Albuquerque (1889). The curriculum of these institutions was widely shaped by Anglo-American and European standards.

> The U.S. system of higher education had formed among English-speaking dwellers of the Atlantic seaboard and the Ohio Valley, not in the deserts and mountains of the Southwest. Their efforts to graft a "classical" European education onto an essentially frontier society based upon traditions of competition and individualism often clashed with the family and community orientation of Native American and Hispano New Mexico. Compounding this was the ambiguous premise of American higher learning: to uplift the masses in a democratic society while ensuring a meritocracy of the learned and, some would say, the privileged. (Welsh 1996, pp. 107–108)

Subsequently, the New Mexico Military Institute was established in Roswell (1891); the Teacher Colleges (later transformed into New Mexico Highlands University and Western New Mexico University), in Las Vegas and Silver City (1893); the New Mexico Institute of Mining and Technology, in Socorro (1898); and Eastern New Mexico University, in Portales (1927). However, an expansion of higher education in terms of increasing numbers of students did not come about until the second half of the twentieth century (see Fig. 5.4).

> New Mexico experienced a continuing expansion in the school and university infrastructure, which can be referred to as a period of mass education. The main aim was equal opportunity in education in order to integrate all parts of society. At the same time various new institutions offering new types of higher academic and professional degrees evolved and attracted many students. At the University of New Mexico, founded in 1889 in Albuquerque, enrolment rose from 5,000 students in the 1950s to 25,000 students on the main campus and another 5,000 students on branch campuses in Fall 2000. A similar increase occurred in the New Mexico State University in Las Cruces, New Mexico's second-largest university with 15,000 students on the main campus and another 8,000 students on branch campuses in Fall 2000. (Freytag 2003b, p. 193)

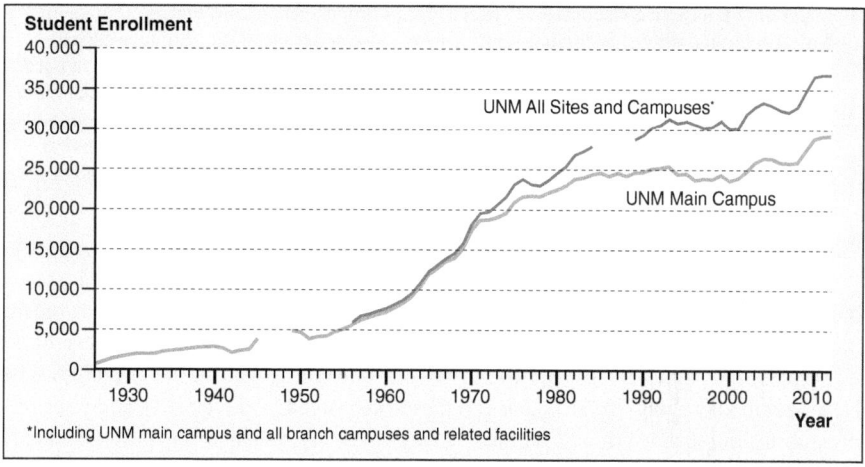

Fig. 5.4 Student enrollment at the University of New Mexico (UNM), fall 1926–2012 (Data from *UNM Factbooks and Official Enrollment Reports, 1926–2012*)

Throughout the second half of the twentieth century, higher education in New Mexico attracted increasing numbers of students and continued to diversify. The wider context of these trends was shaped by the overall dynamics of urbanization, economic restructuring, and the rise of advanced producer services and the knowledge society. To support ethnic and cultural minority students, the educational expansion in the state included the restructuring of curricula, which offered new programs such as Latin American studies and, more recently, the Chicano Hispano Mexicano studies program. All in all, however, the higher education system has largely remained an Anglo-American domain. That orientation prevails today despite the design of the UNM campus, which features a series of buildings incorporating traditional Hispanic and Native American architectural elements, notably the Adobe style. As a result, UNM may have a particular Hispanic ambiance created by its architecture, a relatively high share of Hispanic students, and the profile of specific programs of study, but it remains an institution of predominantly Anglo-American style.

Hispanic students and Hispanic teaching staff thus sometimes find themselves struggling between two conflicting cultural norms and value systems, This tension was pointed out by one of the interviewees:

[W]e tell our students a university setting is a learning institution and it must be reciprocal. This university must continue to learn from us as we continue to learn from that system. We agree that a certain amount of acculturation must take place where we must learn the system here. You have to take some hours [for] this and you have to pass it with a certain grade and got to listen to what the professor says because he is the one that is going to grade you, you know, all of that we need to learn definitely. We need to be opening up to take classes in the classics that—very, very important to me—maybe you were not raised with, but at the same time retain your language, retain your culture, retain your tradition and most definitely pass it down to the younger ones in our community. So I think that mentoring is extremely important for us here at the university because of the young generations that are coming up. Now I feel that we are becoming a much more educated population, formally educated population and I feel that the younger are doing a much better job of trying to look at our youth and support them but unfortunately we are losing our language and we are losing a lot of our culture and traditions because of that. It takes a very special person to try to put them both on that same level. It has to be one or the other. Either we push education and we lose our culture or we push our culture and we can't fit into education. (Interview no. 17, Freytag 2003a, p. 185)

The rather conflictual situation that can be perceived at UNM contrasts with the seemingly much more sensitive and accommodating response in New Mexico's public primary, intermediate, and secondary schools. This difference shows up in the proportion of high school faculty members accounted for by Hispanic teachers; it is much higher than the proportion of Hispanic teaching staff at the universities in New Mexico (Freytag 2003a). One can therefore consider universities and research institutions in New Mexico as eminently Anglo-American islands within the educational landscape of this U.S. state. Consequently, the transfer from high school to college or university can be particularly challenging for a Hispanic student, who may have to make the transition from a familiar cultural context into another world.

Conclusions and Perspectives

The attainment of an academic degree should not be seen as good *per se* or as a universal norm. If Hispanics are less represented among highly skilled and well-educated persons than other ethnic groups are, it may be that the transaction costs of succeeding in education are comparatively great for Hispanic students and that it does not necessarily make sense for all Hispanic students to put priority on acquiring higher academic degrees. However, the broad assumption is that access to education should be equal and open to all ethnic or cultural groups.

This chapter has explored the sociocultural and geographical embeddedness of educational inequalities with a focus on Hispanic students and professors at UNM. I have shown that educational institutions, professors, and students appear to be embedded in a specific sociocultural context that has been shaped over time. One can thus take educational inequalities in this context to be a product of, or at least something reinforced by, difference that a student perceives between him- or herself, an educational institution, and its teaching staff. The particular importance of cultural identity and embeddedness is reflected in the recurrent difficulties of Hispanic students and is underlined by Hispanic professors who try to find a place for or improve the integration of Hispanic culture in their professional environment.

The findings from this study hold not only for Hispanic students but, presumably, for other ethnic and cultural minority groups as well. One can assume that a similar pattern of cultural distance exists among Native American students, for instance. By the same token, the widely praised educational success of a model Asian minority—an achievement that should not obscure the substantial educational disparity among the

Asian populations in the United States (Chou and Feagin 2008)—is arguably a result of socioculturally produced and embedded norms and expectations.

Focusing on the social constructions of ethnicity and cultural identity, which tend to suggest a given homogeneity, only belies the fact that individual paths of education and professional career are involved. Educational success and educational inequalities should be regarded neither as merely resulting from one's ethnic or cultural belonging nor as simply depending on the meaning and significance attributed to educational attainment in an overarching concept, such as social class, socioeconomic status, life-style, or gender. In fact, educational attainment arises from a complex interplay of various parameters operating at the individual, collective, or structural levels. Given the complexity and interrelatedness that academic achievement entails, I find the concept of intersectionality useful for the analysis of educational inequalities.

It is important to be aware of existing barriers and restrictions and to take into account the educational opportunities that can emerge from the various relevant parameters at work in education. The insight that the sociocultural and geographical embeddedness of educational institutions can be instrumental in fostering educational inequalities provides a suitable background for conducting critical research in education and for advancing engaged educational politics. The latter pursuit constitutes an important field in which Hispanics and other ethnic or cultural minority groups can actively help negotiate the content and structure of the curricula, shape specific support programs for minority students, recruit teaching staff, and modernize the general framework and vision of education, especially of higher education, in and far beyond New Mexico.

References

Bourdieu, P., & Passeron, J.-C. (1970). *La reproduction. Eléments pour une théorie du système d'enseignement* [Reproduction in education, society, and culture]. Paris: Éditions de Minuit.

Bourdieu, P., & Passeron, J.-C. (1985). *Les héritiers. Les étudiants de la culture* [The inheritors: French students and their relations to culture]. Paris: Éditions de Minuit.

Chou, R. S., & Feagin, J. R. (2008). *The myth of the model minority: Asian Americans facing racism*. Boulder: Paradigm.

Collins, P. H. (2000). It's all in the family: Intersections of gender, race, and nation. In U. Narayan & S. Harding (Eds.), *Decentering the center: Philosophy for a multicultural, postcolonial, and feminist world* (pp. 156–176). Bloomington: Indiana University Press.

Crenshaw, K. (1989). Demarginalizing the intersection of race and sex: A black feminist critique of antidiscrimination doctrine, feminist theory, and antiracist politics. *University of Chicago Legal Forum, 140*, 139–167.

Freytag, T. (2003a). *Bildungswesen, Bildungsverhalten und kulturelle Identität: Ursachen für das unterdurchschnittliche Ausbildungsniveau der hispanischen Bevölkerung in New Mexico* [*Education, educational behavior, and cultural identity: Causes of the below-average education in the Hispanic population in New Mexico*]. (Heidelberger Geographische Arbeiten, Vol. 118). Heidelberg: Selbstverlag des Geographischen Instituts der Universität Heidelberg.

Freytag, T. (2003b). Mission schools, modernization, and mass education: Historical perspectives on a changing institutionalized education in New Mexico. In J. Nemes Nagy (Ed.), *Frontiers of* geography: *A selection from the wide range of geographical topics* (pp. 189–198). Budapest: Eötvös Loránd Universität.

Freytag, T., & Jahnke, H. (2015). Perspektiven für eine konzeptionelle Orientierung der Bildungsgeographie [Perspectives for a conceptual orientation of the geography of education]. *Geographica Helvetica, 70*, 75–88.

Gallegos, B. P. (1992). *Literacy, education, and society in New Mexico, 1693–1821.* Albuquerque: University of New Mexico Press.

Granovetter, M. (1985). Economic action and social structure: The problem of embeddedness. *The American Journal of Sociology, 91*, 481–510. doi:10.1086/228311

Herrnstein, R. J., & Murray, C. (1994). *The bell curve: Intelligence and class structure in American life.* New York: Free Press.

Holloway, S., Hubbard, P., Jöns, H., & Pimlott-Wilson, H. (2010). Geographies of education and the significance of children, youth and families. *Progress in Human Geography, 34*, 583–600.

McCall, L. (2005). The complexity of intersectionality. *Journal of Women in Culture and Society, 30*, 1771–1800. doi:10.1086/426800.

Meusburger, P. (1998). *Bildungsgeographie: Wissen und Ausbildung in der räumlichen Dimension* [Geography of education: Knowledge and education in the spatial dimension]. Heidelberg, Germany: Spektrum Akademischer Verlag.

Moyers, R. A. (1941). *A history of education in New Mexico.* Unpublished doctoral dissertation, George Peabody College for Teachers, Nashville.

Polanyi, K. (1944). *The great transformation: The political and economic origins of our time.* Foreword by Robert M. MacIver. New York: Farrar & Rinehart.

Ryan, C. L., & Siebens, J. (2012, February). Educational attainment in the United States: 2009. Retrieved 12 Nov 2013, from http://www.census.gov/prod/2012pubs/p20-566.pdf

Sánchez, G. I. (1940). *Forgotten people: A study of New Mexicans.* Albuquerque: University of New Mexico Press.

Seyfried, J. E. (1934). *Illiteracy trends in New Mexico.* Albuquerque: University of New Mexico Press.

U.S. Census Bureau (2010). 2010 Census. Profile of general population and housing characteristics: 2010 Demographic Profile Data. Retrieved 28 July 2014, from http://factfinder2.census.gov/faces/tableservices/jsf/pages/productview.xhtml?src=bkmk

Weber, D. J. (Ed.). (1973). *Foreigners in their native land: Historical roots of the Mexican Americans.* Albuquerque: University of New Mexico Press.

Welsh, M. (1996). Often out of sight, rarely out of mind: Race and ethnicity at the University of New Mexico, 1889–1927. *New Mexico Historical Review, 71*, 105–133.

Chapter 6
Local Cultural Resource Knowledge, Identity, Representation, Schooling, and Education in Euro-Canadian Contexts

George J. Sefa Dei

This chapter presents a discussion of ways in which African people's local knowledge of cultural resources can help enhance the schooling and education of young African learners. In the following pages the local knowledge (subjective voices) that high school and university students have with regard to their schooling experiences is used as legitimate knowledge in order to understand the challenges and possibilities of schooling in Euro-Canadian contexts. The discussion is grounded in the lessons of my past and ongoing schooling research on schooling and in the prism of Indigeneity and anticolonial thought helps re-envision schooling and education. I begin by reconceptualizing local cultural resource (local Indigenous knowledge) and the relevance of understanding current competing tasks and possibilities of education (broadly defined) in pluralistic contexts. Specifically, I theorize the link between identity, knowledge, representation, and schooling in the particular case of Black and minority youth education in Euro-Canadian contexts. I adopt an anticolonial perspective to argue that the epistemologies of Black and marginalized youth, particularly the ways in which these learners come to know and act within schools, offer interesting insights into why and how the questions of identity, representation, and social difference are critical for educational success. I see this chapter as an attempt to situate the local ways of knowing among young learners, minority parents, and educators at the center of theorizing and seeking ways to improve schools in response to the needs and concerns of a diverse body politic. We educators and all others who read this chapter can affirm learners' views and accounts of their schooling experiences while remaining critical and avoiding a reification of student voices.

Some people may argue that a river never returns to its source. But I like to work with the idea of a return to the source. After all, I maintain that every river contains

G.J.S. Dei (✉)
Department of Social Justice Education, University of Toronto,
252 Bloor Street West, Toronto, ON M5S 1V6, Canada
e-mail: gdei@oise.utoronto.ca

© Springer International Publishing Switzerland 2016
P. Meusburger et al. (eds.), *Ethnic and Cultural Dimensions of Knowledge*,
Knowledge and Space 8, DOI 10.1007/978-3-319-21900-4_6

the water from its original source. As Adjei (2010) insists, a river carries the source of its water with it (p. 2). The present is constitutive of the past, and people must learn from that history if they are to understand the present and contest the future. Seeking to reclaim students' agency and power in their own voices is actually something that has been around all the while, and the goal is to bring it to the foreground or surface. An important matter in rethinking education and schooling in Euro-Canadian contexts is what to do and how to re-engage with voices and experiences of Black, minority, and dominant-culture students from multiple vantage points. Drawing on anticolonial theory, I assert that, in order to reconceptualize schooling by using students' experiences, it is important to move beyond a simple narration of these experiences to their theorization (see also Butler and Scott 1992, pp. xiii–xvii). The affirmation of such knowledge is not just a search for an authentic voice, it is a recognition of the contestation of knowledge from different positions (see also Battiste 1998; Battiste and Youngblood Henderson 2000, pp. 35–56; Nakata 2007). It is also a recognition of legitimate voices that can no longer be discounted if one is to understand youth's experiences of schooling. The experiences that marginalized youth have of schooling as narrated by the learners themselves do not necessarily denote an "authentic experience." Research is required to bring a panoptic and critical view to such experiences and knowledge and to explore the limits and possibilities of understanding schooling from the voices of minority students (Dei 2010b; Dei and Kempf 2006; in other contexts see also Bhavnani and Davis 2000; Ernst et al. 1994; Fine and Vanderslice 1991).

I come to the topic informed by my extensive research on issues of minority youth marginality, disengagement, and resistance in schools of the Canadian education system, specifically that in Ontario. I have also done research on exemplary practices of inclusive schooling[1] that are known to promote academic success among youth. Critical research in this field has always embraced a "cultural politics of schooling" to acknowledge the intersections of race, ethnicity, gender, class, sexuality, ability, language, politics, religion, and their impact on the outcomes of educating young people (see Apple 1986, 2004; Apple and Weis 1983; Dei 1996; Fine and Vanderslice 1991; Giroux 1983, 2003; McCarthy 1980; Weis and Fine 2005; Willis 1977, 1983). In working with a nuanced understanding of educational success, I write this chapter to use students' narratives of schooling experiences to shift the conversation beyond a conventional, or dominant, educational agenda. Rather than employ research merely to "generate knowledge about a group of students" (Ernst et al. 1994, p. 2) in order, for example, to understand the academic experiences of successful students or to see these students as objects of knowledge, educational researchers need to focus on youth discourses of resistance and alternative visions of schooling and education that are richly informed by the interplay of history, culture, identity, and the politics of knowledge production (Dei et al. 1997).

---

[1] Inclusive schooling generally refers to practices in the conventional school system. By contrast, inclusive *education* refers to areas of education beyond the schools, including off-school sites.

Theory and Method

My theorization of the "anticolonial discursive framework" (Dei and Asgharzadeh 2001) borrows from the early ideas of Fanon (1963), Memmi (1965), Cesaire (1972), and Foucault (1980), among many others. As argued elsewhere (Dei 2000; Dei and Asgharzadeh 2001; Dei and Kempf 2006), the anticolonial discursive framework is a theorization of colonial and recolonial relations and the implications that imperial structures have for (a) processes of knowledge production, interrogation, validation, and dissemination; (b) the understanding of Indigeneity and the rise of an Indigenous consciousness; and (c) the pursuit of agency, resistance, and subjective politics by learners. An anticolonial reading offers a critique of how conventional schooling provides education that fails to help young learners develop a strong sense of identity, self, and collective agency and empowerment to community-building. Anticolonial theory is an inquiry into the power relations of schooling and the knowledge production processes that help perpetuate internal colonial hierarchies. The anticolonial prism affirms the role and power of local cultural knowledge to subvert such internalized colonial hierarchies of schooling by centering values such as social justice, equity, fairness, resistance, and collective responsibility in the education of the learner.

I contend that part of the colonial and recolonial relations of schooling surfaces in what is deemed valid or invalid and legitimate or illegitimate knowledge, that is, in the ways particular experiences of students are counted differently. Students' identities receive differential recognition, response, and validation. The merit badges of schooling are about privileging certain bodies, experiences, and identities. Through the anticolonial prism one comes to understand the nature and extent of social domination and how multiple places of power and the relations of power work to establish dominant and subordinate relations of schooling. The anticolonial prism offers critical scrutiny and deconstruction of dominant discourses and epistemologies and raises key questions about educators' practices of exclusion and marginalization of certain voices, identities, and knowledge systems.

The framework also highlights what ought to be—analyses and exploration of the contexts and alternatives to colonial relations. There is a particular governing ideology of schooling that privileges particular bodies of knowledge, experience, and history at the expense of others. Such ideologies end up with exclusive notions of belonging, difference, and inferior or superior ability to sustain hierarchies and systems of power. For example, colonial practices of knowledge production uphold versions of reason, authority, and control while scripting the colonized (disadvantaged or racialized) bodies as a violent "Other." These historical relationships of the dominant to the subordinate inform both the contemporary subject's identity formation and knowledge production in school. Dismantling colonial schooling relations and practices (i.e., popular colonial ideologies and mythologies of whiteness) requires a politics of resistance and of affirmation of the cultural knowledge and voices of the marginalized and oppressed youth. It requires engaging in practices of inclusion to ensure that all students have a sense of belonging and connectedness to their schooling and are able to exercise power to reclaim their agency and subjective voices.

In rethinking schooling and education in the Euro-Canadian context. I therefore first want to affirm the ways in which oppressed bodies relegated to the status of racial minorities come to claim a sense of intellectual and discursive agency as well as ownership and responsibility for their knowledge about everyday schooling. I see this recognition as part of the search for multiple ways of knowing about schooling and education from multiple vantage points. Second, re-envisioning and transforming education is about bringing to the fore subjugated voices—histories and experiences of marginalized students. Third, I regard the re-envisioning of education as being about legitimating the link between identity, schooling, and knowledge production and about affirming the idea that practice and experience are the contextual basis of knowledge. The process of creating a decolonized space entails the ability of learners and educators to acknowledge the link between identity, schooling, and knowledge production.

In evoking local cultural ways of knowing, my primary academic project is to examine the role, importance, and challenges of local cultural knowledges in the promotion of multicentric paradigms for schooling and education. Indigenous, or local, cultural resource knowledge is about bodies and representation. I affirm local cultural knowledge in order to assert the intellectual agency of marginalized subjects. I conceptualize and operationalize local ways of cultural knowing as more than simply non-Western forms of knowing. It is about the everyday knowing of local peoples, about their cultural resource base. It is knowledge that speaks to the ways learners from nondominant racial, ethnic, cultural, and religious backgrounds are schooled in the sociocultural and philosophical paradigms of their communities (see also Yankah 2004; Semali and Kincheloe 1999). Thus, validating such local ways of cultural knowing is about the search for "home-grown Indigenous perspectives steeped in culture-specific paradigms" (Yankah 2004, p. 26). Every knowledge producer and user must be grounded in their own local modes of thinking in order not to be alienated from his or her social world. This knowledge emerges from an awareness of the nexus, or interface, between society, culture, and nature and is based on experiencing a particular social world.

It is important to theoretically flesh out some of the tensions, contradictions, and paradoxes of knowledge production in the Western academy by highlighting the efficacy of Indigenous knowledge for schooling in the Euro-Canadian setting. The problem with claiming Indigeneity is not the concept per se but rather the manner in which it is used, evoked, or both. It is necessary to reconsider the complicated understanding of the word *Indigenous* as static, frozen in time and space, and confined to an exclusive domain of a particular group. However, I posit that the notion of static traditionalism sometimes exists only in the imagination of the critic of Indigenous knowledge. The concept of *Indigenous* is evoked in its myriad understandings as legitimate knowledge originating and associated with the land and its long-term occupancy as a place (see also Fals-Borda 1980). Indigeneity can also be a form of political evocation to resist the dominance of Eurocentric knowledge masquerading as universal knowledge. When claims to knowing the other are resisted, Indigenous knowledge is presented as a legitimate way of knowing that highlights the epistemologies of Indigenous and local peoples.

In this chapter I draw on my past and present longitudinal research work in Canadian schools to highlight the relevance of local cultural resources for youth education. Through research grants from the Social Science and Humanities Research Council of Canada (SSHRC), I have worked since the 1990s with graduate students at the University of Toronto to delve into the question of minority education, the challenges of inclusive education in comparative contexts, and the potential innovations that critical and inclusive education offers for redressing persistent inequalities and enhancing social development. For example, we examined the issue of school dropouts from the narratives of students who had dropped out of school and from those still in school (Dei et al. 1995). The sample represented diverse racial and ethnic backgrounds, mostly African-Canadian. The learning objective was to acquire a sense of the structural, sociopolitical, and human dimensions of the problem of leaving school prematurely. Through in-depth interviews and focus-group discussions with students, educators, and administrators, key research questions were taken up: What do youth like about school and why? What are the experiences that young people have with curriculum, classroom teaching, and instructional practices? Why do some students leave school prematurely? And what ought to be done in the context of the structures for teaching, learning, and educational administration to address the problem of youth dropping out of school? Field research using in-depth interviews and focus-group discussions centered on youths, parents, educators, and administrators to examine their interpretations of experiences with schooling. My research team and I also conducted ethnographic observations of school activities and classroom interactions (see Dei et al. 1995, 1997). Within 3 years we had interviewed 150 Black students from four Toronto high schools; over two dozen students randomly selected from metropolitan Toronto schools; 21 school dropouts (some of whom have since returned to school); 7 students designated as being "at risk"; 55 Black parents, caregivers, and community workers; 41 school teachers, including administrators; and 59 non-Black students (mainly White). School dropouts were contacted through our community ties; the so-called at-risk students were identified through a provincially funded summer jobs programs. We also reviewed school board documents, which included information (where available) on the racial, ethnic, and gender composition of the teaching and administrative staff and of the student population as well as a breakdown of Canadian-born and immigrant children and the latter group's countries of origin.

The study documented the ways in which the processes of teaching, learning, and the administration of education fostered youth disengagement from school (e.g., low teacher expectations, differential negative treatment by race, lack of curricular sophistication, absence of a diverse teaching staff, and constant struggles over identity among the youth). The study showed that an adequate exploration of the questions of class, gender, race, ethnicity, power, and history in the discussion of dropping out, particularly of students' lived experiences and social reality, reveals a complex of factors that lead to racial minority youth disengagement from school. The narratives of Black youth pointed out that the search for an understanding of the school dropout dilemma must be rooted in the institutionalized policies and practices of exclusion and marginalization that shape public schooling and characterize the out-of-school environment of many students.

Subsequent research my team and I conducted focused on identifying some of the exemplary practices of inclusive schooling (Dei et al. 2000). The learning objectives of this research were to pinpoint what inclusive educational practices foster successful academic and social outcomes for all students, particularly for those of racial and ethnic minorities. Again through in-depth interviews and focus-group discussions with students, educators, administrators, and parents, we asked what makes schools inclusive in terms of actual classroom practices, curriculum, pedagogy, instruction, school policies, and administrative practices. We supplemented interviews with ethnographic studies and reviews of documents on practices of inclusive education. The study examined best practice that makes for genuine inclusion of *all* students by addressing equity issues and promoting successful learning outcomes, especially among students of racial or ethnic minorities (see also Dei et al. 2002). In addition to inspecting actual educational practices, we asked students, parents, community workers, and educators to speak to us about aspects of schooling (pedagogy, instruction, and texts) that ensured a sense of inclusiveness for all learners. The study provided an opportunity for these interlocutors to share their reflections not only about inclusive schooling but also about what they felt to be some of the possibilities and limitations of current educational practices. The study has made it possible to learn about the programs and educational practices that schools, teachers, families, and local communities have initiated to enhance youth learning. There are, for instance, many educational initiatives outside school that seek to address the needs of diverse student bodies.

In a study that began in late 2005 (Dei 2010b), I focused on what makes some students succeed in schools whereas others fail. High academic achievers from diverse racial and ethnic backgrounds were interviewed for the varying conceptions they had of school success and failure. I also inquired into how students see the role that education and academic success play in enhancing social opportunity, social mobility, and individual life chances. The focus was on students whose academic excellence had earned them prestigious scholarships from the school or the community. These students had made the Dean's list and had come from a cross-section of the racial and ethnic backgrounds represented in the Ontario school system. I also interviewed school administrators, teachers, community educators, and parents. The study probed views the subjects had on how their academic performance and achievement motivation might be explained. The study documented, from their vantage points, the individual strategies (such as learning styles or instructional methods) that had contributed to their success and that might be adopted by others to enhance their learning and school performance. The study also investigated the linkage between equity and excellence by connecting matters of education with those of social difference (see also Dei 2015). Recognizing that relational aspects of difference are closely involved in schooling and knowledge production, I encouraged high achievers to articulate how they understood and interpreted the multiple identities, subjectivities, and histories they brought to their education and how social difference had influenced their definitions and attainment of academic success.

Minority Youth Narratives and Relevant Research Findings

In this section I focus on narratives chosen from the published research on youth, parents, and educators from diverse racial and ethnic backgrounds and what they have said about the school system. My intellectual goal is to highlight the relevance and implications of such voices as legitimate sources of cultural resource knowledge that help inform the processes of educational delivery—teaching, learning, and the administration of education. I have captured voices from the different phases of this longitudinal and ongoing research on schooling to convey understandings that local subjects have of schooling processes and to explain why these voices need to be heard in the search for genuine educational options for all learners. Clearly, the students have their own understanding of the schooling process and how one can account for their marginalization and social exclusion. Yet the students affirm the power of their subjective knowledge and seek to employ these ways of knowing as a form of resistance.

History, Local Knowledge, and the Curriculum

Students, parents, and community educators bring their local cultural resource knowledge and understandings to a discussion of what schooling and education ought to entail in order to engage diverse learners. In two studies (Dei et al. 1995, 1997) Black youth raised the question of school knowledge and curriculum and how culture and pedagogy can and do connect to ensure that all students identify with the Canadian school system. Denise, who dropped out of school but later returned to complete her education, observed the lack of curricular sophistication in schools as a major reason for her having become disengaged.

> The curriculum...was one-sided, especially when it came down to history. There was never a mention of any Black people that have contributed to society....I mean, everything, it's the White man that did. History is just based on the European Canadian that came over.... There was no mention of the Africans that helped build a railway, that ran away from the South and came up to Nova Scotia and helped work and build Canada too;...no mention of that. (File F08: lines 316–332)

The curriculum was seen in this case as the teaching of a narrow Canadian history. To Denise the problem of the one-sided nature of the school curriculum contributed to her eventual disaffection with the school system. The contributions of African-Canadians to Canadian and world history were never appropriately recognized, neglect that added to the sense of Black students' invisibility in schools.

Denise was not alone. Amma, a mother and community worker, agreed. She called for a more globally oriented curriculum. But she cautioned that such an approach must ensure that all learners are central to the making of this global history: "I think it's necessary to have that sort of world vision, but definitely you have to be in that world" (File CWO1: lines 819–822). This sort of identification with history ensures a sense of self- and collective worth and a valuing of all knowledges. Educators do best to avoid parochialism in their teaching and to refrain from problematically insisting on a dominant worldview that is not even holistic (on the concept of holistic education, see Lee 2005, pp. 52–69). Otherwise, most learners will be either left behind or forced to work with the prism that is already established as the Eurocentric norm. Nowhere is this imposition more apparent than when an educator counsels new immigrant students simply to shed their old cultures, identities, and habits and take on the Canadian way or the host country's worldviews and perspectives. Such amputation is even more difficult than the process of coming to know about the host country's outlooks as though that country were itself homogeneous. Yet insensitivity to that reality persists in the Canadian school system. Referring to a number of colleagues participating in the research by Dei et al. (1995, 1997), Andrew, a local school teacher, observed that

> [some teachers] just want to [teach]: "This is Canada, and it's English and French. And if you come from Afghanistan or you come from Uganda, this is Canada. You learn our way. We don't have to learn your way back [home]. We don't even know your way!" (File T01: lines 243–249)

Learning needs to be reciprocal. It must be an exchange of ideas so that all students recognize the multiple and collaborative dimensions of knowledge. All knowledge and historical and lived experience has something to teach the learner. Barbara, another school teacher interviewed in the research (see Dei et al. 1995, 1997), shared Andrew's critique by attesting to the limited stances of some of her colleagues: "They say, 'We live in Canada. We teach Canadian history', or 'We should teach them the Canadian way'. I'm still having difficulties defining what the 'Canadian way' is, and I am a born Canadian!" (File T21: lines 1116–1136). Barbara argued that the Canadian way needs to be inclusive. She spoke of the negation of the contributions, experiences, and knowledges of the different groups that make up the complexity of what Canada is. She pointed out how such complicated understandings of the world can only benefit the learner.

Students, parents, and educators who acknowledged that exemplary practices of inclusive schooling exist also stressed the relevance of educators who make a connection between culture, history, curriculum, and knowledge. Nathalie, a parent and community worker actively involved in the school system for years, openly admitted there were some exemplary practices of inclusive schooling that educators can learn from. In our study on exemplary practices of inclusive schooling (Dei et al. 2002), Nathalie told us what she would like to see more of:

> I think [there should be] more…programs that are more accommodating to students, identify cultural backgrounds and pressures that youth are being faced with, whether, if they're new Canadians, as far as barriers, as far as their English. I think there should be more programs that would focus on identifying their needs and not trying to push them into the mainstream, [programs that would allow] them to get comfortable first, sort of a specialized program. (M/07/05/97)

Schooling should also be about equipping students to solve problems rather than about creating a sameness in which difference fails to be recognized as a strength. Nathalie's narrative upholds the connections of culture, curriculum, and knowledge production in schools to ensure success for all students. Students coming from diverse backgrounds bring significant assets to schools. They include a wealth of local cultural resource knowledge, a diversity of experiences and histories, and lessons on strategies of educational survival.

Learners' Identities, Social Identifications, and Knowledge Production

The relevance of identity, identifications, and representations in schooling is that they point to particular embodiments of knowledge production. A school system that fails to tap into youth identities and identifications as valuable sources of knowledge shortchanges students. Students are neither disembodied learners nor empty vessels simply waiting to be filled with knowledge. They possess a way of knowing. How do schools make use of the rich histories, cultures, and community knowledges of their different bodies to understand the complexity of the world today? The failure to draw on this reservoir of knowledge is particularly disturbing given minority students' desire to reclaim their own knowledges and cultural identities in the school system. The emotions with which a student expressed these feelings in a focus-group discussion with high schoolers in Toronto were not lost on my research team and me:

> I want to be myself and show everybody who I am, but you can't, you can't be yourself too much. You have to hide some of your Blackness when you are around White people because then you become like an alien to them. (Dei et al. 1995, 1997, File 010: lines 691–704)

In other words, this student found it necessary to conceal part of his nature in order to be accepted among Whites. He had a feeling of not being himself in certain spaces, a sense that bodies were being monitored or regulated. This estrangement is not surprising, for some educators would frequently and hypocritically downplay the importance of racial identity by claiming to be color blind. But often this posture is only theoretical. To engage in color blindness is to insist that students negate or amputate a part of themselves, that is, who they are. Yet many students are proud of reclaiming these racial identities in the midst of the pressures to assimilate at the expense of their culture, history, and ancestry. Pitted against society's forces of push and pull, the individual discovers the cost of going against the grain. These issues gradually compound the processes of youth disengagement from school. As reasoned by Richard (a student considered by educators to be at risk of dropping out),

> [y]ou couldn't express it…because you have to assimilate and you have to be like everyone else. So, you had to be like they be. If you didn't, then you're a troublemaker, or you're the strange kid that's got this, you know, psychological problem. But even nowadays when you say it's more expressed, even still, you're looked upon as a troublemaker because you want

to learn your identity and something other than [in a] class[room] where you feel…wrong. You don't feel it. And, you bring it up and…you're a troublemaker now because you're going against the school. So even though they say you can express yourself,…it's not the same still. To me, it's all a big puff of smoke. (File A08: lines 761–782)

To Richard, insisting on who one is means being prepared to face the consequences (including punishment) of claiming a racialized identity. One is labeled a troublemaker, and there is a particular gaze trained on the student. In a way this person's experience shows the powerful linkage between identity, schooling, and knowledge production. The experiences of marginalized minority youth in schooling can be racially coded, and it is intellectually hypocritical for any educator to deny this possibility. No matter how much this connection between identity and schooling may be downplayed, the fact that students, parents, and some community educators make this connection is an important source of knowledge for schools to engage with. Perceptions emerge from actual experiences. They are not manufactured out of the blue. Educators must inquire into how these perceptions are arrived at. As Richard has found, reclaiming identity must be connected to schooling processes because there are rewards and punishments in schools. In a study of high academic achievers, Dei (2010a) found that students did not articulate a relation with the identity of teachers in terms of what teachers know. Rather, that relation was expressed in terms of recognition that the background of an educator (as with that of a learner) is relevant in making sense of knowledge, teaching, and learning. People all speak from particular positions, subject identities, and identifications, experiences, and histories.

Being unable to make racial connections with one's educators does indeed impose limits, as perceived by Mila, an 18-year-old African-Canadian female student who completed high school in a predominantly Black religious school. Comparing that experience with her earlier schooling, she noted that such inability did not necessarily present obstacles per se but that it did preclude some possibilities.

[In my high school] I found that it was different having teachers who were all Black because in my former public school there was only one Black teacher in the whole school. It was just because of the area and location of the school. I felt that I could relate to the teachers a bit more and that they could understand my background because most of the Black teachers were West Indian [Caribbean]. You kind of felt like they were looking out for your best interests on a more personal level and you felt that you could talk to them about whatever. It's more personable. It was more of a family feeling. There was competitiveness, but there was still a great group. (File U16: text units 46–48)

Mila highlighted a degree of connectedness, identification, and understanding that a learner acquires when she or he can relate to teachers who share her or his cultural, racial, ethnic, gender, and sexual backgrounds. It does not mean that only certain kinds of teachers can teach and teach well. Nor does it mean that only teachers of a particular background can provide the much needed nurturing environments. It means that other possibilities can be ruled out. It is essential for all learners to have teachers with whom they can identify. There is a qualitative value of justice in students seeing teachers who look like them in their schools. It is inspiring to young learners (see Solomon 1997). The worn-out argument that students simply need a good teacher neglects the fact that good teachers are equally distributed in a given population. There is something fundamentally and morally wrong when students go through the educational system without being taught by educators who share their cultural, racial, ethnic, sexual, and gender backgrounds. Mila's way of knowing cannot be dismissed when she attributes her educational success in part to the identification with the school environment and the feeling of comfort.

Therefore, teachers and their identities have to be connected to schooling, especially given the critical role educators play in molding the futures of young learners. As stated by Bimpa, a Canadian university student of Nigerian descent in Dei (2010a), "The role they [teachers] have cannot be underestimated. Just having some kind of approval[,]…that kind of affirmation goes a long way. That means a lot to the students" (File U62: text units 165–167). An educator's high expectations of learners do much to motivate students to excel. Receiving a seal of approval from a teacher means a great deal to struggling students and motivates them to aim high academically. Successful students all attest to how important it is to have had teachers who held them to high expectations (see Dei 2010a).

Schools and educators' rich and complex knowledge, when tapped into, means that all learners benefit from an understanding of the complete history of ideas and events that have shaped and continue to shape human growth and development. That kind of knowledge base addresses the insularity of the learning experience. It also enables learners to develop a complex and complete understanding of themselves, their identities, and their histories and to appreciate how histories, identities, and experiences are intertwined and interconnected. In Dei et al. (2000) an 18-year-old female student named Marsi identified her origin as Ethiopian, although she had been born and raised in Montreal. In our discussions she superbly made links between culture, ancestral heritage, and identity. She insisted that she had a rich heritage stemming from her cultural diversity and the complexity of her own lived existence. She opined that she needed to learn about other peoples' histories as much as students from other backgrounds needed to learn about her history and culture. As she emphatically argued,

> I have to say that I would like schools to talk more about our history. For example, I live in a Portuguese neighborhood, and I have grown up in a White community and don't really have much understanding of my ancestry or my heritage,…[I]n that sense I would like schools to have more of a broad subject [base],…not like Black history month. Just to have a month dedicated to Black people isn't quite enough, specially for students like myself who are born in Canada.…I am more aware of my history and my background, and it just boosts my self-esteem. I have something to be proud of instead of the negative aspects I see on the news and stuff like that. (F/12/10/97)

History is about affirmation and reinforcement. Acknowledging ethnic and racial identities strengthens the sense of self and the collective and helps the learner navigate the challenges and possibilities of schooling. School instruction should be complex enough to allow students to engage with the multiplicities and complexities of their existence. Students' voices and knowledge on these matters are helpful in informing everyday classroom teaching and pedagogical practices. Creation of a genuine educational option for youth—one that embodies a comprehensive re-envisioning of schooling—calls for working critically with many different (and sometimes contending) voices of students, parents, and community educators to appreciate what teaching, learning, and administration of education can be for diverse learners.

This conviction does not imply that students necessarily know best. But certainly no one should devalue or diminish the experiences, histories, and cultural knowledges that students bring to the schooling processes. Kambi, a high school student of mixed race, pointed to the dangers of muting or silencing certain voices and knowledges:

> [V]aluing students' opinions is important. Treating them as more than just a student or just a teenager. I think teenagers face a lot of ageism, like, we're represented as this group that sits around and screw things up....We really are. And I think that's what is stupid because... adults are really in charge of the world and the world is not in such a good state,...so it can't just be us that are screwing things up. That's the attitude[:]...treating us like we are people we have to respect. (File HS30: text units 79–87)

Kambi noted that the voices of students are often dismissed, subjected to negative perceptions, and seen as problematic (see Dei 2010a). Yet students are mature in terms of the knowledge gained from their daily experiences. Kambi was asking for respect and an acknowledgment that students' voices can be central to ways of rethinking schooling and education. If schooling is to address the needs of those who are marginalized, then a positive way must be found to affirm and reinforce the local cultural resource knowledge, particularly that of subordinated groups informed by their racial, ethnic, gender, and sexual contexts and identities.

Race, Class, Language, Gender, Sexuality, and Physical and Socioeconomic Disability as Legitimate Dimensions and Sources of Inclusive Schooling and Educational Excellence

Students, parents, and minority educators bring different and multiple readings to social difference and the implications for schooling. It has long been established that schooling outcomes are affected by race, class, gender, sexuality, and disability (see Apple 1986, 2004; Apple and Weis 1983; Fine and Vanderslice 1991; McCarthy 1980; Weis and Fine 2005; Willis 1977, 1983). The categories of race, class, language, gender, disability, and schooling involve both the processes and structures of educational delivery. Social difference is about identity, power, and relations. Social difference can also be an important dimension motivating students either to do well or to disengage from school. In the aforementioned study with high academic achievers (Dei 2010a), some students regarded race, gender, and class as impinging on schooling in direct and indirect ways. Nadra, a Canadian female university student of Middle Eastern background, was a Rhodes Scholar. She saw social-class language as having effects on schooling. She argued that social class had an indirect effect on her schooling, an experience that may have been different for other students.

I don't see it directly having an effect. I see it having an effect indirectly through motivation. So I think that there are a lot of people who, like my family,…are a family of immigrants, and I think there's a lot of motivation from your parents to become a professional, to do well in school…. On the other hand, I think some of the people in higher social classes that I know, also all of their family are doing well…. They're all professional doctors, lawyers, and so on. So there's a lot of pressure to do well there. So motivation as a vehicle affects how you do in school as an indirect consequence of social status. But also the converse of that could be true. Like I know several people who come from well-off families in Toronto, and they have absolutely no motivation to do well because they know that they are assured a good job at the end of it. So I don't think social class directly affects performance but maybe indirectly through how it affects motivation. I think language is a big obstacle. I couldn't imagine studying in Germany, for instance, so I don't know how people do it. Definitely, it would be very difficult. I haven't noticed any real patterns that I could point to for differences in performance correlating with race…. Again, I think social class is compounded with that, given the nature of Toronto as a diverse city. (File U07: text units 59–80)

The pressure to succeed given her family socioeconomic background and professional history motivated her to do well in school. Although claiming that the effect of race was unclear to her, she conceded that others may find that coming from a rich background actually hinders the desire to do well in school.

Language, too, can be an "obstacle," and Nadra admitted it can slow a student's progress, especially if the person is studying in a different context or country. But apart from learning dominant languages such as English and French, minority students experience the relevance of the language of their heritage. Heritage language can be situated in discussions of social difference insofar as questions of identity, culture, history, and schooling collide and connect. Some minority students and their parents see studying heritage languages as an asset for the learner. As noted by a Chinese mother, Olga, who stated that her "children struggled through school when they first arrived [in Canada]," the study on exemplary practices of inclusive schooling (Dei et al. 2000) also dealt with a struggle over language competency. But it is actually an asset to retain heritage language:

I think what I hear when they go to school, where the majority is Chinese, their language is not that good because they're used to speaking their native language instead of English. But it's O.K., I think. It's good to learn both languages. (PF/22/08/97)

In general, the acquisition of multiple languages allows learners to understand their communities better. It also contributes to the breadth of the learner's knowledge.

Other high-achieving students who participated in our study (Dei 2010a) also alluded to the importance of gender in schooling. Anda, a 28-year-old Aboriginal student, was a single parent in law school. She spoke about how female students are often steered into careers that represent gender-based role norms. She argued for schools to be inclusive and to engage with social difference, and she speculated what that call means for working with an understanding that female students are capable of excelling in all subjects and that school subjects are not gender specific.

Probably [it entails] the sensitivity towards gender differences. Probably I think that it's a huge thing that, unfortunately, girls are still not encouraged to take like math and sciences. We are still geared towards the nice, nice things, and music, and social sciences, and family,

and those just are not, um, springboards to the quote endquote more prestigious degrees.
I think if you have an inclusive school, girls should be encouraged on [an] equal footing.
(File U38: text units 129–132)

Students are aware of the cultural knowledge of schooling and how particular
schooling practices engage with social difference in ways that either facilitate learning or hinder the academic growth of students. The narrative of gender-role norms
afflicts schooling in powerful ways and can be disrupted only by a simultaneous
affirmation of difference as a dimension of power and resistance.

Just as with social class, language, race, and gender, disability, too, has implications for schooling. Student voices make it clear that inclusive education is about
making schools accessible in multiple ways to all students. Twenty-four-year-old
Morley was a white male student who had won many prestigious scholarships at the
time of the study (Dei 2010a) and who was studying at a top Canadian university at
the time of our research. He stated enthusiastically that one of the biggest challenges to promoting educational excellence is how to make schools physically
accessible to students—all of them.

> I think it's pretty straightforward. It's things like…in terms of inclusiveness. I think the two
> major concerns…would be physical accessibility like you've described, and socioeconomic
> accessibility, and I think you have to be very aggressive in your approach to both. Meaning
> that you should level the playing field as much as possible….It's making sure every build-
> ing in every classroom is physically accessible, whether it's students who are blind, deaf,
> physically handicapped in wheelchairs, whatever the case is. (File U06: text units 64–90)

To make schools into a level playing field, educational resources and facilities
that enhance learning must be accessible to all students. This necessity is why one
cannot speak of social difference and schooling only in terms of race, ethnicity,
class, and gender. For many students the cumulative effect of the combination of
race, ethnicity, class, gender, language, disability, and religion is either enhancement of or detraction from their performance in and outside the classroom. Thus,
social difference can at least partly explain why some students do well and others do
not. It is crucial to be aware of students' knowledge about social difference in order
to understand the specific impacts it has on learning and the school experience.

The two studies on the matter of school dropouts (Dei et al. 1995, 1997) are
instructive reminders that minority students in particular are able to make a powerful link between identity, social difference, and educational excellence and how that
link becomes working cultural knowledge (capital) in schools. Cheryl, a twelfth-
grader, asserted that Black counselors are able to provide an environment in which
the experiences of minority students are understood.

> I can relate to a Black counselor more and maybe I can talk to her about things that I can't
> talk to my White counselor about. Maybe because, what if I have a personal problem that I
> don't want the White counselor to know about?…But like, maybe I'll feel better that the
> Black counselor knows….She can relate to you…like you're on an even basis. (File W11:
> text units 553–567)

There was a presumed sense of shared understanding based on racial solidarity,
which helped encourage students to communicate. This knowledge rooted in an

awareness of shared history, experience, and culture could not be dismissed as merely fictive. The students realized that shared identities and experiences were not necessarily singular; the attendant shared understanding was a form of local cultural resource knowledge that minority students worked with in schools. Beth, a classmate of Cheryl, added that such local cultural affinity may be an easily assumed, but powerful, inducement for students to engage with the school system

> [when] there is a problem, when there's, you know, something culturally based that they'd be able to understand more than someone who was White. I don't want to have to go into an office and, first of all, explain my culture to someone and then explain my problem. I want to be able to go in there and say, "This is the problem. Can you help me with it?" (File 004: text units 1145–1161)

In effect, the identities of race, class, gender, disability, and language become powerful markers with which to build trusting relations and relationships in schools among some students, teachers, and administrators. The dimensions of difference become important markers for students' strategies of survival. Students use race and the relational aspects of difference to disrupt the centrist hegemonic knowledge systems in schools and validate other ways of knowing. For example, Sandy, also a twelfth-grader, was known to be actively engaged in Black identity issues in her school. My research team interviewed her in the study on school dropouts (Dei et al. 1995, 1997). She discussed a learning technique that is as much a revelation about strategies of survival as it is a critique of knowledge production in the school system:

> [W]e were going to write an exam....I think it was a parenting or sociology exam. All the Black students were at one place studying together, even though we didn't really know each other that much. And some of them were saying, "Oh no, I don't know if I'm going to pass." And then some of the Black students said, "Just think White. Think of the way White people raise their kids, and you'll pass the exam because that's what's going to be on the exam." And we felt a closeness because we know that we're raised differently. (File R06: text units 2312–2329)

Noting her racial identification with the other test-takers, Sandy spoke about a feeling of togetherness, but she also pointed out how powerful meanings are embedded in the creation of communities as additional spaces for surviving school. They constitute relevant knowledge for everyone interested in transforming schools to serve the needs of all students.

Local Cultural Ways of Knowing and the Education of Marginalized Youth

As argued elsewhere (Dei 2009), the history of Black and specifically African and minority youth education in the Euro-Canadian context is replete with sacrifices made by exceptional teachers, parents, and community workers committed to the educational welfare of youth. It abounds with growing resistance to dominant processes of schooling that consign these youth to failure. In re-envisioning schooling and education to promote educational excellence for youth, we educators must closely examine the colonizing relations of schooling that produce academic failures. We must connect dominant perspectives of schooling to the subordinate perspectives of young learners, minority parents, and local communities in order to understand the contradictions of education.

This commitment necessarily means not only validating epistemological conclusions derived from social and schooling experiences of youth but also reinvoking their local cultural resource knowledges. There needs to be a recognition that students coming from diverse backgrounds experience different educational realities not easily comprehended when presented as obstacles and problems. It requires a shift away from the understanding that "students, not their schools or classroom circumstances,…are the sources of unequal educational attainment" (King 2005, p. 201). Most of the current neoliberal policies and practices pertaining to testing, standards, "school improvement and effectiveness," and "school choice" constitute "compensatory" approaches to schooling and do not in reality speak to the structural and institutional problems of minority youth success or failure in schools. Well-meaning educational policies and practices to transform education are often knee-jerk reactions to emerging problems; they are frequently short-sighted responses and not far-reaching in their effects. Many educational practices and policies are developed or couched in ways that constitute temporary expedients and recycled measures without producing truly transformative effects.

In the context of youth of African descent, Akoto (1992), Asante (1988), Hilliard (1995), Karenga (1993), Lomotey (1990), Madhubuti (1990), and Shujaa (1994) have all theorized the African-centered pedagogy and cultural funds of knowledge for educating youth. Many of their ideas allude to the significant political, cultural, historical, and spiritual interconnections in the educational experiences of peoples of African descent. These connections go beyond a commonality of experience. They evoke visionary possibilities seated in the affirmation of the specificities, uniqueness, richness, and complexities of the Black and specifically African existence. This awareness is an important source of local cultural resource knowledge relevant for the education of the youth. A host of African Indigenous philosophies of education allude to a community's shared responsibility to work out solutions to its own problems (see Abdi and Cleghorn 2005; Fafunwa 1982; Sifuna 1992; Shujaa 1994).

As is clear from the aforementioned voices of students and educators, the schooling identities of learners are shaped by experiences, history, cultures, and socialization processes. There are links between these identities—such as race, class, gender, sexuality, and disability—and the manner in which learners and educators come to know and engage in schooling. Schools that ensure success for Black and specifically African and minority learners utilize student knowledge to initiate discussions. This fact in itself is recognition of the power of multiple ways of knowing. Ensuring success for African learners will also require that educators and schools understand and appreciate the role of students in their own learning. It means listening to their voices and finding out what they want to see in their schools and what engages them most in their learning. Learners learn best when engaged and when teachers have high expectations.

In re-envisioning schooling and education to ensure success for ethnic minority learners, a whole new conception of schooling is required, one that defies much of what conventional schools currently do. Apart from sustained curricular, textual, and instructional initiatives, a new vision of schooling is about fundamental changes in the organizational lives of the school, including its environment, culture, and climate. This vision is about an educational philosophy, an educational worldview that answers questions about the why, how, and what of education. A new vision of schooling and education is not about integrating ideas into an already existing school structure. Genuine educational options for youth center the learner in her or his own culture, experience, histories, and complex identities as a way to begin engaging broad knowledge critically. Countervisions of schooling work with the concepts of culture, centeredness, and a rootedness in history and community. Such a form of education uses the social values of community, responsibility, interdependence, interrelationships, solidarity, spirituality, and complex identity consciousness as pedagogical and instructional tools for providing education. This form of education and schooling is about cooperative learning where success is broadly defined to include social and academic success. It is education without hierarchies. It is a philosophy for educating the learner more broadly than is presently the case in myriad subjects. Sustained and direct community involvement and participation in the running of the school is equally important. Community members become Elders who share their rich cultural resource knowledge about schooling and education and who provide alternative visions of schooling, which are required in order to widen current debates about fostering genuine educational options for youth (see Akoto 1992; Asante 1988, 1991; Hilliard 1995; Karenga 1993; Lomotey 1990; Madhubuti 1990; Shujaa 1994).

In articulating the need for radical, incisive, Indigenous African thought, I am also heralding calling for the use of local resource knowledge, a turn to culturally centered pedagogy, and a political engagement of parents and communities in concrete and meaningful action to produce educational change for youth. It is necessary to revisit students' cultural knowledges that used to inform traditional African schooling and education. Community knowledge constitutes a priori cultural resource knowledge that the learner has before entering the formal school system. The engagement and practice of culture and community is contextualized in the understanding of cultural and communal knowledge (e.g., extended family ties and practices, rituals and social networks associated with the understanding of the self and collective spiritual development, and the use of local languages and orality as cultural expressions of identity and belonging). This cultural knowledge constitutes cultural reservoirs of knowledge that are important for educational practices to draw on as assets and possibilities; it does not constitute deficits or limitations in the education of Black and specifically African youth (see also Lee 2005, p. 64; Dei et al. 2010). The cultural practices of schooling must tap into community cultural knowledge in order to create conducive learning environments for all youth. For example, the cultural sources of knowledge that Black and specifically African and minority students have within them on entering the school system show that the students bring an important contribution to the learning process. Educators must tap this wealth of knowledge to enhance learning for all. This approach to schooling and education for Black and specifically African and minority learners entails a socio-political and intellectual shift of paradigm, one that works with the nexus and interface of society, culture, and nature to promote a holistic education for the learner.

Acknowledgements I would like to thank the many students of the Ontario Institute for Studies in Education of the University of Toronto (OISE/UT) who have assisted me in the various research projects in the Ontario school system. They include the many students I have written or worked with as researchers. I am grateful to the many students, educators, parents, and community workers who have shared with me their views and rich experiences about the Canadian school system. I say thank you also to Mairi McDermott and Aman Sium of the Department of Humanities, Social Sciences, and Social Justice Education (formerly Sociology and Equity Studies), OISE/UT, who commented on drafts of this chapter. The text was initially presented as a keynote address at the Eighth Annual International Symposium on Ethnic and Cultural Dimensions of Knowledge, held at Ruprecht-Karls-Universität (Heidelberg University), Germany, October 7–10, 2009. I offer deep thanks and appreciation to the organizers of the symposium for the invitation to share my ideas. The occasion also marked my first trip to Germany. Lastly, I am grateful to the Social Science and Humanities Research Council of Canada (SSHRC) for the funding that made these studies possible.

References

Abdi, A., & Cleghorn, A. (Eds.). (2005). *Issues in African education: Sociological perspectives.* New York: Palgrave Macmillan.

Adjei, P. (2010). *Personal communication.* Toronto: University of Toronto, Department of Sociology and Equity Studies, Ontario Institute for Studies in Education of the University of Toronto (OISE/UT).

Akoto, K. A. (1992). *Nationbuilding: Theory and practice in Afrikan-centered education.* Washington, DC: Pan African World Institute.

Apple, M. W. (1986). *Teachers and texts: A political economy of class and gender relations in education.* New York: Routledge & Kegan Paul.

Apple, M. W. (2004). *Ideology and curriculum.* New York: Routledge Falmer.

Apple, M. W., & Weis, L. (1983). *Ideology and practice in schooling.* Philadelphia: Temple University Press.

Asante, M. K. (1988). *Afrocentricity.* Trenton: Africa World Press.

Asante, M. K. (1991). The Afrocentric idea in education. *Journal of Negro Education, 60,* 170–181.

Battiste, M. (1998). Enabling the autumn seed: Toward a decolonized approach to aboriginal knowledge, language, and education. *Canadian Journal of Native Education, 22,* 16–27.

Battiste, M., & Youngblood Henderson, J. (2000). *Protecting indigenous knowledge and heritage.* Saskatoon: Purich.

Bhavnani, K.-K., & Davis, A. Y. (2000). Women in prison: Researching race in three national contexts. In F. W. Twine & J. W. Warren (Eds.), *Racing research, researching race: Methodological dilemmas in critical race studies* (pp. 227–246). New York: New York University Press.

Butler, J., & Scott, J. W. (Eds.). (1992). *Feminists theorize the political.* New York: Routledge.

Cesaire, A. (1972). *Discourse on colonialism.* New York: Monthly Review Press.

Dei, G. J. S. (1996). *Anti-racist education: Theory and practice.* Halifax: Fernwood Publishers.

Dei, G. J. S. (2000). Rethinking the role of indigenous knowledges in the academy. *International Journal of Inclusive Education, 4*, 111–132.

Dei, G. J. S. (2009, May). *Educational policies and the search for academic excellence for Black/African learners: Towards a new policy framework?* Invited featured address delivered at the DuBois-Nkrumah-Dunham International Conference on: 'Academic Achievement in Africa and its Diaspora: Challenges and Solutions', University of Pittsburgh, Pittsburgh.

Dei, G. J. S. (Ed.). (2010). *Fanon and the counterinsurgency of education.* Rotterdam: Sense.

Dei, G. J. S. (2015). Indigenous philosophies, counter epistemologies and anti-colonial education. In W. Lehman (Ed.), *Reader in sociology of education.* London: Oxford University Press [in press].

Dei, G. J. S., & Asgharzadeh, A. (2001). The power of social theory: Towards an anti-colonial discursive framework. *Journal of Educational Thought, 35*, 297–323.

Dei, G. J. S., & Kempf, A. (2006). *Anti-colonialism and education: The politics of resistance.* Rotterdam: Sense.

Dei, G. J. S., Holmes, L., Mazzuca, J., McIsaac, E., & Campbell, R. (1995). *Drop out or push out? The dynamics of Black students' disengagement from school.* Toronto: Report submitted to the Ontario Ministry of Education and Training.

Dei, G. J. S., Mazzuca, J., McIsaac, E., & Zine, J. (1997). *Reconstructing 'drop-out': A critical ethnography of the dynamics of Black students' disengagement from school.* Toronto: University of Toronto Press.

Dei, G. J. S., James, M., James-Wilson, S. V., Karumanchery, L., & Zine, J. (2000). *Removing the margins: The challenges and possibilities of inclusive schooling.* Toronto: Canadian Scholars Press.

Dei, G. J. S., James-Wilson, S. V., & Zine, J. (2002). *Inclusive schooling: A teacher's companion to removing the margins.* Toronto: Canadian Scholars Press.

Dei, G. J. S., Butler, A., Charamia, G., Kola-Olusanya, A., Opini, B., Thomas, R., & Wagner, A. (2010). *Learning to succeed: The challenges and possibilities of educational development for all.* New York: Teneo Press.

Ernst, G., Statzner, E., & Trueba, H. (Eds.). (1994). Alternative visions of schooling: Success stories in minority settings [Special issue]. *Anthropology and Education Quarterly, 25*(3), 200–394.

Fafunwa, A. B. (1982). African education in perspective. In A. B. Fafunwa & J. Aisiku (Eds.), *Education in Africa: A comparative study* (pp. 14–22). London: George Allen & Unwin.

Fals-Borda, O. (1980). La ciencia y el pueblo: nuevas reflexiones sobre la investigación-acción [Science and people: New thoughts about research activity]. In Asociación Colombiana de Sociología (Ed.), *La sociología en Colombia: balance y perspectivas. Memoria del Tercer Congreso Nacional de Sociología 20–22 de agosto 1980* (pp. 149–174). Bogotá: Asociación Colombiana de Sociología.

Fanon, F. (1963). *The wretched of the earth.* New York: Grove Press.

Fine, M., & Vanderslice, V. (1991). Qualitative activist research: Reflections on methods and politics. In F. B. Bryant, J. Edward, R. S. Tindale, E. J. Posavac, L. Heath, E. Henderson, & Y. Suarez-Balcazar (Eds.), *Methodological issues in applied social psychology: Social psychological applications to social issues* (pp. 199–218). New York: Plenum.

Foucault, M. (1980). *Power/knowledge: Selected interviews and other writings, 1972–1977* (trans: Gordon, C., Ed., Gordon, C., Marschall, L., Mepham, J. & Soper, K.). New York: Pantheon Books.

Giroux, H. A. (1983). *Theory and resistance in education: Towards a pedagogy for the opposition.* South Hadley: Bergin & Garvey.

Giroux, H. A. (2003). *The abandoned generation: Democracy beyond the culture of fear.* New York: Palgrave Macmillan.

Hilliard, A. G. (1995). *The maroon within us: Selected essays on African-American community socialization.* Baltimore: Black Classic Press.

Karenga, M. (1993). *Introduction to Black studies.* Los Angeles: University of Sankore Press.

King, J. E. (2005). *Black education: A transformative research and action agenda for the new century*. Mahwah: Lawrence Erlbaum.

Lee, C. D. (2005). The state of knowledge about the education of African Americans. In J. E. King (Ed.), *Black education: A transformative research and action agenda for the new century* (pp. 45–71). Mahwah: Lawrence Erlbaum.

Lomotey, K. (1990). *Going to school: The African-American experience*. Albany: State University of New York Press.

Madhubuti, H. R. (1990). *Black men: Obsolete, single, dangerous? Afrikan American families in transition: Essays in discovery, solution, and hope*. Chicago: Third World Press.

McCarthy, C. (1980). *Race and curriculum: Social inequality and the theories and politics of difference in contemporary research on schooling*. London: Falmer Press.

Memmi, A. (1965). *The colonizer and the colonized*. Boston: Beacon Press.

Nakata, M. (2007). *Disciplining the savages: Savaging the disciplines*. Canberra: Aboriginal Studies Press.

Semali, L., & Kincheloe, J. (Eds.). (1999). *What is indigenous knowledge? Voices from the academy*. New York: Falmer Press.

Shujaa, M. J. (1994). *Too much schooling, too little education: A paradox of Black life in white societies*. Trenton: Africa World Press.

Sifuna, D. N. (1992). Diversifying the secondary school curriculum: The African experience. *International Review of Education, 38*, 5–20.

Solomon, P. (1997). Race, role modelling, and representation in teacher education and teaching. *Canadian Journal of Education, 22*, 395–410.

Weis, L., & Fine, M. (2005). *Beyond silenced voices: Class, race, and gender in United States schools*. Albany: State University of New York Press.

Willis, P. (1977). *Learning to labour: How working kids get working class jobs*. Farnborough: Saxon House.

Willis, P. (1983). Cultural production and theories of reproduction. In L. Barton & S. Walker (Eds.), *Race, class and education* (pp. 107–138). London: Croom Helm.

Yankah, K. (2004). *Globalization and the African scholar*. Accra: University of Ghana, Faculty of Arts.

Chapter 7
The Knowing in Indigenous Knowledge: Alternative Ways to View Development, Largely from a New Guinea Highlands' Perspective

Paul Sillitoe

After two decades or more, the Indigenous Knowledge (IK) in Development Initiative has not apparently had the success that some commentators anticipated (Antweiler 1998; Brokensha et al. 1980; Kloppenburg 1991; Purcell 1998; Sillitoe 1998b). The initiative, which attempts to increase the prominence of local voices and practices in development contexts, has arguably had its brief fashion moment on the back of the participatory movement, passing out of favor as the shortcomings of participation become evident, due in part to vested interest manipulations (Cooke and Kothari 2001; Mosse 2005). The IK approach has not had a chance, in other words, to show its potential, for development agendas controlled by outside agencies distort and mask it. The question is how those communities subject to development interventions might reduce the power imbalance inhibiting their effective contribution.

Although there are several political dimensions to the limited up-take of the IK initiative, this chapter explores the implications of the possibility that ideas about what it is may vary between populations, particularly what is understood by *knowledge*, notably in context of the Papua New Guinea Highlands. Confusion and disagreement about the meaning of the term doubtless contributes in part to the stalling of the IK initiative. Having already tackled the matter of defining IK in development contexts (Sillitoe 1998a, b, 2000a, 2002c) and taken the opportunity to demonstrate anthropology's "useful knowledge" (as termed by A. M. Strathern 2006) and consultancy potential, I turn in this chapter to exploring what an alternative understanding of IK might look like. I offer one defined from within, not imposed, and ask what it could imply for alternative views to capitalist development.

P. Sillitoe (✉)
Department of Anthropology, Durham University,
Dawson Building, South Road, Durham DH1 3LE, UK
e-mail: paul.sillitoe@durham.ac.uk

© Springer International Publishing Switzerland 2016
P. Meusburger et al. (eds.), *Ethnic and Cultural Dimensions of Knowledge*,
Knowledge and Space 8, DOI 10.1007/978-3-319-21900-4_7

Arguing Over Words

Advocates of IK find themselves in an awkward position, simultaneously both supporting and subverting development. On the supportive side, they seek to focus on issues from the perspective of development agencies whatever their complexion—national or international, NGO or government, bilateral or multilateral. They do so on the grounds that this strategy is effective at getting local views and practices on the agenda, whereas criticism from the outside is not. This cooperation with business-as-usual development includes valuable work, even if it is currently restricted to the margins for the most part. Examples are contributions to local farming in programs for improving food security and to local medicine in programs for improving health (Bentley and Baker 2005; Shepherd 2004, 2005). It features the advancement and refinement of effective methodologies intended, for instance, to further the potential of participation and to demonstrate the value of process, as opposed to blueprint, approaches to development. The cooperation I refer to addresses several things, among them: making sociocultural aspects more accessible, delivering relevant research promptly while maintaining ethnographic integrity, overcoming prejudicial science versus local knowledge discriminations, and advancing interdisciplinary work.[1] Regardless of some hard work on the methodological front (Emery 2000; Grenier 1998; IIRR 1996; Sillitoe et al. 2005), there has been no significant breakthrough to new ways of working with development. People doing practical IK work have largely tried to relate local practices and ideas to whatever technological-cum-economic intervention agencies think will advance development aims and have advised on the implications of associated political and social changes. For those who wish to see an improvement in the standard of living of poverty stricken families, such endeavor has a degree of merit, whatever the problems—and there are several. It is the only way that most western-funded agencies, from major international organizations promoting high-tech interventions to small NGOs advancing low-tech ones, can see of proceeding with development.

Other possibilities and perspectives are needed. Introducing them relates to the subversive dimension of IK work. The dominant narrow capitalist idea that development amounts to technical fixes, market integration, and good governance hampers IK work, which has to challenge this hegemony to meet its potential. The implication is that IK work engages with development on the terms set by agencies and that it does so with a view to changing those stipulations, creating space for alternative local ideas of development in the future. It is difficult to confront such entrenched and powerful interests and simultaneously encourage a positive change of outlook. In order to begin, imagine what shape such alternative stances on development might take. To a great extent, effecting this change of view is for people whose lives are affected by interventions. There is a need to distinguish between furthering understanding of the political concerns involved and engaging in dubious politicking

---

[1] It does so notably by facilitating meaningful communication and promoting a collaborative atmosphere (see Sillitoe 2004; Strathern 2006 underlines communication).

on others' behalf. One way to enter on this task legitimately is to question the use of the term *indigenous knowledge*, its various synonyms, and the assumptions that they prompt, sidestepping rationales about the propriety of becoming involved in development at all.

Despite disagreement about the most appropriate term—and difficulties defining exactly whatever words are used amid ongoing diffusion and continual change (Ellen and Harris 2000)—the assumption is that one finds something akin to IK everywhere, whether in the New Guinea Highlands, the floodplains of Bangladesh, or the Durham dales of North-East England. It is equivalent to assuming that all humans have subsistence regimes, technology, and language; politick over power; acknowledge kinship relations; entertain supernatural ideas; and so on. I am, namely, talking about the long-standing presumption underpinning ethnographic inquiry: that there are certain universal attributes subject to varying sociocultural interpretations. The presumed universality of IK has contributed to its featuring on the development agenda in the first place as agencies seek generic solutions to global poverty. It could possibly contribute to answers if only advocates of IK can figure out how to harness it to the dominant technological and market view. But what if IK itself is a culturally relative concept? After all, anthropologists know that categories such as economy, kinship, and the supernatural, which inform and structure ethnographic enquiry, are relative. The possibility indicates the way to potential alternative perspectives on development.

Although a clash with agency positions is unavoidable if one of the goals is to suggest alternative ideas about development, I need to proceed cautiously and not appear too off the wall to be taken seriously. Agencies seek approaches that they can apply globally—such as integrated rural development, structural adjustment, and sustainable livelihoods—and one needs to beware of putting them off (as those with power in development contexts) when saying that there may be no culturally universal idea of IK, let alone a generic IK approach. One of the problems long encountered in applied work is that IK, whether a universal category or not, varies widely from place to place, change that in part prevents it from contributing to general answers to development problems as opposed to suggesting the need for specific culturally tailored ones. It is IK's inability to play a role in formulating universal solutions that has inhibited its impact on development.

It all comes down to language and what is meant by the words used—in this case, *indigenous* and *knowledge*. Although I do not subscribe to the strict interpretation of cultural relativity according to the Sapir-Whorf hypothesis,[2] namely, that the language used by speakers determines their understanding of the world they live in (Whorf 1956), I do think that its vocabulary and the way those speaker use it are instructive indicators of what they think they know. The words are indubitably an aspect of the users' understanding as communicated to others. A much cited, albeit disputed, example of how the words people use help to determine their knowledge

---

[2] Although referred to as the Sapir-Whorf hypothesis, the views that Sapir and Whorf had on language and culture were not identical. Whorf was interested primarily in grammatical categories and their influences on thought (see Lucy 1992, for a critical review).

and understanding is Inuit snow categories.[3] A contender in the New Guinea Highlands is the plethora of terms for sociopolitical exchange transactions (Sillitoe 1979), against which our gift vocabulary is insignificant. These transactional terms refer to an intriguing stateless polity and way of interacting with fellow human beings, to which the distinction between gift and commodity, a bland dichotomy widely imposed on the ethnography, does scant justice (Godelier 1999; Gregory 1982; Strathern 1988; Weiner 1992). It is necessary to master the complex exchange vocabulary to come anywhere near understanding what people are doing, for while the western mercantile tongue can subdivide the commodity side into several transactions (such as cash sale, lease, installment plan, check, and credit card), it does little more than talk of gifts, which is the reverse of native languages spoken in the New Guinea Highlands.

Debating Appropriate Adjectives

Opinions on the correctness of the adjective *indigenous* and suggested alternatives such as *traditional* and *local* differ. The development community makes wide use of IK and seems to have fair agreement about what it means. It is used in this chapter preferentially to signal engagement with agencies on their terms. But some persons, and certain academics in particular, are unhappy at the use of this adjective, as reflected by a furious debate recently over the propriety or otherwise of employing it in any context, development or otherwise (Barnard 2006; Guenther et al. 2006; Kendrick and Lewis 2004; Kuper 2003). According to detractors, history shows that it is a difficult, if not meaningless, word to define. No one is truly indigenous to anywhere. There is, for example, no such thing as a thoroughly native Englishman because those who inhabit that part of Britain called England are a mix of immigrants and invaders, including Celts, Romans, Angles, Saxons, Vikings, and Normans, with repeated influxes of Irish, Scottish, Jewish, and Huguenot. More recently, this mélange has been expanded by various citizens of the Commonwealth or of other areas. The result is a cosmopolitan, if not mongrel, culture. There is anxiety from a liberal perspective that use of the term *indigenous* may encourage xenophobic, even racial, emotions. Also, unsurprisingly, some colonially founded states

---

[3] According to some linguists (Martin 1986; Pullum 1991), speakers of the Eskimoan family of languages do not have a strikingly larger number of words for snow than, for example, English does, whose vocabulary includes *snow*, *slush*, *sleet*, *hail*, *freezing rain*, *blizzard*, *drift*, *white-out*, *flurry*, and *powder*; compounds such as *snowstorm* and *snowflake*; and terms particularly familiar to skiers, such as *cornice*, *mogul*, *run*, *avalanche*, and *hardpack*. According to critics, many of us have swallowed the hoax of Inuit snow categories because it plays to the exotic view of these nose-rubbing, blubber-eating people (Pullum 1991, p. 162). By contrast, the fact that printers have dozens of words covering a range of fonts and typefaces for what the rest of society simply calls letters arouses scant interest. But this lack of awareness does not detract from the point that the vocabulary people have at their disposal informs their understanding to some extent, certainly as communicated to others.

such as the United States of America and Australia, given their history of domination, ethnocide, and even genocide, do not wish to draw attention to indigenous issues and rights (e.g., Battiste 2000; Four Arrows 2006; Povinelli 1998; Wolfe 1994).

Although some intellectuals warn against the use of *indigenous*, numbers of people use it to describe themselves. Indeed, attempts to discourage its use could arguably inflame tensions because some people fear loss of identity or of the right to think of themselves as native—that is, belonging—to a place where they can relate to the values and way of life. Millions of people think of themselves, for instance, as English indigenes. Some populations are increasingly marking themselves off as native to certain regions in their fights for their rights and interests, of which national governments seek to dispossess them (see, for instance, the debate in *Anthropology Today*—Bowen 2000; Colchester 2002; McIntosh et al. 2002; Rosengren 2002; also Asch 2004; Niezen 2003). Many of these people refer to themselves as tribal, and they have long fascinated and even until recently largely defined anthropology, the only discipline to esteem their ways of life highly enough to study and understand them. Various indigenous movements are currently seeking a voice for their views internationally, sometimes holding alternative world summits (Sillitoe and Bicker 2004). There are also many locally rooted NGOs, some of which constitute self-styled IK networks seeking to represent such interests in development (Sen et al. 2000; Shepherd 2005). Although the known history of the Papua New Guinea Highlands may not extend back far, archaeological evidence is expected to show considerable intermixing of populations over the millennia. Yet people do not currently style themselves as indigenous in the way that others do (such as those displaced by Indonesian colonial expansion in the western half of the island of New Guinea), perhaps reflecting that their rights are not subject to threats found elsewhere in the tribal world. Granted, the dangers of xenophobic or "tribal" tensions are evident elsewhere in the Melanesian region (e.g., conflicts in the Solomon Islands, recent military coups in Fiji, and calls to expel from Morobe Province in Papua New Guinea all those not native there).[4] The nearest that New Guinea Highlanders are likely to come to being indigenous to their area is in talking about belonging to or coming from someplace, with the implication that those living there speak the same language and follow the same traditions.

The discussion turns now to the adjective *traditional*, which some commentators prefer to use to describe the knowledge of interest in this chapter. But *traditional*, too, faces disapproval (Douglas 2004; Hobsbawm and Ranger 1983). To critics, archaeology and history show that all cultures change, albeit some apparently faster than others, so an interest in tradition, conceived of as static custom passed between generations, is wrongheaded, for again there is no such thing. Although this stance may be correct from a western historical perspective, it does not apparently stop some people—again often ethnic minorities of tribal origin—from talking about tradition and their wish to keep it intact, the way they think it has always been for them. They express a need to protect it from extinction in the face of current global pres-

---

[4]The *wantok* (Tok Pisin, literally "one talk," speaker of the same language) system in urban areas is another manifestation of these strains.

sures, and some participants in this discussion even search for *lost*, often suppressed, traditions. Not everyone, it would appear, is enthralled by the idea of globalization (for many, code for American world domination). These forces are ironically push-ing some groups in the opposite direction, seeking to reaffirm their traditions and associated indigeneity, for they see what they value increasingly threatened by nomi-nally democratic capitalist domination. It is common in the Pacific, for example, to hear people talk in Pisin (the regional creole) about looking out for *kastom* 'custom', and in Papua New Guinea to hear reference to *pasin bilong tumbuna* (or *bilong ol* or *mipela*) 'ways of the ancestors (or them or us)' (Keesing 1992; Keesing and Tonkinson 1982; Sillitoe 2000b, pp. 240–258). In the Wola vernacular known to me, they talk about *shumbaen bismiy uwp* 'our ancestor's way of doing things' and *naun maen* 'our customs'. These expressions all signal ideas of the traditional, doubtless sharpened with the arrival of the outside world setting their ways off as different, although people have long noted differences with neighbors (The Wola, for instance, acknowledge that the Foi and Huli have different languages and customs).

Another adjective used to label the knowledge of interest in this chapter is *local* (Bicker et al. 2004; Pottier et al. 2003). This word, too, has its critics who hold that the forces of globalization and migration make definition of the local problematic. The position is similar to that for the previous two adjectives. Many people seem to have no problems talking about the local, identifying strongly with particular regions and even local communities within them. Although some residents may move away and blur ideas about what constitutes local by maintaining social and emotional links with their place of origin, the majority living there throughout their lives experience little confusion. They readily identify with their place and those who live there, likely regarding themselves as indigenes who have traditions. There is a clear idea of the local in the New Guinea Highlands, defined as a place where one has rights to land through kin connections to current and previous inhabitants. In Pisin people talk about their *as ples* 'place of origin', and Wola speakers refer to *ninau suw* 'my land' if asked where they come from (Sillitoe 1996). They have strong emotional associations with particular locales in their regions. As mentioned, these associations are as near as they come to the idea of indigeneity as deployed overtly in political struggles elsewhere.

Escaping the Adjectival Mire

The interest in keeping up local traditions suggests that not all people are enthusias-tic about the prospect of change (often glossed optimistically as development) but wish to preserve some aspects of life much the same from generation to generation. It brings to mind the distinction between so-called hot and cold societies or, more aptly in relation to the passage of time, linear versus circular ones. The critique of adjectives such as *indigenous*, *traditional*, or *local* is of a piece with anthropolo-gists' shift of interest away from their *traditional* circular-tribal subject matter and toward their own society's linear concepts and commercially driven preoccupation with change. Although ideas of circular tradition may reflect the local view in some places, such a synchronic focus (with associated notions of structure and questionable ethnocentric present) is, as history shows, an error. It is an overstatement, even misrepresentation, as occurs in many a radical change of perspective.

The debate over the propriety of the adjectives *indigenous*, *traditional*, and *local* indicates one reason why the IK initiative has not blossomed as some expected, mired as it is in assertions smacking of political correctness that inhibits frank discussion. Criticisms of these adjectives appear to have more to do with western intellectual reservations than those of people to whom the words are regularly applied and who happily use these labels for themselves. When they use them, they may characterize unique attributes of their epistemologies, particularly when associated with the idea of knowledge and contrasted with western scientific, rationalist epistemology. It seems that they wish to, and can in their minds, distinguish the indigenous, traditional, or local categories that substantially overlap with such anthropological ones as culture, custom, and society, used widely until theoretical fashions alighted on humanity apparently merging into a global gloop. Even the UN, arbiter in the global political sphere, sanctions the use of the term *indigenous*, as in the International Decade of the World's Indigenous People and concern for Indigenous Peoples' Rights (Minority Rights Group International 2006; United Nations 1993). Although no institution has the authority to legislate on such things globally, it is at the UN Permanent Forum on Indigenous Issues and in related international bodies such as the Convention on Biological Diversity (e.g., Article 8(j): Traditional Knowledge, Innovations and Practices) and the World Bank (e.g., Bank Procedure 4.10: Indigenous Peoples, July 2005) that interested parties have negotiated a workable international definition.

The case against the use of such terms may be well intentioned, reducing the *we* and *them* distinctions that often characterize misuse of power, also possibly offensive racism, by encouraging people to consider themselves all equal members of humanity with the same rights. But opposing their use, while presumably holding attendant ideas in one's head, comes close to sounding like the colonial authorities so roundly condemned by and large. The line of thought seems to be 'we know better than you how dangerous such terms can be in fostering nasty sentiments, so let's stop using them'. In other words, 'trust us, we know best', when recent history probably makes the users of the suspect vocabulary understandably distrustful of outsiders. These labels often feature in the fight that many tribal peoples wage against historical wrongs, their rights having been grossly violated from a liberal perspective (The point is highlighted regularly by NGOs such as Survival International and Minority Rights Group; see, for example, the ongoing struggle of the San people in Botswana—Minority Rights Group International 2005; Survival International 2006). They surely have a right to use the expressions in relation to themselves if they so wish, if they think it will help their interests. Also ironically, accord that such adjectives are inappropriate for distinguishing between communities implies agreement with modernization approaches to development. After all, if it is illegitimate to distinguish between human beings because everyone is the same, then top-down solutions to underdevelopment presumably apply across the board (Sillitoe

2002b, p. 133). Few of the critics advising against the use of *indigenous*, *traditional*, and *local* would probably accept this let-modernization-rip reasoning. But if one is not to distinguish between people, or is unable to do so, then, logically, the same *solutions* should work everywhere, which they patently do not, as six decades of development demonstrate.

It is possible that humankind will eventually develop a harmonious global culture in which common solutions to problems will be appropriate. But current evidence—an aspect of which is the way people keenly defend their identities as indigenous, traditional, and local—indicates that that eventuality is a long way off. Belief in this process relates to another opinion of why interest in IK has not taken root in development, namely, that it has no future, either for inside custodians or outside researchers. In development-as-modernization, the forces of globalization will drive it to extinction. The evidence suggests otherwise. Since the time of anthropology's Victorian founders, observers have warned about the disappearance of cultures and the need for salvage ethnography. Yet irrespective of significant lifestyle changes in many small-scale societies, hunter-gatherers continue to hunt and gather; shifting cultivators, to shift and cultivate; communities, to believe in local deities; and clan obligations, to inform social life. Local ways, albeit prone to major change, persist against the apparent odds. They are stubbornly resistant to the blandishments and threats of global capitalism and modernism. It is increasingly realized, for example, that local knowledge of natural resources is an integral aspect of any environment, notably in biodiversity management and conservation, where its demise may be as damaging as the loss of species (Posey 1999).

The problem of which adjective to use to describe the knowledge of interest in this chapter appears irreconcilable to the interests of all parties and in some ways reflects the current indeterminate state of anthropology spread-eagled between sociology, geography, biology, history, and recently emerged cultural studies (Cohn 1980; Macfarlane 2013; Wade 1997; White and Tengan 2001). Other terms—such as *people's* or *citizen's knowledge*—are too woolly to mean anything. For example, who are the people meant? Do citizens necessarily share much knowledge? Perhaps the most apt adjective to describe the knowledge that I am discussing is the adroit neologism *glocal* (Robertson 1995), with its clever play on the local affected by global forces. But it has yet to catch on.

Despite the interminable debate, all the adjectives place their users in the same grid square, if not exactly at the same map reference, giving a fair idea of what they are talking about. It is unhelpful, even possibly ethnocentric, for privileged, globetrotting, cosmopolitan academics to foist their worries on others by holding that certain ways of talking should be avoided because they are divisive, especially when the concerns of these populations are pressing and involve basic human-rights issues, even daily survival. The research community and policy-makers need to get back to ethnography to further understanding of why and how people use such terms, particularly in development contexts—that is, there is a need to listen to these locutions nonjudgmentally. In summary, contentions about indigeneity scarcely relate to ideas that Wola people have about knowledge or development, for the members of that community touch on it only through their understandings of the local and the traditional.

What Is Meant by Knowledge?

During an IK conference, a prominent anthropologist broke into a discussion over the appropriateness of different adjectives and asked, "what is knowledge?" There was an audible groan, implying that everyone knows the answer. Someone whispered, "more clever-sounding rhetoric," and the adjectives debate resumed. On reflection I can see some merit in the question; enough to prompt this chapter. Can western academics assume that other indigenous/local/traditional ideas of knowledge will be similar to their own? If not, how might those ideas differ, and what are the implications? Anthropologists need to go beyond debating adjectives to look also at the noun *knowledge* as a word potentially opening up a glimpse of an alternative idea of development. Although adjectives are useful for furthering comprehension, it is necessary to have a firm understanding of the noun to which they refer. To talk about gents and ladies and unisex bicycles (according to the angle of the crossbar), or sports, racers, roadsters, BMX, or mountain bikes is all very well, but one needs to know what a bicycle is for any such adjectives to have import.

A vast philosophical literature deals with the topic of what constitutes knowledge. The externalist theories, for example, relate to causality, reliability, and counterfactual matters. There are also internalist theories, such as foundationalism, coherentism, and subjective awareness (Honderich 1995, p. 447). It is not necessary, even if space permitted, to engage in a review of these tangled epistemological debates. The crux is that the justification of true belief, as the philosophers put it, yields knowledge. In other words, knowledge relates to the idea of justified belief or "truth," something of significance to New Guinea Highlanders such as Wola speakers. This chapter consequently focuses on the empiricists' theory of knowledge, drawing on philosophers such as Locke, Mill, Berkeley, Hume, Peirce, and Ayer. Induction has a prominent role in this view, with secure knowledge of truth derived from experience (Russell 1948), which relates directly to what Wola language construction suggests is important to speakers. According to Hume (1739/1896), for instance, ideas derive from impressions of sense: Granite is harder than chalk, as one can demonstrate with the fingernail. This stance relates to the veracity of facts, those fragments of information that are a significant component of knowledge and which a person assesses as either true or false. As Ayer (1971) comments, "To say that I know that something is the case, though it does imply that I am sure of it, is not so much to report my state of mind as to vouch for the truth of whatever it may be" (p. 17). Wola speakers of the Southern Highlands Province of Papua New Guinea routinely assess the reliability of knowledge imparted to them. They realize—as do most, if not all, humans—that knowledge is often partial or based on unreliable evidence. They customarily signal this awareness in their verb constructions. Knowledge also has to do with understanding, both one's own and that of others, and there needs to be some way of assessing the reliability of the information that features in it.

Furthermore, people know that the truth may be disputed hotly, whatever the evidence; witness those who deny the holocaust regardless of the gruesome physical evidence, distressing photographs, and testimony of thousands. The postmodern turn has exacerbated the sense of uncertainty by promoting subjectivity as the only truth, particularly in the arts and social sciences, where it is deeply unfashionable in some quarters to talk about facts. The consequences have been unfortunate for anthropology, with insufficient attention being paid to ethnographic data, even their eschewal as never reliable by some scholars who favor self-conscious accounts of personal experiences. An onion is not a cow, and confusing the two hinders understanding, as evident in the Kalahari debate, where one finds some significant and humorous contributions to the propriety-of-indigeneity argument (Barnard 2006; Lee and Guenther 1991). Indeed, anyone regularly making such mistakes is unlikely to survive for long (or will end up in a lunatic institution for everyone's safety), for as the philosophers put it, that individual will not be someone whose belief is justified and who does well in this life.

There is a danger of confusing facts with imagination. Facts relate to reality (something that exists out there independently of the mind, something such as granite and chalk), as opposed to imaginary ideas (products of the mind only, such as ghosts and fairies). The stuff of the one category can be shown to exist using objective evidence, whereas the stuff of the other depends on faith. Any interpretation of facts involves imagination and possibly abstraction. Although people may concur that facts are true, the interpretation of them in a wider context may be moot, assuming that the reasoning about them is valid and consistent. One of the key points of the postmodern critique is that human interpretation and arrangement of facts is subjective, not that facts are imaginary or unimportant. Politics enters as adherents of one interpretation try to make it prevail over others (see the much-cited work of Foucault 1988; see also Crick 1982, pp. 303–307; Rabinow et al. 2001). An example is the process of determining what constitutes development.

Any interpretation also depends not only on the meaning and veracity attached to the words used to express it but also on the ordering of the words into coherent utterances. It is necessary to go beyond debating the meaning of one or two words to exploring the possibility that the structure of the language people employ to arrange words may critically inform their understanding of the facts that feature in any communication. It is necessary to pay "attention to the way that epistemic modality works in the grammar of non-Western languages" (Howard et al. 2002, p. 4). After all, the way we humans construct the communications that we make with others, and not only the vocabulary we use, has some implication for the intelligence that we share and how we convey what we know. Such structuring is a main aspect of the interpersonal communication by which we primarily turn private thoughts into something socially distributed. It conditions our comprehension and assessment of knowledge passed on, our knowledge of knowledge itself.

This approach to knowledge, which takes its cue from the concerns of the New Guinea Highlanders of interest in this chapter, differs from that more usually taken in anthropology, which commonly centers on the social context and symbolic implications of knowledge.[5] For example, Barth (2002) talks of investigating "how bodies of knowledge are produced in persons and populations in the context of the social relations that they sustain" (p. 1). Such an interpretation, he says, goes beyond questions of "truth and rationality" to the "interleaved phases of its construction, representation, distribution, and reproduction and the uses made of it by positioned actors" (p. 9). Indeed, it nearly goes so far as to embrace the broad and nebulous anthropological notion of culture, as some commentators on his paper observe.[6] It is an attempt to respond to Mannheim and van Vleet's (1998, p. 341) injunction that it is necessary for "ethnographers to work with linguistic form as central to understanding social life."

New Guinea Highlander Expressions of Knowledge

Westerners agree that whatever knowledge may be, humans store and process it in the brain, whereas for the Wola cognition occurs in the chest. It is there that an individual's *konay* resides, which is the mind, intellect, conscience, will, even personality. It resides in the *hibiy-aumuw* 'liver–heart' region but is not physiologically visible. It relates to the *konem*, which is the *mind* in action manifest in thoughts, opinions, understanding, decisions, desires, and intentions that emanate from the *konay*.[7] The formulation of the *konem* with respect to particular matters involves acts of *nendsay*, which is to ponder, think, reflect, or recall something, and involves the *konay* in operation so to speak. It is possible for one's *konay* to *okhemay* 'forget' things that it knew or understood previously. The Wola refer to the *konay* and *konem* as belonging to individuals and anticipate that they may differ from one person to the next. All have their own *konay* located in their chests, and when persons pursue certain courses of action, people may talk of it as *obun konem* 'her or his wishes, intentions, idea' (depending on context). Although they have their own *konay* and may act on its promptings and calculations, these responses are not isolated. Other people influence and shape one's *konay* through interaction, and kinsfolk are particularly influential. Individual experiences also mould one's *konay* throughout life. Another abstract feature of the person is the *wezow*, which everyone has by definition, as it relates to being alive. It is the life force, self-consciousness,[8] or spirit that

---

[5] This point is evident, for example, in Crick's (1982) review of the anthropology of knowledge. He deals there largely with the social dimensions of knowledge, comments on cognitive issues, and concludes with reflections on reflexivity and the nature of anthropological knowledge but omits any consideration of philosophical questions relating to the nature of knowledge and their relativity.

[6] As an aside, this chapter differs in another way, repudiating Barth's (2002) assessment of IK as "not the most felicitous way" to investigate knowledge (p. 2).

[7] These ideas are similar to the Melpa one of *noman* (see Stewart and Strathern 2001; Strathern and Stewart 1998, 2000, pp. 64–66).

[8] The use of the term *self-consciousness* here does not imply the negative connotations of being self-conscious.

animates every human from birth. It is also one's shadow. Unlike the *konay* it does not grow and change through life. One has it at birth. It leaves the body at death, at which time it transmutes into a *towmow*, an ancestor spirit, ghost, spectral presence, or apparition that people believed malevolent, which may attack living kin, causing sickness and death.

The Wola talk of knowing something as *henday* 'seeing' or *pangay* 'hearing' it. This form of expression is reminiscent of the English 'Oh, I see' when one understands something, or 'I've not heard of that' if you do not know a thing. What you know is stored in your *konay*, where it may *sit* (*obun konay bort beray*—literally, 'his or her mind in sits'). But there are degrees of trust that one can have in what sits there, in what you think you know. By encoding for what linguists term *evidentiality* on the verb, the Wola language obliges its speakers go to great lengths to specify the faith one can have in the truth of statements (Aikhenvald 2004; Epps 2005; Willett 1988). It is one of many non-Austronesian New Guinea languages, which are well known for the complexity of their verb constructions, a feature of which has to do with evidentials that indicate veracity (e.g., Foley 1986, pp. 165–166; on languages of neighboring highlands, see Draper and Draper 2002; Franklin 1971; Lang 1973). In the Wola language the verb can indicate not only tense, person, and number but also seen and unseen aspects of actions, that is, direct and indirect evidence (Tipton 1982, pp. 78–81).

The ubiquitous verb *bay* 'to do or undertake' something, which also features as an auxiliary verb in talking about a range of activities (Sillitoe 2006a, p. 137), illustrates the conjugation of past tenses (Table 7.1; see also charts 4 and 5 in Tipton 1982). People use these tenses to talk definitively about what they and others know. There are two time frames, the recent and distant past. It is *niy baenda*, for example, if you talk of doing something recently and *niy bundis* if you did it longer ago. The measure of time in this regard is relative, depending on the periodicity associated with the act discussed. If you are talking about having eaten something, the period between recent and distant past will be a few days; if you are referring to the seasons, it may be months or years. This dual time frame is an indication of reliability. The more recently something occurred, the more likely the speaker is to recall it accurately; people have more faith in short-term than long-term memory.

The Wola language goes considerably further in indicating the trust one can have in the knowledge on which persons draw in conversation. At one extreme the speaker may talk about something that both speaker and hearer know about for certain—something, say, they have witnessed or undertaken together; such as *ibiy bisiybindis* 'you two did [as I saw] some time ago'. At the other extreme the speaker may refer to something that is only hearsay, which neither party has witnessed; such as *obuw bisesa* 'she or he did [we are told] long ago'. The *neither-speaker-nor-hearer-witnesses* evidential signals unconfirmed intelligence that one has to take on trust because neither party to the conversation can vouch for or check it. It is the form often used when telling myths, many of which continue to have currency even though people have nominally turned to Christianity. It may explain in part why Wola speakers gave up their old supernatural beliefs and the corresponding activities so amazingly quickly, taking only a decade or so to become nominal Christians.

Table 7.1 *Past Tense Evidential Conjugations of the Verb bay 'to do' in the Wola language*

Personal Pronoun	Both speaker and hearer witness		Either speaker or hearer witnesses		Hearer did not witness but heard of previously		Speaker did not witness		Neither speaker nor hearer witnesses	
	Recent past	Distant past	Recent past	Distant past	Recent past	Distant past	Recent past	Distant past	Recent past	Distant past
I (*niy*)	b **aenda**	bu **ndis**	bau **wa**	bu **wa**	bu **waenda**	bo **wasaenda**	buw **asol**	buw **asol**		
You (*njay*)	b **aenda**	bi **sindis**	b **ai**	b **isiy**	bae **saenda**	bi **sisaenda**	ba **sol**	bi **siysol**		
She, he (*obuw*)	b **aenda**	bi **sindis**	bo **no**	b **iso**	be **naysaenda**	bi **sorsaenda**	be **naysol**	bi **sorsol**	be **naysa**	bi **sesa**
We (2) (*yau*)	ba **baenda**	bi **subindis**	ba **ba**	b **isba**	ba **basaenda**	bi **sbasaenda**	ba **basol**	bi **sborsol**		
You (2) (*ibiy*)	bai **baenda**	bi **siybindis**	ba **biy**	b **isbiy**	ba **biysaenda**	bi **sbiysaenda**	ba **bisol**	bi **sbiysol**		
They (2) (*mbiybom*)	bai **baenda**	bi **siybindis**	ba **biy**	b **isbiy**	ba **biysaenda**	bi **sbiysaenda**	ba **bisol**	bi **sbiysol**	ba **bisa**	bi **sbisa**
We (*nau*)	ba **maenda**	bi **sumindis**	ba **ma**	b **isma**	ba **masaenda**	bi **smasaenda**	ba **masol**	bi **smorsol**		
You (*onyiy*)	bai **maenda**	bi **siymindis**	ba **miy**	b **ismiy**	ba **miysaenda**	bi **smiysaenda**	ba **misol**	bi **smiysol**		
They (*mbinyom*)	bai **maenda**	bi **siymindis**	ba **miy**	b **ismiy**	ba **miysaenda**	bi **smiysaenda**	ba **misol**	bi **smiysol**	ba **misa**	bi **smisa**

Note. The tense and evidential suffixes added to the verb stem are shown in boldface (variably *ba, be, bi*, and *bu* for *bay*, depending on vowel harmony). The two columns beneath the heading "either speaker or hearer witnesses" display tense suffixes of only the recent and distant past, respectively. All the other columns have evidential suffixes added (recognizable if one takes vowel harmony into account). Design by author.

When faced with the technical superiority of outsiders and their consequently irresistible political power, combined with aggressive missionary proselytizing (extending to the smashing of cult objects), the Wola soon concluded that their ancestors had gotten things wrong. These descendents could no longer trust what had been passed on to them. The *neither-speaker-nor-hearer-can-vouch-for-this-knowledge* tense arguably made it easier to reach this decision and discard their previous beliefs.

In between these two evidential aspects—*both speaker and hearer witness* and *neither speaker nor hearer witnesses*—are aspects that indicate one of the parties to the conversation knows first-hand about the topic. They may also indicate possible indirect types of support, such as hearsay or inferred evidence, rather than the absence of performative, sensory, or other direct evidence. The speaker may not specify which one of them knows about the matter, as in *mbinyom bismiy* 'they did some time ago [as one of us knows]'. Or the speaker may indicate that she or he did not witness the event, as in *onyiy bamisol* 'you all did recently [so I understand]'. Or the speaker may indicate that the listener did not witness it, as in *yau bisbasaenda* 'we two did long ago [as you may have heard]'. The apparently anomalous existence of the first person (singular and plural) in the former of these last two conjugations and second person (singular and plural) in the latter further shows the lengths to which the Wola go to indicate the state of the interlocutors' knowledge of what is under discussion. They allow speakers to talk of events that they or the listener understand they participated in but cannot remember doing so. So *niy buwasol* has the sense 'I did lately [although I cannot remember it]', and *ibiy babiysaenda* has the sense 'you both did recently [although you cannot recall it]'—intimating that the speaker might be putting words in their mouths. Although *okhemay* to 'forget' something implies having once known it, persons will not use a tense suggesting firm knowledge (such as the *indis* suffix) if, when reminded of the thing, they cannot recall it.

In addition to these various suffixes indicating tense, person, number, and seen/direct and unseen/indirect aspects, the Wola language employs other affixes to modify verbs further, such as by temporal aspect. They include *mba* and *pa* to indicate completed concrete actions (i.e., a house thatched) or completed nonconcrete behaviors (i.e., something learned), in ways similar to the English past perfect tense, and *aka* to indicate that the speaker has undertaken some action for someone else (i.e., cooked food for them). There are modifying final suffixes as well, such as *sha* to indicate a fully finished action or task (where either or neither speaker or hearer witnessed it), as in *nau bismasha* 'we did it entirely' and *munk* to indicate a prior condition, as in *obuw bisomunk* 'she or he took part previously'. The language also has query markers that indicate the conditional or interrogative nature of an utterance based on imperfect knowledge. They include the weak *ye*, which is similar to 'perhaps', as in *obuw ngo kiso ye* 'perhaps she or he said that', and the strong *be*, which functions like a spoken question mark, as in *obuw ngo kiso be*, 'did she or he say that?'

These various grammatical arrangements relate to the trust one can have in any intelligence referred to. Their meaning ranges from complete confidence (so far as one can rely on one's senses anyway) because one has seen or experienced it for oneself, to the dubious (in that neither speaker nor hearer have first-hand experience of the topic at hand). The use of the word *henday*, literally 'to see', for 'to know' underlines this point with verb distinctions in that it suggests that if you *know* something for certain you will have *seen* it for yourself; you will have been present and will have witnessed it. The phrase for 'I don't know' is *nae henday*, literally 'I haven't seen it'. In addition, when it comes to abstract notions and things that no one could ever demonstrate definitively, the Wola regularly talk about them as depending on individuals' *konay*. They realize that abstract notions can vary widely. They regularly express express themselves in such matters by using the *neither-speaker-nor-hearer-witnesses* tense because what they say is not factually verifiable. The truth criteria apply to inferred knowledge, too, albeit one step removed so to speak (Russell 1948, cited in Barth 2002, p. 2) in that one's faith in any inference will depend on the faith one has in the intelligence upon which one bases the inference. It is fairly certain that the tree seen as a leafy crown in the forested distance will have bark on its trunk like all the other trees one has seen in life. Less sure is the inference that the booming sound a relative has spoken of hearing in the vicinity indicates the presence of a cassowary nearby.

Implications for Assessing Knowledge

What might such evidential verb morphology suggest about speakers' stances on knowledge? It indicates that the concern for assessing the veracity of knowledge imparted in any communication, that is, the grounds for justified belief, surpasses that apparent in English. Although all human languages allow speakers to signal evidentiality, just as with time and place, my point is that when evidentiality is integral to the grammatical system, it obliges speakers routinely to address the likely veracity of communications. English does not do so. To signal the reliability of statements, English speakers rely on interaction during discourse, including verbal cues such as optional lexical devices (e.g., adverbs such as *apparently* or qualifying expressions such as 'they say') and nonverbal cues (e.g., shoulder-shrugging or eye-rolling).[9] The consequent looseness with respect to evidence allows speakers scope to make obscure statements, either intentionally, as politicians do regularly, or unintentionally, as social scientists do sometimes, a problem compounded in English by the passive voice. One may contend that the British political system would not be held in such popular disrepute and social science would not be in such disarray if English grammar obliged MPs and writers to give more direct statements. It would compel social scientists, for instance, to face the implications of expressing

---

[9] Linguists think that evidentiality, as a grammatical feature, has lexical origins in that where evidentiality is expressed both grammatically and lexically, lexical forms (such as the verbs *to see, to hear*, and *to say*) are slowly absorbed into verb forms generally, first by clitization and subsequently as verb endings (Epps 2005; Willett 1988).

themselves in obscure, jargon-littered prose so as to avoid charges of naivety and the appearance that they attribute action to an imagined idea, namely, the concept of society. It is revealing to search for the subject in such passages, for it frequently appears to be society. In other words, this imagined concept perplexingly remains the heavily disguised active agent. Directly addressing this issue could stop much trite thought masquerading as profundity.

This interpretation of evidentiality, relating it to the veracity of knowledge, does not preclude other more sociological interpretations noted earlier, such as those of linguists who see it as a cultural expression of politeness and responsibility for information (Beier et al. 2002, p. 133; Epps 2005; Hill and Irvine 1993). In small communities, such as those in New Guinea, people may avoid taking responsibility for information, for trouble can ensue if the information proves incorrect and others act on it. Hence, speakers have recourse to prudent strategies that include specifying responsibility for information together with indirection and vagueness, particularly when persons may base accusations of sorcery or other antisocial behavior on seemingly innocent remarks. It is safer to say "I [*speaker-did-not-witness* form] heard that Ongol went to the garden yesterday" than "Ongol was [*speaker-witnessed* form] at the garden yesterday."

These grammatical arrangements may relate in part to the contrast between oral traditions and literate ones (Finnegan 1992; Finnegan and Orbell 1995; Goody 1987). In oral traditions a person's knowledge can build on what he or she hears and experiences in the course of life. Whereas people can rely on what they directly experience, the faith that they can have in what others tell them varies. Outright liars exist everywhere (*kaega kay* 'untruth say' in Wola), persons make mistakes (*korob kay* 'wrong say'), and there is often no sure way to check on them. (In this context the Wola may speak of *duwn kay* 'straight talk' and *aengal henenj* 'true words'.) The literate traditions have, in addition, a sizeable accumulation of written knowledge, which people tend to take on trust and build on in advancing their own understanding—just think how often one hears persons say, "it is true because I read it in a book." The written word can carry an authority that may not extend to the spoken; people tend to accept the idea that it represents the considered testimony of an expert who knows best. For some historical events, for instance—the written word is all the information that exists.

This literate effect is apparent in anthropology. I regularly lecture on regions where I have no first-hand experience, trusting in the ethnography I consult and expecting students to do likewise. The consequences of doubting the trust that people can have in such written knowledge are increasingly evident. Observers in whose minds postmodern critique has debased the currency of enthnographic facts no longer have faith in them and think that the discipline is in crisis. The limited opportunities available to check the veracity of ethnography and discount fraudulent records[10] make it a ready target for such criticism, a vulnerability compounded by the relative poverty of much existing ethnographic data by which to assess the truth

---

[10] An example is the controversy over Mead's (1928) Samoa account. See Brady (1983), Freeman (1983), and Holmes (1987).

of statements. The implications for the knowledge that we ethnologists think we have are great. If some of the ethnographic data available are fictions—often suspiciously convenient in their support of the ideology of the moment—such invention may partly explain why comparative studies have not proceeded far in uncovering any consistencies in human behavior. This critique has further undermined IK enquiries in development, portraying them as unavoidably biased and partial. Oral traditions may by definition be more skeptical and questioning of claims to knowledge. Anthropology perhaps has something to learn in this regard from these societies, which have constituted much of its subject matter. There is a need to focus on making the documentation of ethnographic facts as accurate as possible and on allowing them to drive interpretations—"ethnographic determinism," as I call it (Sillitoe 2003, p. 336). We ethnologists have to adhere to methods that will restore trust in the ethnographic evidence, to rebuild the discipline's confidence in itself and its creditability. We must, for instance, exploit the increasing opportunities to record substantial amounts of data electronically. After all, in order to assess the plausibility of any intellectual interpretation, it is necessary to know the facts on which it rests. We need reliable ethnographic data so that we can trade with one another.

Colleagues who kindly read this chapter in draft suggested that it would give credibility to my discussion of reliability and trust to address the research context and the methods used to collect the linguistic data I present and analyze and would enable the reader to assess their credibility and accuracy. Having addressed this topic in various other publications, I had taken it for granted. The information comes from some 35 years of acquaintance with the Was valley region of the Southern Highlands Province of Papua New Guinea, during which time I have learned the Wola language. Occasionally, I have made notes on grammar, seeking to understand the language (with which I was helped by language material graciously made available to me by missionaries working in the Nipa and Mendi regions). They have expanded along with my competence (on evidentiality, for example). The illustrative words, phrases, and sentences chosen for this chapter are what occur to me as a speaker to be representative of and appropriate to points made. I did not, for instance, collect them from key informants, for I have never worked with such persons, preferring to interact with the entire community where I have stayed. Various friends have explained facets of the language to me, such as the meanings of words and phrases. In short, I did not set out to collect the linguistic evidence with a view to writing this chapter or any other. I came by it as a consequence of learning the language in order to conduct other enquiries (for more information on my fieldwork, see Sillitoe 1979, 1996, 2003; Sillitoe and Sillitoe 2009). It is in the course of that work that I formulated the aforementioned approach known as ethnographic determinism.

The prominence accorded to the role that experience plays in constituting what individuals know in oral traditions raises issues pertaining to tacit knowledge—or what some writers call embodied knowledge (Csordas 1994; Gibbs 2006; Polanyi 1967; Strathern 1996; Weiss and Haber 1999). Many people, including Wola speakers, emphasize such enskilled knowledge, which individuals learn to act on without necessarily seeking to express it in words (Ayer 1971, p. 13). It relates to knowledge of a different order from that framed linguistically. It prompts some scholars, such as Hirst (1974), to distinguish between two types of knowledge: that in which the "objects of knowledge are true propositions and that in which the objects are practical performances of some kind" (p. 65). Only with another practitioner can one meaningfully discuss the performing of some aspect of such experiential knowledge in such a way that it falls into the both-speaker-and-hearer-witness class by definition; that is, one knows something only if one can do it. This point underscores the idea that first-hand experience is an important component of knowledge, particularly knowledge that is securely known. This practical aspect of knowledge, which can be a significant part of what one knows in an oral tradition such as that of the Wola, poses problems for people who trade in words, and mostly written words at that. Such knowledge cannot be reduced to words, certainly not ones that nonpractitioners can appreciate. As Crick (1982) comments, "For some reason which may be no more than cultural bias, we closely link 'knowledge' and 'language,' but knowledge is not only coded in linguistic systems" (p. 300). To explain practical behavior in language is "to let the left brain explicate right brain formulations. . . . Since we can perceive and know without language, this semiotic transmutation whereby language is the privileged system for explanation will certainly distort the phenomena being explained" (p. 300).

Humans acquire such tacit knowledge by doing it, not talking about it (Collins 1974). The capturing of that knowledge, which features prominently in domains where development is intended to intervene, poses a particular methodological challenge to IK enquiries. A frequently cited example is the act of riding a bicycle. Although a mathematician can work out and write down the various physical forces involved (e.g., gravitational force, kinetic energy, and vector velocities) and even express them in equations, they will not teach anyone how to ride a bike (though they may enable a person to design an ergonomically more efficient cycle). You have to cock your leg over the saddle and peddle to learn, for only experience can teach you the skill (Bloch 1991, p. 187; Borofsky 1994, p. 342, cite driving a car, with the latter author referring to such knowledge as dependent on implicit memory). Many activities in Wola life feature the mastery of such competencies. They range from the sedentary (e.g., rolling bark fiber string, which, with my hairy legs, I have never mastered and am still unsure of exactly how people do it) to the energetic (e.g., clearing forest and establishing neat gardens, tacit aspects of which also remain a mystery to me). Another aspect of the trust warranted by such knowledge once it is mastered is that it belongs to the person who has acquired it. It is beyond surreptitious manipulation by any authority, beyond control without the skilled person's knowledge. The knowledge is part of that individual's very person.

The Wola verb constructions also draw attention to the individual variability of knowledge (Borofsky 1994). What human beings know differs from one individual to the next. Sometimes the topic of conversation is known equally to speaker and hearer (both are witnesses) and not to others (either only one or neither can vouch for it). It is a commonplace that individual culture bearers do not replicate each other's knowledge but vary in what they know, yet this everyday fact is not always sufficiently acknowledged. It relates to another problem facing anthropology, which has tended, under dominant sociological theory derived from French social philosophy, to treat any society as a homogeneous entity structured by shared morals and norms. Although there is mounting criticism of the attendant muting of other ways of understanding human behavior and interaction—notably its individual dimensions (Bloch 2005; Rapport 2003)—the real challenge will be to devise methods to capture such real-life variety. The acquisition of such knowledge is another methodological challenge to IK enquiries when development, in which the interest is in generic solutions, causes entire regions, not only communities, to be seen as homogeneous. A possible way to advance the understanding of knowledge variation between actors is to imagine them constituting extended transactional networks, as in my proposed spheres-of-knowledge model (Sillitoe 2002b). This perspective not only affords a way to handle differences between knowledge traditions, such as the local and scientific, but also variability within and across communities. It can even accommodate conflicts of view—previously swept under the normative carpet, especially with advances in computer modeling and e-science—by using agent-based models and flocking algorithms.

The Authority of Knowledge

Questions of knowledge variability take on a particular significance in stateless polities, such as that of the Wola, where no authority exists to standardize what is known and to adjudicate who is right when persons are at odds with each other (Sillitoe 2002a). The only sure knowledge, that on which no authority is recognized to rule what is true, pertains to issues and events of which both speaker and hearer have agreed, possibly shared, experience. It contrasts with state orders where people acknowledge authorities who pronounce on what is true. Speakers of the Wola language would, ironically enough, appreciate the Royal Society's motto: *nullius in verba* 'take nobody's word for it' (which underlines belief in empirical experiment to further understanding of the natural world).[11] Again, trust is a focus. It is necessary to the acceptance of such authoritative knowledge declarations (Cook 2005; Hardin 2002; Rothstein 2000). It is central to the authority that experts claim for their knowledge. Nonexperts have to trust that such persons (e.g., scientists) know their fields and are correct. Global science severely sanctions fraudulent work because it threatens to undermine the entire scientific edifice by sapping confidence in the authorities and their expert representatives, which is central to the acceptance of their knowledge pronouncements as justified or true.

The consequences of mistrust are evident whenever the *populus* senses that science is somehow wrong, as it currently does with respect to ever-changing advice on diet and lifestyle. The fickle guidance suggests, for example, that whatever peo-

---

[11] The Society coined the motto when science was emerging in a society dominated by a church that demanded unquestioning faith, that is, before the learned organization had become part of the establishment's authority and expected nonscientists to believe its word.

ple eat or do threatens their well-being (e.g., recent rat experiments by Colman et al. 2009, recommend a starvation diet to increase longevity). Other dubious matters include unnecessary experiments on animals, faulty convictions about sudden crib death (sudden infant death syndrome, SIDS), abuses of Big Brother surveillance technology, the connection between environmental pollution and worrisome climate change, and genetic engineering's interference in life itself through fears about freak crops and animals. The current commercialization of research, notably in universities, is exacerbating the distrust, with the perception that companies put money-making before scientific truth (see Clift 2007, on consequences of the Bayh–Dole Act of 1980 in the United States). All these trends offer further opportunities for other beliefs to dispute the veracity of scientific knowledge above the spiritual, a challenge exemplified by Christian and Islamic fundamentalist arguments for creationism against evolutionary theory. It is not necessarily that scientific know-how itself is incorrect. Its authority is being undermined instead by a lack of faith that humans have the wisdom to use the knowledge appropriately, a realization that returns this discussion to political aspects. One has only to consider nuclear technology proliferating horrendous weapons of mass destruction, worryingly in the hands of fallible, even deluded politicians.

A key point is that, in states, we humans collectively cede to others authority over aspects of our lives and what we know, even when we mistrust some of their pronouncements, whereas in stateless contexts this partial surrender does not occur. In an acephalous polity, such as that of the Wola, one finds no recognition of authority (or expert) that can direct people. By definition there is no acknowledgment of authoritative offices, no expectation that one will put particular trust in some persons' statements above others. Indeed, the idea that some individuals have special knowledge beyond others challenges the core egalitarian values of the Wola. Why should some know more than others? Furthermore, how can one rely on such persons to use such knowledge in your interests and not to further theirs at your expense and extend power over you? Although people may differ to some extent in what they know (variability as discussed above), such differences do not translate into one section of society systematically monopolizing knowledge of domains important to the well-being of others and using it to dominate them politically. There is thus no scope for the political manipulation and cheating found in states (e.g., the misrepresentation of the scientific evidence to justify the illegal invasion of Iraq by the United States and the United Kingdom).

A stateless context featuring egalitarian values does not mean that persons are all the same. As I have written in detail elsewhere (see Sillitoe 2006a), it does not conflate equality with sameness. Knowledge is inevitably unevenly distributed, but this variability does not translate into differences in authority that may threaten the sovereignty of individuals. Although persons have rights, particular kinds of knowledge are not treated as secret. For instance, only certain groups claim corporate control of land and have oral traditions that justify these claims based on origins of ancestors, but stories that refer to ancestors are not restricted by the owners solely to adult group members. Nor is the passage of such knowledge between generations marked by any initiations linked with wealth, power, fertility, or other attributes, as in some regions of New Guinea (e.g., see Timmer 1998). Individual ability is acknowledged and sought in practical and ceremonial spheres, with some persons being recognized as more knowledgeable and skilled than others in particular ways—such as in hunting, making certain objects, or conducting rituals—but any influence they have extends only to these limited spheres. Although there are individuals who, for instance, had knowledge of certain rites and spells over others, they use this knowledge in the service of relatives and friends when called upon to do so (e.g., to cure illness), usually in return for a payment (Sillitoe and Sillitoe 2009). They behave in much the same way as *misin boi* 'pastors' do today in relaying the biblical message to people. Beyond manipulating certain ritual effects, they cannot offer more authoritative explanations of illness and other states of the person or world. They cannot exploit such knowledge to exert power over others. If such an alien idea should ever occur to them, they would find their kin turning to others with such knowledge for help, being free to consult whomever they wish and to heed their advice or not. Similarly, Wola society's nominal big men—or *ol howma* 'men of the clearing'—derive no authority from the transactional skills on which their social status depends, being no more than transitionally first among equals (Sillitoe 1979). Likewise, differences between women and men, which to many writers are evidence of inequality, do not, I assert, translate into one person being able to dominate the other. The subtleties of the division of labor between men and women (Sillitoe 1985, 2006a), which ensure a gender-defined focus on different domains, promote equal partnerships.

The structure of the Wola language reflects the political dimensions of their decentralized approach to knowledge, with trust a prominent facet of the foregoing discussion of verb tenses and conjugations. If persons have no direct experience of an issue, their language prompts them to mark their knowledge as conditional. They use postmodern tenses. They implicitly grant the critique that what humans know is largely subjective, without engaging in the torturous academic discourse that characterizes this point of view. They not only question the trust one can have in others' knowledge but also the use to which holders may put it, signaling that "true facts" are those known from personal experience, not those received by word of mouth or read in books.

This attitude is discernible elsewhere in Wola life, as in the variability evident in plant and animal classifications (Sillitoe 2002a). It is also apparent in the approach to child socialization, where light parental authority reflects the wider stateless environment. Childhood experiences, too, lead youngsters to ask what trust they can have in any intelligence imparted to them. Older siblings and adults often play tricks, seemingly for amusement, to judge from the laughter that can result from a child's perplexed response or resulting mistaken actions (see Goldman 1995, 1998, on this aspect of play among the neighboring Huli).

Alternative perspectives on knowledge and, consequently, other ideas about development and modernity exist elsewhere in the world as well (Sahlins 1999), albeit to the frustration of agencies that unrealistically seek one-size-fits-all global solutions to poverty. On the other side of the Pacific, the Andean and Amazonian regions of Latin America also have languages that place some importance on specifying the soundness of the evidence that informs any communication of knowledge (Aikhenvald and Dixon 1998; Beier et al. 2002), on assessing the "epistemological status of the information being imparted" (see Howard 2002, p. 42). Quechua speakers, for instance, use a combination of the tense/aspect system and evidential suffixes to indicate the source of knowledge and the faith that speaker or hearer can have in its reliability. In their language, intelligence of which the speaker has direct personal knowledge is distinguished from that dependent on indirect knowledge or hearsay—the so-called *drunken past tense*[12] (Howard-Malverde 1988, pp. 127–133; Mannheim and van Vleet 1998, pp. 337–340). It relates to the idea that the past is ever latent in the present and is constantly reformulated in the light of present experiences. The notion finds expression in a cosmology where "the past is located in front of the viewer" such that "what you can see out there is your past and that behind you is the unseeable or the future" (Rhoades and Nazarea 2007, p. 249). The implications are significant for development intended to manipulate the future seen as opportunities ahead. These grammatical structures also relate to the idea that time and space are interdependent, which, in turn, informs Andean ideas of learning as nurturance, featuring an ethic of regeneration that involves a continuous dialogue with nature—between people, plants, animals, stars, and deities (Bourque 2002, p. 195). These nurturing relationships are reciprocal: Social relations and interaction with nature merge, and, intriguingly, learning does not feature a hierarchical relationship between knowing teacher and ignorant student, for all participants teach and learn.

The Andean concept that the future lies behind its beholders suggests an inherently conservative and sustainable perspective centered on nurturing relations with mother earth. It survived the brutal Spanish conquest and centuries of Catholicism and deserves space to express itself. It brings to mind again the idea of tradition and circular time. Nearer to New Guinea, the Australian Aborigines express sophisticated ideas of time–place interconnectedness and circularity in the *Alcheringa* ('Dreamtime' in Aranda languages), a supernatural sanctioning of knowledge (Sackett 1977, pp. 159–160; Smith 2007, p. 85) that underpinned an innately sustainable way of life. It is sacrilege to interfere with this supernaturally ordained relationship between humans and nature, to speak of capitalist-style development. Although it is hazardous to generalize for the 200 or more Aboriginal languages mentioned by Blake and Dixon (1979, pp. 1–2), it is noteworthy that these authors talk of "the transparent character of Australian languages" (p. 9). Reminiscent of the Wola language, the high number of affixes that verbs bear can result in long words that dominate sentences (Blake and Dixon 1991, p. 19), sometimes also distinguishing between recent and remote past (p. 19; Dixon 1980, p. 283). Particularly relevant to this discussion are the postinflectional clitics added to the first word of sentences to indicate that what is said is "certainly true, or possibly true, or that the sentence reports what the speaker was told by someone else, and so on" (Dixon 1980, p. 284). In addition there are perfect and nonperfect relative clause markings

---

[12] The term derives from the conjugation's function of marking an action in which the speaker has not participated or was not fully conscious (Mannheim and van Vleet 1998, p. 338).

(p. 380) and irrealis for events that may have happened (Blake and Dixon 1991, p. 19).[13]

Alternative Knowledge, Alternative Development

The IK initiative's aim of increasing the prominence of the local voice and practices in development contexts is clearly not going to be straightforward if people differ in their ideas about what constitutes knowledge (even leaving aside for a moment the topic of political power). The importance of the anthropological tenet of holism is evident, that people need to see knowledge in a wide cultural context (with the "extraneous bits"—A. M. Strathern 2006, p. 84) and not as something dissolvable into bytes of information that can be matched up to the narrow concerns of the development moment, such as improving the productivity and sustainability of specific agricultural regimes. When those working in development talk about comparing indigenous, local, traditional, or whatever-kind-of knowledge with capitalist scientific technological knowledge, it is not a clear-cut matter of matching up "our" and "their" ideas and practices within the western paradigm of development, the exercise that agencies currently demand. Although few observers would expect technical interventions appropriate to highly mechanized farming in England to apply to New Guinean swidden cultivation, why should anyone expect that taking the farmers' knowledge into account should imply similar assumptions about what constitutes justified belief?

If a native population has an idea of knowledge different from that current in development agencies, how can its IK assist in development? The question suggests that the population's participation under current arrangements is unlikely to result in its knowledge significantly informing interventions but rather in people talking past one another. There is a need for anthropologists and others who assist with business-as-usual development to try and ensure that different knowledges inform that effort to some (even minimal) extent, but such engagement is only to secure a toe in the development door. Given the current structural and political arrangement of agencies, it may be the only possibility for those who wish to contribute now in some way to alleviating poverty. It may be this option or nothing if IK is to feature in development. It is necessary to look further ahead, to force the door open to allow in alternative envoys of development. It is widely accepted that the capitalist-informed position on development, as a primarily technological and market-centered matter, is not the only one (Apffel-Marglin and Marglin 1996). To judge by the vast resources wasted in such efforts, it is not apposite for many parts of the world. In short, trying to match up aspects of local practical knowledge to global scientific knowledge may assist with development interventions, but it only touches the possible implications of involving IK in development. Going further and taking alterna-

---

[13] I am grateful to Rosaleen Howard and Bob Layton for discussions on Quechua and Aboriginal languages, respectively.

tive views of knowledge into consideration points the way to alternative developments.

What different ideas about development do such different perceptions of knowledge suggest? An initial challenge may be to bring the provisional and fragmentary Highlands idea of knowledge and practice into a meaningful balance with the unified and systematic knowledge of global science and technology that informs development policy. The trick in doing so, however, will be to leave space for another development policy to emerge. In the acephalous New Guinea Highlands, it will be one that favors equality over capitalism's creation of a few rich winners and many poor losers. In relation to such power issues, the structure of the Wola language resists the idea of conceding superiority to another party. We inhabitants of states take much knowledge on trust, whereas people native to stateless contexts are not so inclined, the political implications making them distrustful of extending such an expert mandate to anyone else. Instinctively, culturally speaking, they defend themselves against any such authority. They therefore differ in their perception of what constitutes knowledge, at least in assessing the faith they can have in any intelligence (ranging from that based on first-hand experience to that of which neither party to a communication has direct experience). The Wola way of knowing, as signaled in their verb forms, would certainly make it far more difficult for western politicians and their PR representatives to engage in spin or fabrication of truth. The practice of querying the reliability of information in the repeated negotiations that New Guineans have with mining companies, that inclination to distrust what they are told about agreed arrangements, is also an aspect of their transactional outlook on the world, where reciprocal relations involve endless negotiations. Furthermore, one can appreciate how the negotiators, mostly young persons who speak English and Tok Pisin, are arguably disadvantaged by having to do so in foreign languages that lack evidential markers by which they customarily assess statements.

Trust is a concern generally in development contexts, where poor populations are expected to have faith in western technology and solutions to improve their lives even though they have good reason to be suspicious and not have confidence in them. Yet trust is central to the integrity of social capital (Fafchamps 2002), something worth noting with regard to the current sustainable-development approach. The colonial experience and subsequent events have in many parts of the world undermined trust in the capitalist, democratic system and its solutions, which seem to mire them in ever deeper debt and trouble. The promised developments never seem to come through for the poor, as opposed to the wealthy. A system that prioritizes profit-making ultimately corrodes trust, encouraging some to outsmart and even swindle others, as seen in the problems attending the aforementioned commercialization of science and the current crisis in banking and finance.

The colonial interlude was something of an abnormality in the New Guinea Highlands, where people recognize no political authority and are not used to doing what others direct. Initially, they were in awe of the European newcomers, particularly their astounding technology—unsurprisingly for people who moved from stone-age to satellite technology in a generation—and were soon subdued by their superior firepower (Sillitoe 2000b, pp. 13–57). They subsequently experienced political authority, ceding willingly on the whole, in part believing that it would improve their lives, that it would lead to development by giving access to the newcomer's technology. Yielding also contributed to some extent to the Highlanders' rapid conversion to Christianity, though beliefs have subsequently taken off in a distinctly local way with some syncretism. The people of the New Guinea Highlands first worked on clearing airstrips and building roads and planted new cash crops such as coffee and pyrethrum but gained only limited access to manufactured goods. In their minds they became poor, until then an unheard of idea. The colonial authorities withdrew rapidly from the region, just as the native inhabitants were beginning to doubt the efficacy of what the Europeans had told them. The Highlanders started to distrust the newcomers' advice because it did not seem to bring anticipated benefits. The national government that replaced the colonial officials seeks to exercise force from time to time but is quite ineffective in the region. What education the New Guinea Highlanders have received has not supplied solutions but rather mystified things further. They are no longer in a mood or, in the Southern Highlands Province, a position to listen to outside experts. Although they continue to admire and covet manufactured goods such as canned food, clothing, radios, vehicles, and firearms, they are unsure how to expand their access to them and improve their material standard of living.

The ideas that have emerged relating to economic development illustrate further the gulf between local perceptions and western assumptions. They show the absence of unanimity about what constitutes development and what qualifies as sound knowledge. The two most often locally heard words relating to economic development are the Tok Pisin terms *developman* 'development' and *bisnis* 'business'. Their different connotations are informative about the applicability or inapplicability of the idea of economic development (Sahlins 1999, pp. ix–x). When the Wola talk about *developman*, they have a cargolike image in mind of a massive outside intervention that will dramatically improve their material situation with little or no effort on their side. This understanding stems from the establishment of mines in other regions of the New Guinea Highlands, particularly the Kutubu, Moran, and Hides petroleum and gas field to the south as of the 1980s, where local populations have received unheard of sums in royalties, compensation payments, and so on (Sagir 2004). Having seen exploratory drilling since the arrival of the oil and gas installations, the inhabitants of the Wola region, with its oil seeps, have been anticipating similar windfalls. *Developman* will come with the long-awaited appearance of the oil company. It is *bisnis* when they work to produce things and supply services to earn money. When these people talk about *bisnis*, they have in mind to run small-scale enterprises, such as trade stores and public motor vehicles (PMVs), activities that depend on their efforts and at which they expect to toil in the economic sense. So far, these enterprises have been short-lived with little sustained business growth in the region and scarce linkage to a wider market to foster it.

The attitude to *developman* chimes in with expectations of access to wealth to transact in sociopolitical exchanges, which are now widely believed to be a key feature of the egalitarian polity of the Highlands (Feil 1984; Lederman 1986; Sillitoe 1979; Strathern 1971, 1988). These exchanges require persons to work at transacting for wealth, not producing it. In that political economy one obtains wealth through transaction, with important consequences for the egalitarian polity and the negation of hierarchy (Sillitoe 2006b). To many New Guinea Highlanders development implies access to more wealth to use in such transactions, as indicated by the investment of ever more cash in the exchange system (Nihill 1989). It is puzzling that Europeans have not chosen this practice of generously distributing their obviously superior wealth, for it is normal behavior and the way to achieve social renown. According to the current sustainable-livelihoods type of development, the inhabitants of the Wola region prioritize investment in social capital, whereas economists tend to give precedence to physical infrastructure and financial capital (Carney 1998). As a result, the Wola wait for development to occur, the herald of much wealth to strengthen their sociopolitical transactional position, albeit with inflationary consequences if events elsewhere are any guide (Sillitoe 2006b). They also expect it to give them the means to satisfy their currently limited demand for consumer goods, which is the aspect that appeals particularly to women not greatly engaged in the exchange sphere.

The two contrary local concepts of development and the extent to which they differ from those of western agencies suggests that the Wola had no idea of development before they encountered the intercultural space populated by white colonial administrators and missionaries, multinational mining companies, and international development agencies. It is the same intercultural space in which they might come to recognize their own indigeneity for the first time. This divergence further underscores the local character of IK understandings, which render people vulnerable to external forces such as impositions by a capitalist state. The issue from a development perspective is that the local view alone is insufficient and has to be supplemented by other vantage points and knowledge. This need inevitably implies education, without which there is no role for development, which the Wola themselves want to see in some form in that they wish to share the material benefits of industrial technology. But they want it on their terms, a stance that implies negotiation. Perhaps academic IK debates and publications are unlikely to have a directly traceable influence on this process; indeed, any such effect would almost amount to a top-down approach. Instead, they might be expected to stimulate politicians, technicians, and administrators to leave a space for local knowledge and to avoid precluding options by denigrating it, for although IK is more circumscribed than scientific knowledge, it serves people well. It often matches and sometimes betters science-based understandings of, for instance, managing human use of land and natural resources (Sillitoe 2007). It is on this strength that much IK research focuses. In that sense, this chapter's opening pessimism about the IK initiative may be misplaced. There are many NGOs, funding bodies, and governments committed to a role for IK, as shown, for instance, by the "endogenous sustainable development" movement (Haverkort 2006) and the consideration that various international organizations give to IK interests.

A major challenge for the future is to have the agenda include such alternative ideas of what development might be and work out their implications. In the New Guinea Highlands it is *bisnis* activities that agencies are likely to recognize as meriting development assistance, not *developman* interventions. Yet with some people profiting more than others, it is difficult to see how such commercial growth can occur without conflict with local egalitarian values. This presents a significant test for any future alternative views of development intended to achieve material advancement with equity. On the other hand, experience elsewhere shows that extractive industry, if that *developman* does occur, is likely to inhibit *bisnis* activities (Banks 2005). Recipients of royalties, becoming wealthy beyond their dreams of avarice, lack incentives to engage in *bisnis*, whereas their neighbors remain dependent on subsistence livelihoods. The consequences are worrying, not only with respect to tensions with those left out but, looking ahead to the closure of mines and wells, also with a generation unfamiliar with subsistence livelihood and characterized by scant investment in alternatives beyond the receipt of royalty checks (Gilberthorpe 2006). Papua New Guinea's Oil and Gas Act of 1998 requires oil companies to take some account of local arrangements and knowledge through social impact analysis and social mapping—work that anthropologists sometimes undertake. The legislation is better than nothing but is largely designed to forestall disputes over land and royalties and comes nowhere near penetrating the sociocultural world of the people affected by companies' activities, let alone taking it into account. It is in subsequent misunderstandings and dissention that they reassert their worldview. These clashes are the stuff of local resistance to capitalist hegemonic processes and can turn acrimonious, even violent, to the detriment of companies' interests, an eventuality that makes an appreciation and accommodation of local understandings beneficial to everyone from the outset. IK may feature as an additional mobilizing force in such rights claims and politics of identity affirmation.

Implications of Alternative Development

Language issues alone do not account for the relatively low impact of IK in development. Political relations between outsider agencies and insider actors, often mediated by governments, also feature prominently as mentioned at several junctures, though language-interposed understanding or misunderstanding informs the political dynamic and can contribute to inappropriate interventions from the local perspective when the other party is more powerful and seeks to impose its views. Other cultural factors, too, such as different understandings of time, are likely to play a role in local resistance to western-style economic development. The Wola and other Highlanders, from their worldview, do not see time passing linearly, featuring progress. It comprises spirally arranged cycles (Sillitoe 2006a) with good and bad periods that can be influenced if one ensures that relationships with spiritual forces stay in balance. Any remodeling of development dialogue with the inhabitants of the region should reflect the implications of this cyclical, spiral conception of time, change, and succession. Such translation of development ideas into local idioms is central to the IK initiative.

People will draw on these idioms anyway, so IK is likely to influence development whatever outsiders may think. In this event IK may have an unnoticed effect while itself simultaneously changing. It is part of the interface at any knowledge transfer, being party to the process by which innovation takes place. It does not disappear (except in the limited sense that 25 local landraces, or varieties, of sweet potato might be replaced by 3 high-yielding, introduced cultivars). Rather, IK is continually reconstituted, as demonstrated by several studies (e.g., Arce and Long 2000; Long 2001). It is not that previous ways are necessarily erased in development but rather that people rework and reconfigure whatever ideas, innovations, and institutions are introduced, recreating them to produce their own counterwork that complies with local politics, practices, and realities (Sahlins 1999). Such recasting gives yet another slant on the power imbalances between host group and intervening development agency. One aim of the IK initiative is to make outsiders working in development aware of these trends so that they, too, develop. If the developers are cognizant of local realities, they can more readily accommodate any interventions in them. The precept of this chapter is that what constitutes knowledge and development should not be taken for granted in the formulation of development policies that are intended to include indigenous people.

The view of development that should prevail has to be that of the local population, not only on the ethical grounds that it is the future of those people but also on the grounds of realpolitik in the Southern Highlands region (including the recently established Hela Province), where the authority of the nation-state is weak to nonexistent. Imposing the capitalist idea of development is not an option in the current stateless context. Participation (featuring *bisumindis* the 'we-do, both-parties-witness' knowledge) will be a sine qua non if development initiatives are to have any hope, for Southern Highlanders are unlikely to go along otherwise. They will not heed what someone else claims to be best unless they can *see*—think or know—that it will work for them. Given their propensity to question the necessary validity of others' experience and only fully trusting in their own, paying heed to what they see themselves, the Highlanders are suspicious of experts (with at best *biso*, the 'she-or-he-does, speaker-only-witnesses' knowledge). This tendency has been reinforced by decades of attempting to follow the nostrums of outsiders (e.g., *didiman* 'agricultural officers') without the anticipated returns. The implications that the local view of knowledge has for development are substantial; agencies will have to involve indigenous people and convince them of the effectiveness of outside perspectives and proposals.

The application of knowledge to development may involve political matters to a great extent, but it is not up to academics to engage in politicking on behalf of others. Rather, it is to better understanding of the political considerations. Anything else would patently be inappropriate in a stateless context such as that of the Southern Highlands region, where all have an equal say and there is no recognition of expert opinion. A necessary first step to improve understanding is to agree on the meaning of the words used to communicate ideas and their implications. The ethnocentric assumption that development universally means progress has caused confusion that has led to many problems and a waste of resources. At the same time, it is equally necessary to avoid an academic penchant to debate the meaning of words endlessly—as when discussing the appropriateness of the adjectives *indigenous*, *local*, and *traditional*—while development agencies that could make a worthwhile contribution struggle with quandaries of dire poverty. Hence the advocacy for engagement in development on agencies' terms, as imperfect as they are, if the IK initiative is to help ameliorate problems. All parties involved need to proceed with caution to be heard and to have influence on development policy.

It would be inappropriate for me to conclude this chapter, in search of some climax, by suggesting what might be the contours of negotiated alternative views of development, for the whole premise of the IK agenda is that this task falls to local knowledge holders, not to foreign commentators. To create a genuine place for development players in development discussions, attention must first focus on informing them of what IK comprises and what it has to contribute. But whatever perspective one takes on manufactured goods produced and distributed under the capitalist market regime, it is a puzzle how an acephalous order can improve its access to them without compromising its egalitarian relations and ethic of equality. It is equally puzzling how an acephalous order's captious approach to knowledge might cope with a world that stores and passes on vast amounts of information in *bisesa* 'it-does, neither-speaker-nor-hearer-witnesses' contexts, which demand deference to expert knowledge. Some of the current turmoil in the Southern Highlands region relates to this conflict of interests as people seek to come to terms with such conundrums (Sillitoe 2000b, pp. 219–239).

References

Aikhenvald, A. Y. (2004). *Evidentiality*. Oxford, UK: Oxford University Press.

Aikhenvald, A., & Dixon, R. M. W. (1998). Evidentials and areal typology: A case study from Amazonia. *Language Sciences, 20*, 241–257.

Antweiler, C. (1998). Local knowledge and local knowing: An anthropological analysis of contested "cultural products" in the context of development. *Anthropos, 93*, 469–494.

Apffel-Marglin, F., & Marglin, S. A. (1996). *Decolonizing knowledge: From development to dialogue*. Oxford, UK: Clarendon.

Arce, A., & Long, N. (Eds.). (2000). *Anthropology, development and modernities: Exploring discourse, counter-tendencies and violence*. London: Routledge.

Asch, M. (2004). Political theory and the rights of indigenous peoples. *Canadian Journal of Sociology, 29*, 150–152.

Ayer, A. J. (1971). *The problem of knowledge*. Harmondsworth: Penguin. (Original work published 1956).

Banks, G. (2005). Globalization, poverty and hyperdevelopment in Papua New Guinea's mining sector. *Focaal, 46*, 128–143.

Barnard, A. (2006). Kalahari revisionism, Vienna, and the 'indigenous peoples' debate. *Social Anthropology, 14*, 1–16.

Barth, F. (2002). An anthropology of knowledge. *Current Anthropology, 43*, 1–18.

Battiste, M. (2000). *Reclaiming indigenous voice and vision*. Vancouver: UBC Press.

Beier, C., Michael, L., & Sherzer, J. (2002). Discourse forms and processes in indigenous lowland South America: An areal-typological perspective. *Annual Review of Anthropology, 31*, 121–145.

Bentley, J. W., & Baker, P. S. (2005). Understanding and getting the most from farmers' local knowledge. In J. Gonsalves, T. Becker, A. Braun, D. Campilan, H. de Chavez, E. Fajber, M. Kapiriri, J. Rivaca-Caminade, & R. Vernooy (Eds.), *Participatory research and development for sustainable agriculture and natural resource management: A sourcebook* (Vol. 3, pp. 58–64). Laguna: CIP-UPWARD/IDRC.

Bicker, A., Sillitoe, P., & Pottier, J. (2004). *Investigating local knowledge: New directions, new approaches.* Aldershot: Ashgate.

Blake, B. J., & Dixon, R. M. W. (1979). Introduction. In R. M. W. Dixon & B. J. Blake (Eds.), *The handbook of Australian languages: Introduction, Guugu Yimidhirr, Pitta-Pitta, Gumbay Nggir, Yagir* (Vol. 1, pp. 1–26). Amsterdam: John Benjamins Publishing.

Blake, B. J., & Dixon, R. M. W. (1991). Introduction. In R. M. W. Dixon & B. J. Blake (Eds.), *The handbook of Australian languages: The aboriginal language of Melbourne and other grammatical sketches* (Vol. 4, pp. 1–28). Oxford, UK: Oxford University Press.

Bloch, M. (1991). Language, anthropology and cognitive science. *Man, 26,* 183–198.

Bloch, M. (2005). *Essays on cultural transmission.* Oxford, UK: Berg.

Borofsky, R. (1994). On the knowledge and knowing of cultural activities. In R. Borofsky (Ed.), *Assessing cultural anthropology* (pp. 331–347). New York: McGraw-Hill.

Bourque, N. (2002). Learning and re-learning how to plant: The impact of new crops on the spread and control of new agricultural knowledge in the Ecuadorian Andes. In H. Stobart & R. Howard (Eds.), *Knowledge and learning in the Andes: Ethnographic perspectives* (pp. 194–206). Liverpool: Liverpool University Press.

Bowen, J. R. (2000). Should we have a universal concept of 'indigenous peoples' rights'? Ethnicity and essentialism in the twenty-first century. *Anthropology Today, 16*(4), 12–16.

Brady, I. (1983). Speaking in the name of the real: Freeman and Mead on Samoa. *American Anthropologist, New Series, 85,* 908–909. doi:10.1525/aa.1983.85.4.02a00170.

Brokensha, D. W., Warren, D. M., & Werner, O. (Eds.). (1980). *Indigenous knowledge systems and development.* Lanham: University Press of America.

Carney, D. (1998). *Sustainable rural livelihoods: What contribution can we make?* London: Department for International Development.

Clift, C. (2007). Is intellectual property protection a good idea? In P. Sillitoe (Ed.), *Local science vs global science: Approaches to indigenous knowledge in international development* (pp. 191–207). Oxford, UK: Berghahn.

Cohn, B. S. (1980). History and anthropology: The state of play. *Comparative Studies in Society and History, 22,* 198–221.

Colchester, M. (2002). Indigenous rights and the collective conscious. *Anthropology Today, 18*(1), 1–3.

Collins, H. M. (1974). The TEA set: Tacit knowledge and scientific networks. *Science Studies, 4,* 165–186.

Colman, R. J., Anderson, R. M, Johnson, S. C., Kastman, E. K., Kosmatka, K. J., Beasley, T. M.,…, & Weindruch, R. (2009). Caloric restriction delays disease onset and mortality in rhesus monkeys. *Science, 325*(5937), 201–204.

Cook, K. S. (2005). Networks, norms, and trust: The social psychology of social capital. *Social Psychology Quarterly, 68,* 4–14.

Cooke, B., & Kothari, U. (2001). *Participation: The new tyranny?* London: Zed Books.

Crick, M. R. (1982). Anthropology of knowledge. *Annual Review of Anthropology, 11,* 287–313.

Csordas, T. J. (1994). *Embodiment and experience: The existential ground of culture and self.* Cambridge, UK: Cambridge University Press.

Dixon, R. M. W. (1980). *The languages of Australia.* Cambridge, NY: Cambridge University Press.

Douglas, M. (2004). Traditional culture—Let's hear no more about it. In V. Rao & M. Walton (Eds.), *Culture and public action* (pp. 85–109). Stanford: Stanford University Press.

Draper, N., & Draper, S. (2002). *Dictionary of Kyaka Enga Papua New Guinea.* Canberra: Pacific Linguistics.

Ellen, R. F., & Harris, H. (2000). Introduction. In R. F. Ellen, P. Parkes, & A. Bicker (Eds.), *Indigenous environmental knowledge and its transformations: Critical anthropological perspectives* (pp. 1–33). Amsterdam: Harwood Academic.

Emery, A. R. (2000). *Integrating indigenous knowledge in project planning and implementation.* Nepean: Partnership Publication with KIVU Nature, The World Bank and Canadian International Development Agency.

Epps, P. (2005). Areal diffusion and the development of evidentiality. *Studies in Language, 29,* 617–650.

Fafchamps, M. (2002). Social capital, trust, and development. Retrieved from http://www.google. com/url?sa=t&rct=j&q=&esrc=s&source=web&cd=1&ved=0CDUQFjAA&url=http%3A%2 F%2Fciteseerx.ist.psu.edu%2Fviewdoc%2Fdownload%3Fdoi%3D10.1.1.196.6962%26rep% 3Drep1%26type%3Dpdf&ei=b7ApUZDAHc6GrAf624CoAw&usg=AFQjCNEMaA5u9Ea gom-X9t83iXNxu3K5kg&bvm=bv.42768644,d.bmk&cad=rja

Feil, D. K. (1984). *Ways of exchange: The Enga tee of Papua New Guinea.* Lucia: St. University of Queensland Press.

Finnegan, R. (1992). *Oral traditions and the verbal arts: A guide to research practices.* London: Routledge.

Finnegan, R. H., & Orbell, M. (1995). *South Pacific oral traditions.* Bloomington: Indiana University Press.

Foley, W. A. (1986). *The Papuan languages of New Guinea.* Cambridge, UK: Cambridge University Press.

Foucault, M. (1988). *Politics, philosophy, culture: Interviews and other writings, 1977–1984.* New York: Routledge.

Four Arrows (Don Trent Jacobs). (2006). *Unlearning the language of conquest: Scholars expose anti-Indianism in America.* Austin: University of Texas Press.

Franklin, K. J. (1971). *A grammar of Kewa, New Guinea.* Canberra: Australian National University.

Freeman, D. (1983). *Margaret Mead and Samoa: The making and unmaking of an anthropological myth.* Cambridge, MA: Harvard University Press.

Gibbs, R. W., Jr. (2006). *Embodiment and cognitive science.* Cambridge, UK: Cambridge University Press.

Gilberthorpe, E. (2006). 'It's raining money': Anthropology, film and resource extraction in Papua New Guinea. *Anthropology in Action, 13*(3), 13–21.

Godelier, M. (1999). *The enigma of the gift* (Nora Scott, Trans.). Cambridge, UK: Polity.

Goldman, L. (1995). The depths of deception: Cultural schemas of illusion in Huli. In A. Biersack (Ed.), *Papuan borderland: Huli, Duna, and Ipili perspectives on the Papua New Guinea Highlands* (pp. 111–138). Ann Arbor: Michigan University Press.

Goldman, L. (1998). *Child's play: Myth, mimesis and make-believe.* Oxford, UK: Berg.

Goody, J. (1987). *The interface between the written and the oral.* Cambridge, UK: Cambridge University Press.

Gregory, C. (1982). *Gifts and commodities.* London: Academic.

Grenier, L. (1998). *Working with indigenous knowledge: A guide for researchers.* Ottawa: IDRC.

Guenther, M., Kendrick, J., Kuper, A., Plaice, E., Thuen, T., Wolfe, P.,..., & Barnard, A. (2006). The concept of indigeneity. *Social Anthropology, 14,* 17–32.

Hardin, R. (2002). *Trust and trustworthiness.* New York: Russell Sage Foundation Publishers.

Haverkort, B. (2006). *Moving worldviews: Reshaping sciences, policies and practices for endogenous sustainable development.* Leusden: ETC/Compas.

Hill, J. H., & Irvine, J. T. (1993). *Responsibility and evidence in oral discourse.* Cambridge, UK: Cambridge University Press.

Hirst, P. H. (1974). *Knowledge and the curriculum: A collection of philosophical papers.* New York: Routledge.

Hobsbawm, E., & Ranger, T. O. (Eds.). (1983). *The invention of tradition.* Cambridge, UK: Cambridge University Press.

Holmes, L. D. (1987). *Quest for the real Samoa: The Mead/Freeman controversy and beyond.* South Hadley, MA: Bergin & Garvey Publishers.

Honderich, T. (1995). *The Oxford companion to philosophy.* Oxford, UK: Oxford University Press.

Howard, R. (2002). Spinning a yarn: Landscape, memory, and discourse structure in Quechua narratives. In J. Quilter & G. Urton (Eds.), *Narrative threads: Accounting and recounting in Andean Khipu* (pp. 26–49). Austin: University of Texas Press.

Howard, R., Barbira-Freedman, F., & Stobart, H. (2002). Introduction. In H. Stobart & R. Howard (Eds.), *Knowledge and learning in the Andes: Ethnographic perspectives* (pp. 1–13). Liverpool: Liverpool University Press.

Howard-Malverde, R. (1988). Talking about the past: Tense and testimonials in Quechua narrative discourse. *Amerindia, 13*, 125–155.

Hume, D. (1896). *A treatise of human nature* (3 vols). Reprinted from the original edition and edited by L. A. Selby-Bigge. Oxford, UK: Clarendon Press. (Original work published 1739).

IIRR (International Institute for Rural Reconstruction). (1996). *Recording and using indigenous knowledge: A manual.* Silang: Regional Program for the Promotion of Indigenous Knowledge in Asia (REPPIKA), International Institute of Rural Reconstruction.

Keesing, R. M. (1992). *Custom and confrontation: The Kwaio struggle for cultural autonomy.* Chicago: Chicago University Press.

Keesing, R. M., & Tonkinson, R. (Eds.). (1982). Reinventing traditional culture: The politics of *kastom* in island Melanesia [Special issue]. *Mankind, 13*, 297–399.

Kendrick, J., & Lewis, J. (2004). Indigenous peoples' rights and the politics of the term 'indigenous'. *Anthropology Today, 20*(2), 4–9.

Kloppenburg, J. (1991). Social theory and the de/construction of agricultural science: Local knowledge for an alternative agriculture. *Rural Sociology, 56*, 519–548.

Kuper, A. (2003). The return of the native. *Current Anthropology, 44*, 389–402.

Lang, A. (1973). *Enga dictionary.* Canberra: Pacific Linguistics Series.

Lederman, R. (1986). *What gifts engender: Social relations and politics in Mendi, Highland Papua New Guinea.* Cambridge, UK: Cambridge University Press.

Lee, R. B., & Guenther, M. (1991). Oxen or onions? The search for trade (and truth) in the Kalahari. *Current Anthropology, 32*, 592–601.

Long, N. (2001). *Development sociology: Actor perspectives.* London: Routledge.

Lucy, J. (1992). *Language diversity and thought: A reformulation of the Linguistic Relativity Hypothesis.* Cambridge, UK: Cambridge University Press.

Macfarlane, A. (2013). *History and anthropology.* CreateSpace Independent Publishing Platform.

Mannheim, B., & van Vleet, K. (1998). The dialogics of Southern Quechua narrative. *American Anthropologist, New Series, 100*, 326–346.

Martin, L. (1986). "Eskimo Words for Snow": A case study in the genesis and decay of an anthropological example. *American Anthropologist, New Series, 88*, 418–423.

McIntosh, I., Colchester, M., & Bowen, J. R. (2002). Defining oneself, and being defined as, Indigenous. *Anthropology Today, 18*(3), 23–25.

Mead, M. (1928). *Coming of age in Samoa: A psychological study of primitive youth for western civilisation.* New York: Morrow.

Minority Rights Group International. (2005, March 31). Botswana must amend discriminatory tribal policies says UN Committee. Retrieved from http://allafrica.com/stories/200504080839.html

Minority Rights Group International. (2006, July 18). UN makes move for Indigenous Peoples' rights: MRG urges support of General Assembly. Retrieved from http://www.cilevics.eu/minelres/mailing_archive/2006-July/004691.html

Mosse, D. (2005). *Cultivating development: An ethnography of aid policy and practice.* London: Pluto Press.

Niezen, R. (2003). *The origins of indigenism: Human rights and the politics of identity.* Berkeley: University of California Press.

Nihill, M. (1989). The new pearlshells: Aspects of money and meaning in Anganen exchange. *Canberra Anthropology, 12*(1–2), 144–160.

Polanyi, M. (1967). *The tacit dimension.* New York: Anchor Books.

Posey, D. A. (1999). *Cultural and spiritual values of biodiversity: A complementary contribution to the global biodiversity assessment.* London: Intermediate Technology Publications.

Pottier, J., Bicker, A., & Sillitoe, P. (Eds.). (2003). *Negotiating local knowledge: Power and identity in development* (Series: Anthropology, culture and society). London: Pluto Press.

Povinelli, E. A. (1998). The state of shame: Australian multiculturalism and the crisis of indigenous citizenship. *Critical Inquiry, 24,* 575–610.

Pullum, G. K. (1991). *The Great Eskimo vocabulary Hoax: And other irreverent essays on the study of language.* Chicago: University of Chicago Press.

Purcell, T. W. (1998). Indigenous knowledge and applied anthropology: Questions of definition and direction. *Human Organization, 57,* 258–272.

Rabinow, P., Foucault, M., & Faubion, J. D. (2001). *Power: The essential works of Michel Foucault, 1954–1984* (Vol. 3). London: Allen Lane.

Rapport, N. (2003). *I am dynamite: An alternative anthropology of power.* London: Routledge.

Rhoades, R. E., & Nazarea, V. (2007). Forgotten futures: Scientific models vs. local visions of land use change. In P. Sillitoe (Ed.), *Local science vs global science: Approaches to indigenous knowledge in international development* (pp. 231–256). Oxford, UK: Berghahn.

Robertson, R. (1995). Glocalization: Time-space and homogeneity-heterogeneity. In M. Featherstone, S. Lash, & R. Robertson (Eds.), *Global modernities* (pp. 25–44). London: Sage.

Rosengren, D. (2002). On 'indigenous identities': Reflections on a debate. *Anthropology Today, 18*(3), 25.

Rothstein, B. (2000). Trust, social dilemmas and collective memories. *Journal of Theoretical Politics, 12,* 477–501. doi:10.1177/0951692800012004007.

Russell, B. (1948). *Human knowledge: Its scope and limits.* London: Allen & Unwin.

Sackett, L. (1977). Confronting the dreamtime: Belief and symbolism in an aboriginal ritual. *Ethnos, 42,* 156–179. doi:10.1080/00141844.1977.9981139.

Sagir, B. (2004). The politics of petroleum extraction and royalty distribution at Lake Kutubu. In A. Rumsey & J. Weiner (Eds.), *Mining and indigenous lifeworlds in Australia and Papua New Guinea* (pp. 145–156). Oxford, UK: Sean Kingston Publishing.

Sahlins, M. (1999). What is anthropological enlightenment? Some lessons of the twentieth century. *Annual Review of Anthropology, 28,* i–xxiii.

Sen, S., Angell, B., & Miles, A. (2000). The Bangladesh Resource Centre for Indigenous Knowledge and its network. In P. Sillitoe (Ed.), *Indigenous knowledge development in Bangladesh: Present and future* (pp. 213–218). London: Intermediate Technology Publications.

Shepherd, C. J. (2004). Agricultural hybridity and the 'pathology' of traditional ways: The translation of desire and need in postcolonial development. *Journal of Latin American Anthropology, 9,* 235–266.

Shepherd, C. J. (2005). Agricultural development NGOs, anthropology, and the encounter with cultural knowledge. *Culture and Agriculture, 27,* 35–44. doi:10.1525/cag.2005.27.1.35.

Sillitoe, P. (1979). *Give and take: Exchange in Wola society.* Canberra: Australian National University Press.

Sillitoe, P. (1985). Divide and no one rules: The implications of sexual divisions of labour in the Papua New Guinea Highlands. *Man, 20,* 494–522.

Sillitoe, P. (1996). *A place against time: Land and environment in the Papua New Guinea highlands.* Amsterdam: Harwood Academic.

Sillitoe, P. (1998a). Defining indigenous knowledge: The knowledge continuum. *Indigenous Knowledge and Development Monitor, 6*(3), 14–15.

Sillitoe, P. (1998b). The development of indigenous knowledge: A new applied anthropology. *Current Anthropology, 39,* 223–252. doi:10.1086/204722.

Sillitoe, P. (2000a). Cultivating indigenous knowledge on Bangladeshi soil: An essay in definition. In P. Sillitoe (Ed.), *Indigenous knowledge development in Bangladesh: Present and future* (pp. 145–160). London: Intermediate Technology Publications.

Sillitoe, P. (2000b). *Social change in Melanesia: Development and history.* Cambridge, UK: Cambridge University Press.

Sillitoe, P. (2002a). Contested knowledge, contingent classification: Animals in the Highlands of Papua New Guinea. *American Anthropologist, New Series, 104,* 1162–1171.

Sillitoe, P. (2002b). Globalizing indigenous knowledge. In P. Sillitoe, A. Bicker, & J. Pottier (Eds.), *Participating in development: Approaches to indigenous knowledge* (pp. 108–138). London: Routledge.

Sillitoe, P. (2002c). What is indigenous knowledge? The definition of an emerging field in development. In S. M. Nurul Alam (Ed.), *Contemporary anthropology: Theory and practice* (pp. 365–395). Dhaka: The University Press.

Sillitoe, P. (2003). *Managing animals in New Guinea: Preying the game in the Highlands.* London: Routledge.

Sillitoe, P. (2004). Interdisciplinary experiences: Working with indigenous knowledge in development. *Interdisciplinary Science Reviews, 29,* 6–23.

Sillitoe, P. (2006a). What labour engenders: Women and men, time and work in the New Guinea highlands. *The Asia Pacific Journal of Anthropology, 7,* 119–151.

Sillitoe, P. (2006b). Why spheres of exchange? *Ethnology, 45,* 1–23.

Sillitoe, P. (Ed.). (2007). *Local science vs global science: Approaches to indigenous knowledge in international development.* Oxford, UK: Berghahn.

Sillitoe, P., & Bicker, A. (2004). Introduction: Hunting for theory, gathering ideology. In A. Bicker, P. Sillitoe, & J. Pottier (Eds.), *Development and local knowledge: New approaches to issues in natural resources management, conservation and agriculture* (pp. 1–18). London: Routledge.

Sillitoe, P., & Sillitoe, J. (2009). *Grass-clearing man: A factional ethnography of life in the New Guinea Highlands.* Long Grove: Waveland Press.

Sillitoe, P., Dixon, P., & Barr, J. (2005). *Indigenous knowledge inquiries: A methodologies manual for development.* London: Intermediate Technology Publications.

Smith, B. R. (2007). 'Indigenous' and 'scientific' knowledge in Central Cape York peninsula. In P. Sillitoe (Ed.), *Local science vs global science: Approaches to indigenous knowledge in international development* (pp. 75–90). Oxford, UK: Berghahn.

Stewart, P. J., & Strathern, A. J. (2001). *Humors and substances: Ideas of the body in New Guinea.* Westport: Bergin & Garvey.

Strathern, A. J. (1971). *Rope of moka: Big-men and ceremonial exchange in Mount Hagen, New Guinea.* Cambridge, UK: Cambridge University Press.

Strathern, A. M. (1988). *The gender of the gift: Problems with women and problems with society in Melanesia.* Berkeley: University of California Press.

Strathern, A. J. (1996). *Body thoughts.* Ann Arbor: University of Michigan Press.

Strathern, A. M. (2006). Useful knowledge: Isaiah Berlin lecture. *Proceedings of British Academy, 139,* 73–109.

Strathern, A. J., & Stewart, P. J. (1998). Seeking personhood: Anthropological accounts and local concepts in Mount Hagen, Papua New Guinea. *Oceania, 68,* 170–188.

Strathern, A. J., & Stewart, P. J. (2000). *Arrow talk: Transaction, transition, and contradiction in New Guinea Highlands history.* Kent: Kent State University Press.

Survival International. (2006). *Botswana: Bushman case.* Retrieved from http://www.survival-international.org/news.php?id=1816

Timmer, J. (1998). Lost power, concealed knowledge, and the return of the Kingdom among the Imyan of the Bird's Head of Irian Jaya. In J. Miedema, C. Odé, & R. A. C. Dam (Eds.), *Perspectives on the Bird's Head of Irian Jaya, Indonesia* (pp. 79–116). Amsterdam: Rodopi Press.

Tipton, R. A. (1982). *Nembi procedural and narrative discourse.* Canberra: Pacific Linguistics Series.

United Nations. (1993). UN General Assembly Resolution 48/163. Passed 21 December 1993, and proclaiming the International Decade of the World's Indigenous People. Retrieved from http://www.un.org/ga/search/view_doc.asp?symbol=A/RES/48/163&Lang=E&Area=RESOLUTION

Wade, P. (Ed.). (1997). *Cultural studies will be the death of anthropology*. Manchester: Manchester University Group for Debates in Anthropological Theory.

Weiner, A. B. (1992). *Inalienable possessions: The paradox of keeping-while-giving*. Berkeley: University of California Press.

Weiss, G., & Haber, H. F. (1999). *Perspectives on embodiment: The intersections of nature and culture*. London: Routledge.

White, G. M., & Tengan, T. K. (2001). Disappearing worlds: Anthropology and cultural studies in Hawai'i and the Pacific. *The Contemporary Pacific, 13*, 381–416.

Whorf, B. L. (1956). *Language, thought, and reality: Selected writings of Benjamin Lee Whorf*. Cambridge: Technology Press of Massachusetts Institute of Technology.

Willett, T. (1988). A cross-linguistic survey of the grammaticalization of evidentiality. *Studies in Language, 12*, 51–97.

Wolfe, P. (1994). Nation and miscegenation: Discursive continuity in the post-Mabo era. *Social Analysis, 36*, 93–152.

Chapter 8
Local Knowledge as a Universal Social Product: A General Model and a Case from Southeast Asia

Christoph Antweiler

This chapter deals with the specific ontic character of local knowledge, the epistemic implications of that character, and the proper methods for studying such knowledge. Using an empirical study of urban knowledge from Indonesia, I offer examples of local knowledge in the field of development. Local knowledge consists primarily of cognitive capabilities related to the performance of action. It is usually gained through long experience and has a certain basis in localized cultural contexts, so its distribution is spatially limited. Nevertheless, local knowledge is not necessarily confined to one locale or specific ethnic group. Debates about local knowledge are heavily biased toward issues connected with its potential uses for development or the conservation of biodiversity. Especially since the 1990s, its status as a special form of intellectual property has stood at the center of these exchanges (Agrawal 1998; Anderson 2009; Brush and Stabinsky 1996; Greaves 1994; Posey and Dutfield 1996; Sillitoe 1998c; Schareika and Bierschenk 2004). Basic aspects of local knowledge are sidestepped all too often in these user- or policy-oriented discussions. For instance, scientific and local (or "traditional") knowledge tend to be cast as polarities, often on the assumption that they are incompatible. The resulting inclination is to see both as homogenous and thus to essentialize each and to regard one as more valuable than the other. Local knowledge tends to be associated erroneously with sustainable actions, purity, naturalness, social equality, equity, and political justness; with outsiders; and with powerless or otherwise marginal people or subaltern interests (e.g., Lyotard 1979/1984).

First, it is important not to confound knowledge with information (see Abel 2008; Meusburger 2013). Knowledge is not information but experience. Knowledge in general, not only local knowledge, is action-oriented and often not explicit (or

C. Antweiler (✉)
Department of Southeast Asian Studies, Institute of Oriental and Asian Studies (IOA),
University of Bonn, Nassestr 2, 53113, Bonn, Germany
e-mail: christoph.antweiler@uni-bonn.de

© Springer International Publishing Switzerland 2016
P. Meusburger et al. (eds.), *Ethnic and Cultural Dimensions of Knowledge*,
Knowledge and Space 8, DOI 10.1007/978-3-319-21900-4_8

165

codified) but tacit. Knowledge may be conceived of as purposeful coordination of action (Zeleny 2010, pp. 24–28). The question in this chapter is whether local knowledge might have general structural aspects despite its tremendous diversity across persons, settings, and cultures. I make two propositions. First, local knowledge should be neither equated with scientific knowledge (e.g., as ethnoscience) nor contrasted in an essential way with science (e.g., as wisdom or counterscience). Second, local knowledge should not be automatically associated with indigenous groups or indigeneity. I therefore use the term *local knowledge*, which is also somewhat problematic, but less so than the term *indigenous knowledge*.

Local knowledge and empirical scientific knowledge share certain attributes. But are some of them lost if local knowledge is formalized and delocalized? To address that question, I first present a general model that outlines ten universal features of local knowledge. This kind of model is needed in a search for traits that span the diversity of specific instances of local knowledge. It is also needed especially for systematic comparison of empirical cases. I then demonstrate this model with examples of environmental and migration-related knowledge documented in an ethnographic case drawn from fieldwork in Indonesia. My contention is that local knowledge might be a specific form of knowing and rationality found in all societies worldwide.

Beyond Science Versus Belief: Knowledge, Knowing, and Ignorance

Conceptual problems regarding local knowledge are notorious. Definitions of it differ greatly, ranging from fact-oriented knowledge to beliefs and wisdom. Perhaps the biggest obstacle to understanding local knowledge is simple dichotomies between local knowledge and science. Examples of the conceptual problems created by the oversimplification of concepts surface in a practical field where local knowledge is currently often used: development.

Despite often being called a *knowledge system* (especially among cognitive anthropologists), local knowledge in general is not necessarily a comprehensive system. Empirically, it is usually a patchy form of detailed and systematic knowledge in certain topical areas and sketchy knowledge in others. Local knowledge is not only cognitive; it entails emotive and corporeal aspects, too. It is situated in the present way of life and historically accumulated experience. Local knowledge may thus be best understood as a cultural or social product (Antweiler 1998). It encompasses knowledge in the strict sense of shared information and ways of knowing, ontology, framing reality, being acquainted with a topic, and bodily knowledge. These kinds of knowledge transcend the purely cognitive realm (Borofsky 1994; D'Andrade and Strauss 1992; Farnell 1999; Friedberg 1999, pp. 6–10; Harrison 1995; Kronenfeld 1996, pp. 14–19; Nygren 1999, p. 278; Siverts 1991, p. 308; Strauss and Quinn 1997).

Local knowledge of processes is especially relevant to development measures (Arce and Long 2000; Brokensha et al. 1980; Honerla and Schröder 1995; Lölke 2002; Miehlau and Wickl 2007; Pottier et al. 2003; Sillitoe 2009). Drawing on local knowledge in development should not be restricted to the elicitation of cognitive issues, nor should local knowledge be seen as simply a countermodel to western science (e.g., Agrawal 1998). Contrary to popular views, activities informed by local knowledge are not necessarily sustainable or socially just (Hauck and Kößler 2004; Murdoch and Clark 1994). Local knowledge is frequently geared to real-life practices and may be understood only with reference to the situation to which it is to be applied, that is, only in terms of referential transparency (Quinn and Holland 1987). This point is important in the selection of research methods. Knowledge relating to practices and situations may be called *situated knowledge* (Lave 1993, 1996, p. 90; Nygren 1999, p. 277). Hobart (1993) speaks of *situated practices* (p. 4) and *knowledge as practice* (p. 17). But local knowledge should not be reduced to practical, mundane, everyday, or routine knowledge.

Empirical data on urban knowledge in Makassar, Indonesia, show that local knowledge may comprise fixed and structured knowledge that can be articulated or that local knowledge may, by virtue of its combination with the performance of actions, involve a more fluid process of knowing (Antweiler 2002). Local knowledge should not be contrasted with science but may be conceived of as a continuum between formal science and everyday rationality (Antweiler 2007). Furthermore, knowledge in the sense of *what* is known must be distinguished from knowing in the sense of *how* something is known (Barth 1995; Borofsky 1994, pp. 335–339; Hobart 1993, p. 19).

Local knowledge is made up not only of technological and environmental knowledge but of the social environment as well. Ultimately, one can speak broadly of *knowledge of social systems* (von Cranach 1995). Given that human beings exist in a continuous flux of experiences and practices, local knowledge in the widest sense thus also includes the social management of information, learning, teaching, and decision-making routines. Local knowledge can involve knowing about group peers and their interrelations, a subject matter called *social cognition* in social psychology (Augoustinos and Walker 1995; Fiske and Taylor 2013; Pennington 2000). Social cognition may include the wider social environment (e.g., neighboring communities; see Stokols and Altman 1987). Despite its importance in human societies in general, social cognition in this sense is mostly skipped in debates on local knowledge. An aspect of social cognition related to local knowledge is what a people of one cultural group know about the network of experts in their own and neighboring collectives. Social cognition is especially important for development measures (Boggs 1990; Quarles van Ufford 1993). Other fields worthy of mention include organizational and management knowledge and legal knowledge, such as that pertaining to nonformal organizations (Marsden 1994). The locally available knowledge of development projects, which is still a largely unstudied subject, also falls into this category.[1] Alongside it is knowledge of a more individual, less collective

---

[1] Many underpaid academics in Southeast Asian countries, for example, have an intricate knowledge about the ideals and procedures of northern nongovernmental organizations (NGOs). In Indonesia they use it to acquire money from these sources or to establish their own NGOs (Lembaga Swadaya Masyarakat [LSM] in Bahasa Indonesia), often as an income-generating device.

Table 8.1 Forms and levels of local knowledge

Description	Example
Declarative	
Recognition and naming	Attribution of entities to terms, discreet entities, and diversity
Factual	Traits of animals, plants, temperature, social status, prices, salaries, and administrative levels
Categorical	Classifications of organisms, colors, kinship, and development project types
Procedural	
General processes, rules	Farming calendar, religious calendar, environmental crises, household cycle, and development-project cycle
Specific processes (scripts, schemas, and action plans)	Everyday routines (e.g., greetings and farewells, natural resource management, ritual sequences, project request schema, and nonroutines)
Complex (concepts, belief systems, and knowledge systems)	Cosmology, model of whole society, models of "honor," of "marriage," of "justice," cropping systems, therapies, and decision-making procedures

Adapted from Antweiler (1998), p. 475. Copyright 1998 by Anthropos Institute. Adapted with permission

character, such as how to mobilize joint interests in an action group (if no suitable organizational form exists locally).

Another field of political relevance is knowledge of the local legal system and conflict management. This knowledge is pertinent when someone asks how to assert a claim. This aim may require information of unwritten rules and locally appropriate ways of arguing one's case (see Hutchins's 1980 cognitive study of Trobriand land-use conflicts; Hutchins 1996). These examples illustrate the overlaps between various fields of knowledge. Empirical studies suggest three levels of thought in any knowledge: (a) recognition and naming of discreet entities; (b) classification and sequencing of discontinuity and diversity; and (c) unification of ensembles of living beings, including one's own society. Distinguishing between these levels may be important for eliciting parallels between scientific and local representations and establishing a dialogue between them. The chances are good for an exchange regarding the first two levels of each, but problems ensue with the third (Friedberg 1999, p. 14).

Table 8.1 shows general levels of local knowledge. The examples come from traditional fields and modern contexts (such as development) to illustrate not only that local knowledge is a traditional, rural, or other knowledge but that it results from a universal capacity (Barth 1995; Hauck and Kößler 2004; see Jones 1999, pp. 559–566; Malinowski 1948, p. 196; Nader 1996, p. 7; Salmond 1982; Scott 1998, p. 331). This universal quality challenges the supposed contrast between knowledge of a different kind (nonwestern knowledge or nonwestern wisdom) and western, or scientific, knowledge. Real-life situations (e.g., development exchanges) usually involve more than two parties and two sets of knowledge.

The first level of local knowledge is of a declarative nature. It pertains to discreet entities relating to the natural and social environment, facts relating to neighboring groups or, for example, details on development organizations. But local knowledge may involve categories and classifications, such as of plants, animals, or relatives. The practical value of such classifications, known in anthropology for decades, has become evident in development work only in recent years. Current debates about them revolve around their internal coherence and their similarity to western classifications. The underlying causes of these classifications are also disputed. To what degree does local knowledge represent (a) panhuman perception and intellect acting on natural discontinuities, (b) utilitarian concerns, or (c) reflections of cultural relations (Atran and Medin 2008, pp. 17–25; Berlin 1992; see Crick 1982, pp. 293–298; Ellen 1993, p. 3; Hunn 1982, pp. 839–844; Nazarea 1999, p. 4; Sillitoe 1998b)?

The second level of local knowledge may relate to processes. It might comprise knowledge of rapid changes in the natural environment, in market prices for goods, or in experiences with development projects. Analytically, one can distinguish between knowledge of general and specific processes. An example of the former type is knowledge of a religious calendar. The latter type, specific knowledge, includes knowledge of the precise sequence of steps involved in processes (e.g., in rituals or daily activities). Prototypical, idealized process models embody this kind of knowledge. Stage performers use this resource when they follow a sequence of schematically prescribed steps, ("scripts," Schank and Abelson 1977; Quinn 2005). A classic example of such an everyday script is the sequence of actions a person performs upon entering a restaurant: Open the door, look for an unoccupied table, occupy the table, look for the cloakroom or check in the overcoat(s), sit down at the table, pick up the menu list, order drinks, and so on. These scripts, although not fixed in writing anywhere, constitute cultural rules. They are not necessarily taught but rather are learned by individuals themselves through frequent performance and frequent observation. Consequently, this procedural knowledge usually remains unconscious or at least tacit. Nevertheless, even minor errors in the sequence of actions can cause confusion among actors or can lead to failure. If, for instance, I were to enter a restaurant in the United States and follow the customary European script of looking for a table myself, I would be breaking the rule of waiting to be seated. Conversely, applying the American script in a European restaurant (unless it were an American steakhouse or an exclusive establishment) would, as a rule, mean waiting at the entrance to no avail.

The third level illustrates that local knowledge may be considerably more complex than these scripts suggest. It may include causal knowledge and knowledge of complex systems, such as ways of farming crops or treating environmentally induced diseases (see Kalland 2003; Berkes 2012; cases in Hornidge and Antweiler 2012). It may pertain to relationships in everyday life or information of relationships within the cosmos or to the aetiology of diseases, the creation of humankind, or the origin of a community. This systemic knowledge informs action and should be considered a crucial dimension of development (see Warren et al. 1995). Such complex knowledge constructs may be termed *belief systems*, *knowledge systems*, or *meaning systems* (see van der Ploeg 1989). The word *system* indicates that these constructs are not simply aggregates of isolated concepts but that they comprise interlinked concepts and their constituent elements. Religious worldviews are an example. Within such systems, cosmologies and prototypical concepts play a major role. They are known as *cultural models* or *schemas*. Models of this kind, which may center on ideas such as *honor and shame, the self-made man*, or the *typical American marriage*, are usually shared by the majority of the people within the given culture. Although these schemes or models constitute only loose patterns of association, they are thematically organized and are highly stable both individually and over time (Quinn 2005; Quinn and Holland 1987, p. 24; Rogoff and Lave 1984; Strauss and Quinn 1997). Cultural models structure knowledge-processing, for they link affects and motives with accumulated experiences. They are also linked to linguistic metonymies (*pars pro toto*), analogies, or visual metaphors. A self-reflexive case in point is the conceptions of knowledge in modern talk about science. They often revolve around the metaphor of physical landscape. Knowledge is a *territory*, and scientific activity is a *journey*. Thus, science has a *destination*; scientists try to understand the *world*, to *see* reality. The mind is a *container* and language is a *conduit* transmitting knowledge (Salmond 1982, pp. 67–74).

A General Model of Local Knowledge

Local knowledge always has a history and is synchronically dynamic. The kernel of local knowledge consists of skills and acquired intelligence responding to constantly changing social and natural environments. Scott (1998) describes this assessment with reference to the ancient Greek concept of *mêtis*, translatable as a sort of cunning intelligence, but one specifically responding to constant dynamics (p. 311). The diversity and dynamics of local knowledge need to be understood as part of the cultural system. There are various types of public regarding knowledge distribution. Some information is made manifest to all, whereas other information may remain concealed from the majority. Certain things are known only to women or to men. Only a few specialists may possess in-depth knowledge of a particular field, such as medicine, or an activity, such as cropping expertise. Certain individuals, or either sex, may be excluded in principle from certain fields or topics of knowledge on cultural grounds. Even some agricultural knowledge may be categorized as secret (Pottier 1993, pp. 30–32).

The ability to know something about certain topics (*knowledgeability*, see Lave 1993, pp. 13, 17) varies across members of a culture, and it changes, for it is itself a social product. Expertise may be conceived of as socially sanctioned coordination of action (Zeleny 2010, p. 28). An individual's knowledge changes through time with the situation (Borofsky 1994, p. 334). There may be different or even rival items or forms of knowledge, different versions of the world. As for content, practical relevance, and centrality, knowledge is often differentially distributed between the young and the old, a process that may even lead to virtual parallel knowledge systems (Mersmann 1993). In times of rapid change, one knowledge system may be morally significant yet without practical value. "It is in the nature of interactions within and between 'knowledge systems' to make manifest certain sorts of information and to keep other sorts hidden" (Marsden 1994, p. 43). Knowledge is power. In this sense Thrift (1996, pp. 99–100) differentiates between five kinds of unknowing:

- Knowledge that is unknown because it is spatially or historically unavailable
- Knowledge that is not understood, that is, outside a frame of meaning
- Knowledge that is undiscussed and perhaps taken for granted
- Knowledge that is actively and consciously concealed
- Knowledge that is distorted

It is possible to draw a broad distinction between local specialist knowledge and local everyday knowledge (knowledge of "just plain folks"; Lave 1996, p. 87). Studies in the sociology of knowledge and network anthropology have demonstrated that differential knowledge distribution concerns not only specialist knowledge but also everyday knowledge. It is necessary to determine empirically who within a population at any time has what knowledge. The most intriguing issue is the attitude people themselves have to their concepts and the extent to which they adhere to them. What, for instance, constitutes "factual knowledge" for people as opposed to what they believe "can only be credibly claimed" (Flick 1995, pp. 56–59; Sperber 1982, pp. 98–103)?

How is a particular local knowledge realm to be defined and to what extent does it represent a body of knowledge? The fields of social psychology and sociology of knowledge supply useful definitions of knowledge. One of them characterizes knowledge as "stored information which refers to important structures, processes and functions of the system producing it...and which therefore generates evaluative processes" (von Cranach 1995, p. 25). According to another definition, knowledge is "any and every set of ideas and acts accepted by one or another social group or society of people—ideas and acts pertaining to what they accept as real for them and for others" (McCarthy 1996, p. 23).

A close inspection of many studies of local knowledge (few of whose authors attempt explicit generalization[2]) reveals a number of characteristic features and patterns of local knowledge. Table 8.2 outlines some of them in a general model of local knowledge, which is needed for seeking common traits that cut across the diversity of local knowledge. The model stresses the universal capability and specific situational nature of such knowledge and emphasizes its practical relevance and social significance. It is not knowledge isolated and abstracted from daily life. It encompasses not only information (factual knowledge) as a resource but also capabilities and skills. The knowledge utilized for the exercise of skills most often

---

[2] The following works from different disciplinary backgrounds have informed my understanding of the general characteristics of local knowledge, mostly Atran and Medin (2008); Barth (1995); Berkes (2012); Crick (1982); DeWalt (1994); Ellen and Harris (2000); Friedberg (1999); Geertz (1983); Harrison (1995); Hutchins (1996); Lambek (1993); Lave (1993, 1996); Lindblom (1959); Malinowski (1948); Nader (1996); Nazarea (1999); Nygren (1999); Pasquale et al. (1998); Schareika (2004); Scott (1998); Sillitoe (1998a, b); Strauss and Quinn (1997); Tambiah (1990); and Worsley (1997).

Table 8.2 A general model of local knowledge

Key interrelated features	Aspects
1. Knowledge plus skills	Rooted in a combination of specific factual knowledge and practical, action-oriented skills
2. Adaptation to situational dynamics and variability	Keyed to common, but never precisely identical, features of a particular place; thus adapted to ambiguous, mutable, stochastic, and thus indeterminate issues
3. Empirical local basis and experiential saturation	Generated by local observation and low-cost, low-risk, trial-and-error, and natural experiments; proven by coping over a prolonged period in the laboratory of life
4. Redundancy and holism	Represented parallel in several cultural domains; embedded; characterized by holistic orientation through systemic relations with other aspects of culture
5. Tacit nature of knowledge	Manifested as often implicit, uncodified, intuitive, embodied, nondisciplinary, less verbalized and less susceptible to verbal or written communication than performed procedural knowledge is
6. Informal learning	Experienced through oral transmission, decentralized and piecemeal learning, learning by imitation, demonstration, and apprenticeship more than by instruction
7. Scientific approach	Conducted with at least partially systematic, methodical, parsimonious, empirical-hypothetical, comprehensive, and distinguished by a capacity to generate causal theory
8. Optimal ignorance	Achieved through information only as detailed and accurate as necessary for addressing the problem, no more and no less
9. Evaluation criterion, test	Based on practical efficacy as the yardstick vs. other criteria such as theoretical consistency, parsimony, and elegance (but see 4)
10. Resulting actions and solutions to problems	Arrived at through solutions familiar and thus broadly accepted by local peoples, oriented to "satisficing" and optimizing (vs. maximizing) and the use of local or endogenous resources

Adapted from Antweiler (2004). Reprinted with permission

remains implicit because it is usually won through a process of learning by doing. With all human beings, capabilities and skills tend to be less conscious than factual and formalized knowledge. Instead of giving detailed verbal instructions, experienced persons demonstrate to their students in a few words how to perform actions.

Despite being communicated and transmitted primarily through speech and listening, local knowledge may be highly specific. The performance of complex actions may depend on procedural knowledge that is only partially conscious. This contingency is true especially of actions involving knowledge embodied in movements (see Farnell 1999), rituals (Tambiah 1990), and decision-making (Antweiler 2000; Fjellman 1976; Gladwin 1989; Prattis 1973). Procedural knowledge is less verbalized than formal knowledge. For example, it often happens that a colleague is able to help solve a computer problem after several trials but then cannot explain the steps followed. The question today is less about the ways in which specific items of information find their way into people's minds than about the ways in which individuals constitute communities of knowing by participating in their knowledge (Karamoy and Dias 1982). Furthermore, local knowledge is linked to one or more shared views of the future. They may comprise best worlds, ontologically based theories of the good life, or goals of endogenous development.

The combined qualities of being both a result of universal cognitive capacities and localized, situated performance may explain apparent contradictions that arise if local knowledge is systematized in the global knowledge arena. What are the consequences of isolating, simplifying, documenting, storing, repackaging, codifying, transferring, and commodifying local knowledge (Agrawal 1995; Ellen and Harris 2000, p. 17; Menkhoff et al. 2010; Sillitoe 2009)? These transformations imply delocalization, decontextualization, and depersonalization that may be possible for general know-how, specific expertise, and certain skills. These aspects are especially relevant for nations striving to make a transition into a knowledge society, as in Southeast Asia (Evers 2010). In other dimensions such as locale and the environmental context, however, cultural embeddedness and socially situated character (sometimes featuring specified topographical information) cannot be discarded without distortion. There is a need for systematic and comparative, yet culturally sensitive, methods.

Methods: Systematic Anthropological Techniques

Empirical research on local knowledge is possible with several types of ethnographic data gained from fieldwork. Most often they are based on open interviews or participant observation. Fieldwork methods, being experiential and person-oriented, have their pros and cons. Their strength is that they reveal knowledge in naturalistic settings, giving researchers real-space, real-life, and real-time data. The weakness of fieldwork methods is that the data they yield largely lack representativeness and are difficult to compare with data from other fieldwork. I deal in this section mainly with elicitation and only hint at issues of data analysis.

Problems with Eliciting Cognition

Current data-collection techniques for eliciting local knowledge are torn between two opposing poles. On the one hand, there are standard ethnographic methods, such as structured interviews in cognitive anthropology (Spradley 1979; Strauss 1987; Werner and Fenton 1970; Werner and Schoepfle 1987), which are often very formal and time consuming. On the other hand, ethnologists have the simple tools of rapid and/or participatory appraisal and learning methods (see Chambers 1991; Schönhuth and Kievelitz 1995; Sillitoe et al. 2005). Both approaches have drawbacks. Cognitive anthropological methods are usually used only to conduct detailed inquiry into specific cultural domains (e.g., the classification of animals or soils), not to account for people's understanding and practical use of concepts. Moreover, intracultural cognitive variability, which often derives from talking to knowledgeable specialists, remains largely unexplored (Nygren 1999, p. 277; Sankoff 1971; Sillitoe 1998b, p. 190). These methods, too, cost a great deal of time. Participatory methods are rapid, inexpensive, and generally more inclusive than survey methods. But they pose problems if one tries to scale them up. They are theoretically ungrounded as a result of tinkering, are laden with political agendas (which are often hidden), and generally omit the larger cultural context.

A Call for Simple Cognitive Methods

In this chapter I argue for systematic, systemic, and multifocal approaches (Ellen 1996, p. 459; see Vayda 1983). Specifically, I argue that ethnologists need comparative, yet culturally sensitive, methods for samples larger than those used thus far. Methods such as the Rapid Rural Appraisal (RRA), the Participatory Rapid Appraisal (PRA), and the Participatory Learning Approach (PLA) should be complemented with procedures from cognitive anthropology and clinical psychology. The problem with cognitive methods is that there are many works on data-processing procedures but few on data-collection techniques in cross-cultural contexts. Anthropological experience shows that systematic elicitation of data may be very problematic in a cultural context different from that of the researchers. The assumption made in methods handbooks (e.g., Weller and Romney 1988; Werner and Schoepfle 1987) is that certain methods—systematic interviewing, systematic data collection, and systematic elicitation techniques—may be universally applied.

Cognitive anthropologists who use these methods claim the ability to reveal the insider view (emic perspective); critics maintain that the methods are overly formal, too complicated, and inapplicable to real-life nonwestern settings. They are largely based on U.S. experiences with people who are accustomed to formal tests and whose cultural background is already well known. Reported experiences with such methods in the context of anthropological fieldwork in nonwestern settings are varied. Some colleagues have reported that informants found them funny and interesting; others have regarded them as childish and proposed having coffee, *ouzo*, or a *kretek* (clove) cigarette instead. The gap between bold textbook claims and fieldwork experiences comes across clearly in the following comparison of statements by Weller and Romney (1988) with Barnes (1991) (Table 8.3).

[T]he interviewing and data-collection tasks contained in this volume are as appropriate for use in such exotic settings as the highlands of New Guinea as they are in the corporate offices on Wall Street. (Weller and Romney 1988, p. 9)

[The informants] quickly got stuck, re-sorting the cards as each new name was added, before stopping and declaring the task to be impossible. (Barnes 1991, p. 290)

We social scientists may respond to these problems in three ways. First, we can continue to use the methods of the established participatory approaches regardless of the problems, remaining aware of the limitations of the methods and the fuzziness of the term *participation*. Second, we can dispense with systematic elicitation and resort to less formal ways of collecting local knowledge data. This option is

Table 8.3 Methods for eliciting local knowledge

Method	Examples of themes and aspects
1. Listening, talking to people	Terminology, locally relevant cultural topics

2. Systematic, structured interviewing, formal elicitation techniques
   Free listing; question–answer frame
   Card- or pile-sorting; label-sorting
   Triadic comparison; triad test
   Rating, rating scale
   Ranking, rank ordering
   Sentence frame format, frame elicitation
   Combination of the above (e.g., repertory-grid method)
   Graphic methods (visualization, drawing trees, cognitive maps)
	Domains, topics, propositions
	Dictionary, basic cognitive structure
	Similarity comparisons, taxonomy
	Evaluative comparisons
	Hierarchy of values, coherence vs. diversity
	Logical relationships, causes–effects
	Environmental perception, personal constructs
	Concepts; hierarchies; spatial concepts

3. Observation	
Nonparticipant (e.g., time allocation)	Practices, routines, products
Participant	Procedural knowledge, knowing
4. Documentation and study of documents	
Photos, films, video	Knowledge products (texts, objects)
Mapping	Knowledge distribution
Recording of narrative texts ("orature")	Nature and mode of knowledge, knowledge transfer
Recording of natural discourse	Topics, forms of discourse, implicit knowledge
5. Combinations of the above methods	
Natural decision-making	Procedural knowledge, rules, cultural models
Apprenticeship, teacher–pupil interaction	Implicit knowledge, scripts
Action research	Knowledge acquisition, creativity, implicit knowledge
Participatory methods (e.g., RRA, PRA, PLA)[a]	Participation in the gathering, dissemination, and utilization of knowledge

Design by author

[a]Rapid Rural Appraisal, Participatory Rapid Appraisal, and Participatory Learning Approach

often taken by anthropologists as they collect everyday discourse (gossip). The approach foregoes control over data collection, and results are difficult to compare. Third, social scientists can adapt and simplify the methods of cognitive anthropology. Criticisms of cognitive anthropology mentioned above apply to older, often very formal procedures (e.g., Metzger and Williams 1966). Some modern methods, such as free listings, pile-sorting, and triad tests, are simple and especially suitable if the aim is to know from the outset topics, criteria, and problems relevant to local people (Antweiler 1993; Bernard 2011, Chaps. 9–12, 2013, Chap. 10; Weller and Romney 1988, Chap. 2; Sillitoe 2012). In fact, they are already used in some participatory approaches and can elicit informants' knowledge as well as evaluations and sentiments. They allow for intracultural variation as a consequence of age, gender, network position, social rank, or other factor.

Localizing a Systematic Method

Residential Knowledge and Repertory Grids

During my research on intraurban residential mobility,[3] specific questions arose about knowledge and the sentiments of residential relocation and the subjective relevance of the built environment (residential cognition; Aitken 1990; Saegert and Winkel 1990; see Tognoli 1987): What do people know of the area (knowledge)? How do they know it, that is, what are the ways in which they represent streets cognitively (knowing)? What are their evaluations and the meanings of the built environment? What are the evaluations of neighborhoods or places as potential residential locations?

Environmental psychologists and geographers interested in perception often use techniques of cognitive or mental mapping to understand such issues. Months of living with people in Makassar and exchanges with a research assistant showed that they are not accustomed to using maps. Any results would have been quite artificial. Because of the comparatively few assumptions and the simplicity of implementing the repertory grid method (also called the *repgrid technique*), I turned to that instrument. It elicits cognitive and emotive data via language to reveal so-called constructs (Scheer 1993, pp. 25–36). The method was developed by Kelly (1955; see Fransella 1995/2009) and is classically used in clinical psychology to reveal subjective theories (Catina and Schmitt 1993). Guides to the method include Scheer and Catina (1993), Jankowicz (2004), and Fromm and Paschelke (2010). The background is explained in Fransella et al. (2004, Chap. 1). The technique is also used by organizational scientists, environmental psychologists, and urban geographers interested in neighborhood evaluation and residential choice (e.g., Aitken 1984, 1987, 1990; Anderson 1990; Preston and Taylor 1981; Tanner and Foppa 1996). The repertory grid method is hardly known in knowledge research and seldom used in the field of anthropology (for exceptions see Barker 1980, p. 300; Richards 1980, p. 187; Seur 1992, pp. 124–127).

The simple basic assumption is that humans order their world cognitively by using dual polarities. With respect to everyday decisions, it is assumed that individuals construe several aspects within the diversity of their experiences according to similarities and dissimilarities (Catina and Schmitt 1993). Every person uses many polarities, and, unlike the principal dualism assumed by structuralists, they differ intra- and interculturally. The main assumptions underlying the repertory grid method are:

- Humans do not merely react to events but are in charge of their actions.
- People think in dichotomies, but the content of these dichotomies differs from one person to the next.

---

[3] The research was a case study of everyday rationality, especially of decision-making on intraurban residential moves, in Makassar (formerly Ujung Pandang), the multiethnic capital city of South Sulawesi, Indonesia. The fieldwork took place over 1 year (1991–1992) and in subsequent, almost annual visits (between 1992 and 2010) to that regional metropolis of 1.3 million people. For the main results see Antweiler (2000, 2002).

- Personal constructs consist of psychic polarities. Each pole is relevant only with its counterpart, but they need not be logical poles or contradictions; they need not be rational or precise.
- Personal repertoires are rooted in the biographical framework of earlier experiences and the anticipation of future ones.
- A person has a repertoire of several such constructs.
- Individuals select between several of their constructs according to the given situation.
- Individuals can rebuild their constructs. Alternative views are present anytime, an important point in everyday decision-making.

Repertory Grid Method: Procedure and Advantages

The repertory grid method has two steps:

1. An elicitation of constructs via a comparison of two or three items (dyads, triads)
2. An evaluation of other items according to the constructs elicited in a scaling procedure

Step 1 is the construct question. Two or three items, called *elements*, are compared by the person interviewed. These elements may be words or short sentences written on cards, they may be photographs, or they may be concrete objects such as a plants or an animal specimen. The interviewer asks the interviewee to discriminate between the items on the basis of similarity (*dyad comparison, triad comparison*), without any other inputs. This procedure reveals certain stated characteristics, as in "these two are similar because they are both clean" or "this one is different because it is dangerous." An *initial construct pole* thereby emerges. Asking for the opposite of the stated trait—if not obvious—reveals pairs of contrasts (e.g., *clean–dirty, dangerous–safe*). These pairs of contrasts are called *polarities* or *personal constructs*. It is possible to elicit several constructs per dyad or triad.

In Step 2 the constructs that this exercise elicits as poles on a scale are used for ordering other items presented to the interviewee, who evaluates new items by ranking or rating them on a scale that he or she establishes, not the ethnographer. Only the items for comparison come from the interviewer. This restriction differentiates the procedures of the repertory grid method from those of the *semantic differential* and the *polarity profile*, with which given items are rated or ranked. A matrix can be formed by arranging the constructs (from Step 1) horizontally and the rating values (from Step 2) vertically. It represents the person's cognitive and emotive repertoire regarding a specific theme or domain. The semantic space made up of elements and constructs is called the person's *repertory grid*.

I focus now on the method as a recipe adapted to the Makassar fieldwork setting and explain the adaptation to the local setting—an important step often not mentioned in cognitive anthropology methods texts and RRA, PRA, and PLA handbooks.

Step 1: Elicitation of constructs by means of triad comparison

1. Present three elements (houses, lanes, neighborhoods) as photos A, B, C.
2. Ask, "Which of these two elements are similar, or which one is specific in any respect?"
3. Ask, "Why this one [these ones]?" Note response as one pole on a 10-point scale.
4. Ask, "What would be the opposite of that?" Note reply as the other polarity.
5. Ask, "Which of these two qualities would you prefer?" Note preference with a symbol (X).

Repeat triads several times with the same procedure to obtain several constructs.

Step 2: Assessment of named urban areas with the constructs (elicited in Step 1)

1. Rank eight plastic strips bearing the names of residential areas or streets within the poles of the first polarity profile.
2. Repeat this ranking of the eight strips within the other constructs elicited in Step 1.

The repertory grid method has several advantages over survey-like methods and open interviews. In the media used there are similarities to participatory methods (e.g., PRA), except that the method is more rigorous yet not less participatory. First, it elicits systematic emic data from a sample (same stimuli for everyone) and is thus both qualitative and quantitative. Second, it yields emic cognitive results on individuals (not generalized results) and can document intracultural variation. Third, the researcher gives only stimuli instead of a prearranged polarity profile, allowing for local perspectives (e.g., local classifications or sentiments). Fourth, photographs replace words as stimuli, improving control over stimuli. Fifth, the items are presented parsimoniously rather than in an elaborate procedure and are therefore applicable to real life. Sixth, notation is easy, simple, and thus transparent for the interview partners, all of which takes account of ethical issues. Seventh, the interview procedure is short and thus conducive to further dialogue on emic-related topics.

The method can be modified in many ways. One could, for example, have the interviewees sort or rank items according to desired future states of their residential environment (Aitken 1990, p. 253). The method elicits subjective theories but, by virtue of its comparative nature, may be used to understand intersubjective emic theories (Fig. 8.1).

An Essential Step: Selecting Culturally Relevant Themes and Suitable Media

If participation is the aim of textbook methods, they need considerable modification. The challenge is to simplify and adapt the methodology without localizing it to the point that it prevents comparison and generalization of the data. The localization aspect is often forgotten when PRA methods are applied, for it needs time and a degree of ethnographic grounding, as I now show. I resorted to this method after five months in the field while trying out different approaches, such as cognitive maps. I

Introductory triad (three persons A, B, C)
to sensitize the interviewees for an
evaluative similarity comparison

Notation form for constructs
and ranking (Steps 1 and 2)

Kretek (clove) cigarettes

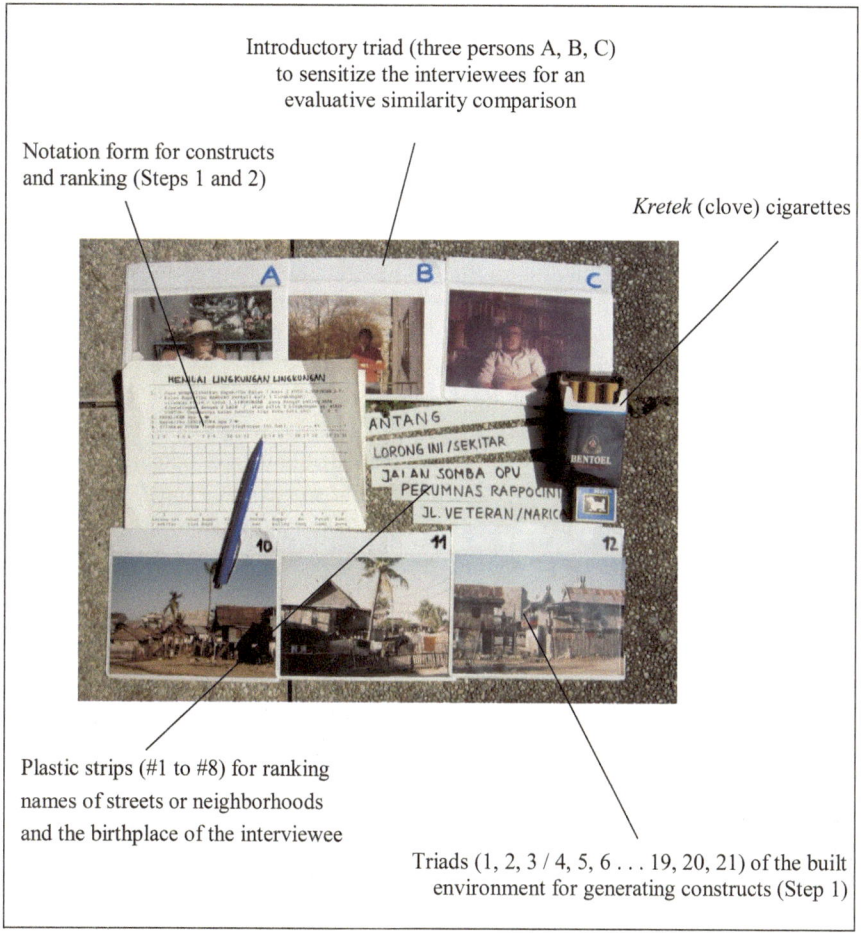

Plastic strips (#1 to #8) for ranking
names of streets or neighborhoods
and the birthplace of the interviewee

Triads (1, 2, 3 / 4, 5, 6 . . . 19, 20, 21) of the built
environment for generating constructs (Step 1)

Fig. 8.1 Set of interview material for repertory grid interview (Photo by author, 2004; design by author)

interviewed a random sample of house owners (30 %; $n=21$) from the neighborhood. My respondents determined the interview locations. Often, we talked in their small guest rooms (*ruang tamu*), on the terrace, or in front of their dwelling while standing at the edge of the small lane there. Interviews lasted from 30 to 90 min, depending on the age and education of the interviewees and on situational factors such as their mood and the presence of neighbors.

Initial experiences confirmed that triads motivate people to think and evaluate far more than dyads do, the latter often being too obviously similar or different. For each triad I selected photographs depicting relatively similar situations to make the comparison interesting but tried to maximize the variance of living situations through the triads. One triad showed three lower-income houses, whereas another one showed three well-organized middle-class neighborhoods. I used photographs (Fig. 8.1) for Step 1 because I had found during the first few months of my stay that people liked to talk about photographs. They kept family albums, and almost every household displays photos. The use of photos made it relatively easy to give the same stimuli to each person interviewed. Furthermore, photographs were suitable for exploring urban environmental knowledge because they revealed details of living spaces that people could compare.

The color photographs showed typical residential areas from Makassar. The selection of this material was important. I did not use photos of the area where the interviewees lived, for the purpose of Step 1 was to elicit general value orientations based on observable traits, not on an evaluation of a known area (Nasar 1998). Nor did I use photos of well-known places, streets, buildings, or billboards. To cover the diversity of living situations and residential areas, I selected from hundreds of images that I had taken while documenting the city during the first five months of fieldwork. Pragmatic and ethical considerations prompted me to show only outdoor situations. First, indoor photos too obviously reflected residents' social status and thus would not have been altogether productive. Second, I could take indoor photos only if I knew the people well. Using indoor photos of households personally known to the interviewees would have been dubious in a society where living conditions are a common topic of daily gossip. For practical reasons I used the locally common print format (10×15 cm, or 4×6 in.) and numbered the photos for easy identification. This practice also facilitated notation of responses and was transparent to interview partners—a significant consideration for the methodology of participatory citizen science.

The comparative evaluation of urban areas within the constructs (Step 2) also needed preparation based on my fieldwork experience. I presented eight urban areas as the stimuli, using their names (not photographs). The neighborhood names were written clearly on plastic strips. They were easy to handle if illumination was dim within a house or in the evening, and they did not become dirty if the interview had to be conducted outside in front of a house or during rain in the monsoon season. I selected six areas. To connect the ranking with the biographical experience of the interviewee, I also included that person's present neighborhood and his or her birthplace. During my fieldwork on residential decision-making, it became apparent that previous experiences with migration from the country to the city were a key factor in the selection of areas for intraurban residential moves. Attention was to be directed to general evaluations, images, and prejudices, not to specific traits. Ranking was used instead of rating because the former is less laborious and implies a comparison of elements.

Systematic Yet Culturally Adequate Interviewing: Experiences with Strengths and Weaknesses

As one might expect, reactions to this method varied among the people within this ethnically and economically mixed neighborhood. Most of the interviewees were interested or amused, but some of them found it was strange. Having been in the field for 5 months, I had visited all interviewees several times before the repertory grid interview, having informal discussions, conducting a household census, and tracing their residential history with them. They knew me and knew that my overall research topic was intraurban residential mobility.

When introducing each person to this specific interview, I stressed that I was not seeking *correct* answers but rather her or his personal perceptions and evaluations. Such understanding was important because my trials with this formal procedure revealed that some people associated it with intelligence tests. In a city full of schools, universities, and offices, many people have encountered such instruments. After some conversation and a few obligatory *kretek* (clove) cigarettes, I would show the interviewee three outdoor photographs: one of me, one of my brother, and one of my mother. I used my family because the interviewees were keen on seeing photos of western people and of families. I then asked the triad question and answered it by grouping the people in the photographs according to gender: my brother and me in one pile, my mother in the other. Then I pointed out that the three of us could also be grouped in the same way but for other reasons. For example, my mother was improperly (*kurang cocok*) dressed for women, whereas my brother and I were properly dressed. The aim was to demonstrate the method to the interviewee by using an everyday topic. I also used gender roles to make them aware of an evaluative comparison in the interview. I deliberately used simple notation—upper-case letters—because of my poor handwriting and the dim lighting. The notation was facilitated by the numbering of the photographs (Step 1) and the plastic strips (Step 2).

The repertory grid method discussed in this chapter is simple regarding theoretical assumptions and the procedure followed, but its use requires reasoning and preparation. Time is needed, for most of this preparatory work has to be done in the field. I drew the following conclusions from the repertory grid interview in my case:

- Even simple cognitive textbook methods are complicated and time consuming.
- A problem of cognitive textbook methods is that they often require many interviews with the same person.
- A short interview is better than a long one in most instances.
- Transparency is to be increased with simple notation that allows for visual sharing.
- It is best to restrict the number of personal constructs elicited.
- Items presented have to be culturally appropriate.
- Photographs are often suitable for comparison because people are interested in them.
- Time to prepare and pretest suitable photos must be planned into the procedure.
- Awareness of the locally relevant topics of discourse is helpful.
- To prepare the interview partners to use the chosen method, the researcher should demonstrate its use with another locally relevant topic.

Pan-Cultural Local Knowledge and Delocalization of Local Knowledges

Research on local knowledge richly illustrates the intimate relation between theory and method in social science and the humanities. The very nature of local knowledge as shown in the theoretical model poses the challenge to study this knowledge empirically with universal methods. The key methodological problem in research on local knowledge is to elicit and analyze data in naturalistic settings systematically. It is especially demanding to elicit emic data in a way that allows one to compare them analytically with those that other researchers have gathered among other peoples elsewhere.

The repertory grid method in the version demonstrated in this chapter exemplifies a method that combines systematic elicitation with local adaptation. It can yield various results, ranging from simple qualitative data on cases to detailed quantitative sample data. They may be used and analyzed in several ways[4]:

- Simple graphic representation to allow visual sharing, which lays a sound basis for direct further discussion and which reveals proposals for evaluation criteria that are not yet recognized
- Analysis through simple sorting by hand (see Raethel 1993, pp. 47–49)
- Use of analytic software tools, such as *Anthropac*[5] (Borgatti 1989), *OpenRepGrid*, and *Sci:Vesco*[®6]
- Use of specific quantitative and graphic data-processing for which there is software available (see Raethel 1993, pp. 53–67)

The results showed that the idiom and perceptions of local residents differed strikingly from the language and concepts employed in official urban planning brochures (Table 8.4). Formal, but simple, cognitive methods have potential in development work, as demonstrated by the use of local urban knowledge to humanize urban planning and enhance its effectiveness. Regarding methods, the universality of cognitive approaches can be maintained if informants are accustomed to formal questioning or if textbook versions are adapted to the local cultural setting. The latter condition requires a certain ethnographic grounding in local culture, which is not normally available in development projects.

---

[4] See Bernard and Ryan (2009) for an exceptional handbook on systematic analysis of qualitative data, combining reflection, hands-on-information, and many examples of ways to combine elicitation techniques with appropriate techniques of analysis (e.g., schema analysis, sequential analysis, analytic induction, and grounded theory).

[5] Anthropac is especially suited for the analysis of cultural domains and offers other analytic tools as well (e.g., for consensus analysis). Anthropac 4.98 and all manuals can be downloaded for free. Retrieved July 31, 2013, from http://www.analytictech.com/anthropac/anthropac.htm

[6] For the latter two resources see http://www.openrepgrid.uni-bremen.de/wiki/index.php?title= Main_Page and http://elementsandconstructs.de/de/products/scivesco/scivescoweb (both retrieved July 31, 2013).

Table 8.4 Selected concepts about residential areas: emic concepts versus official idiom

Everyday concepts		Official idiom (e.g., in maps, planning documents)	
Pole 1	Pole 2		
(Being) alone (*sendiri*)	Populated (*bermasyarakat*)	a	
Like village (*sama kampung*)	Urban (*kota*)		
Still like *kampung* (*masih kampung*)	Already urban (*sudah kota*)		
Plain, simple (*sederhana*)	Luxurious (*mewah*)		
Orderly (*rapi*)	Not (yet) orderly (*belum rapi*)		
Dense (population) (*padat*)	Distantly spaced (*renggang*)		
Dense (population) (*padat*)	Good ventilation (*udara bagus*)		
Dense (population) (*padat*)	Still empty (*masih kosong*)		
Calm (*tenang, sunyi*)	Ado, fuss, gossip (*cencong*)		
Calm (*sunyi*)	Full of life (*ramai*)		
Secure (*aman*)	Disturbed, unsafe (*rawan*)		
Secure, clear (*tenang*)	Insecure (*kacau*)	Same as everyday concept	
Orderly planned (*teratur*)	Not orderly (*tidak teratur*)	Same as everyday concept	
The rich (*yang kaya*)	Ordinary people (*orang biasa*)	Rich (*orang kaya*)	Poor (*miskin*)
Modern (*maju*)	Not (yet) modern (*belum maju*)	Modern (*maju*)	Traditional (*tradisional*)
Dirty (*kotor*)	Clean (*bersih*)	Dirty area (*kawasan kumuh*)	
		Economically strong (*economi kuat*)	Economically weak (*economi lemah*)
		Upper people (*masyarakat tinggi*)	Lower people (*masyarakat rendah*)

[a]Empty cells signify that no complement was observed

The repertory grid method is presented in this chapter as a method that allows the user to elicit information on local cognition and emotion systematically, yet sensitively. In urban situations space and mobility options, knowledge of prices, and unwritten rules of the public sphere and bureaucracy are particularly important in addition to knowledge about dwellings (Burgess et al. 1997; Cresswell 1996; see also Irwin 1995; Knorr-Siedow and Gandelsonas 2004; Nas 1995). Such knowledge could be used both to counter dominant official regulations and to enrich expert knowledge. If local knowledge research were less idealistic and hurried and more systematic, multifocal, and contextually sensitive than it currently is, it would improve the effectiveness and participatory capacity of development, specifically of urban planning. The aim would be to enable communities to take part in knowledge-related decision-making about their environment: a truly citizen science.

Cognitive anthropological methods will be applicable if textbook versions are considerably simplified and adapted to the local cultural setting. Participatory methods often used in development are currently useful but are frequently too quick and mostly lacking in theoretical grounding. Local knowledge research is conducted mostly in the realm of development; it is problem oriented. But even such applied research needs well-defined methodologies, which require a clear theoretical understanding of the phenomenon of local knowledge.

A central question for research on local knowledge in the future is to find cross-cultural patterns. Particularistic documentation of the local specifics will remain a key research area, but it is necessary to go beyond that practice and ask what the general properties are of local knowledge found throughout the varieties of western social milieus and the diversity of nonwestern cultures. These general questions pertain not only to the very content and character of local knowledge. Research on general-process properties of local knowledge must increase after that knowledge has been empirically found. What are general properties of the dissemination, documentation, and conservation (archives, data pools)? What are the general effects of the inevitable standardization and delocalization (decontextualization) involved within these processes? What is gained and what is lost when local, predominantly oral knowledge makes the transition into printed books like this one?

The kernel of local knowledge is a form of cognition and performance that probably exists in all societies. Local knowledge comprises skills and acquired intelligence, which are action-oriented, culturally situated, and responsive to constantly changing social and natural environments. Given that the general properties of local knowledge as outlined in the above model (or a variant of it) will pass additional empirical cross-cultural tests, local knowledge might be regarded as a human universal. People may live differently in this world, but they do not live in different worlds.

References

Abel, G. (2008). Forms of knowledge: Problems, projects, perspectives. In P. Meusburger, M. Welker, & E. Wunder (Eds.), *Clashes of knowledge: Orthodoxies and heterodoxies in science and religion* (Knowledge and space, Vol. 1, pp. 11–33). Dordrecht: Springer.

Agrawal, A. (1995). Dismantling the divide between indigenous and scientific knowledge. *Development and Change, 26*, 413–439.

Agrawal, A. (1998). Geistiges Eigentum und "indigenes Wissen": Weder Gans noch goldene Eier [Intellectual property and "indigenous knowledge": Neither goose nor golden eggs]. In M. Flitner, C. Görg, & V. Heins (Eds.), *Konfliktfeld Natur: Biologische Ressourcen und globale Politik* (pp. 193–214). Opladen: Leske und Budrich.

Aitken, S. C. (1984). Normative views and ordering the urban milieu. *The East Lakes Geographer, 14*, 1–16.

Aitken, S. C. (1987). Households moving within the rental sector: Mental schemata and search spaces. *Environment and Planning A, 19*, 369–383.

Aitken, S. C. (1990). Local evaluations of neighborhood change. *Annals of th American Association of Geographers, 80*, 247–267. doi:10.1111/j.1467-8306.1990.tb00290.x.

Anderson, T. J. (1990). Personal construct theory, residential decision-making and the behavioural environment. In F. W. Boal & D. N. Livingstone (Eds.), *The behavioural environment: Essays in reflection, application and re-evaluation* (pp. 133–162). London: Routledge.

Anderson, J. E. (2009). *Law, knowledge, culture: The production of indigenous knowledge in intellectual property law*. Aldershot: Edward Elgar.

Antweiler, C. (1993). Universelle Erhebungsmethoden und lokale Kognition am Beispiel urbaner Umweltkognition in Süd-Sulawesi/Indonesien [Universal survey methods and local cognition with the example of environmental cognition in South Sulawesi, Indonesia]. *Zeitschrift für Ethnologie, 118*, 251–287.

Antweiler, C. (1998). Local knowledge and local knowing: An anthropological analysis of contested "cultural products" in the context of development. *Anthropos, 93*, 469–494.

Antweiler, C. (2000). *Urbane Rationalität: Eine stadtethnologische Studie zu Ujung Pandang (Makassar), Indonesien* [Urban rationality: An anthropological study on Ujung Pandang (Makassar), Indonesia] (Kölner Ethnologische Mitteilungen, Vol. 12). Berlin: Dietrich Reimer.

Antweiler, C. (2002). Rationalities in Makassar: Cognition and mobility in a regional metropolis in the Indonesian periphery. In P. Nas (Ed.), *The Indonesian town revisited* (Southeast Asian dynamics, Vol. 1, pp. 232–261). Münster/Singapore: Lit Verlag/Institute of Southeast Asian Studies (ISEAS).

Antweiler, C. (2004). Local knowledge. In A. Bicker, P. Sillitoe, & J. Pottier (Eds.), *Investigating local knowledge. New directions, new approaches* (p. 17). Aldershot: Ashgate Publishing.

Antweiler, C. (2007). Wissenschaft quer durch die Kulturen: Wissenschaft und lokales Wissen als Formen universaler Rationalität [Science across cultures: Science and local knowledge as universal forms of rationality]. In H. R. Yousefi, K. Fischer, R. Lüthe, & P. Gerdsen (Eds.), *Wege zur Wissenschaft: Eine interkulturelle Perspektive. Grundlagen, Differenzen, Interdisziplinäre Dimensionen* (pp. 67–94). Nordhausen: Verlag Traugott Bautz.

Arce, A., & Long, N. (Eds.). (2000). *Anthropology, development and modernities: Exploring discourses, counter-tendencies and violence*. London: Routledge.

Atran, S., & Medin, D. L. (2008). *The Native mind and the cultural construction of nature*. Life and mind: Philosophical issues in biology and psychology. Cambridge, MA: MIT Press.

Augoustinos, M., & Walker, I. (1995). *Social cognition: An integrated introduction*. London: Sage.

Barker, D. (1980). Appropriate technology: An example using a traditional African board game to measure farmers attitudes and environmental images. In D. W. Brokensha, D. M. Warren, & O. Werner (Eds.), *Indigenous knowledge systems and development* (pp. 297–302). Lanham: University Press of America.

Barnes, R. H. (1991). Review of Röttger-Rössler 1989. *Zeitschrift für Ethnologie, 116*, 289–291.

Barth, F. (1995). Other knowledge and other ways of knowing. *Journal of Anthropological Research, 50*, 65–68.

Berkes, F. (2012). *Sacred ecology* (3rd ed.). New York: Routledge.

Berlin, B. (1992). *Ethnobiological classification: Principles of categorization of plants and animals in traditional societies*. Princeton: Princeton University Press.

Bernard, H. R. (2011). *Research methods in anthropology: Qualitative and quantitative approaches* (5th ed.). Oxford, UK: AltaMira Press.

Bernard, H. R. (2013). *Social research methods: Qualitative and quantitative approaches* (2nd ed.). Los Angeles: Sage.

Bernard, H. R., & Ryan, G. W. (Eds.). (2009). *Analyzing qualitative data: Systematic approaches*. Thousand Oaks: Sage.

Boggs, J. P. (1990). The use of anthropological knowledge under NEPA. *Human Organization, 49*, 217–226.

Borgatti, S. P. (1989). *Provisional documentation ANTHROPAC 2.6* [Manuscript]. Columbia: University of South Carolina.

Borofsky, R. (1994). On the knowledge and knowing of cultural activities. In R. Borofsky (Ed.), *Assessing cultural anthropology* (pp. 331–347). New York: McGraw-Hill.

Brokensha, D. W., Warren, D. M., & Werner, O. (Eds.). (1980). *Indigenous knowledge systems and development*. Lanham: University Press of America.

Brush, S. B., & Stabinsky, D. (Eds.). (1996). *Valuing local knowledge: Indigenous people and intellectual property rights*. Washington, DC: Island Press.

Burgess, R., Carmona, M., & Kolstee, T. (1997). Contemporary policies for enablement and participation: A critical review. In M. Burgess, M. Carmona, & T. Kolstee (Eds.), *The challenge of sustainable cities: Neoliberalism and urban strategies in developing countries* (pp. 139–162). London: Zed Books.

Catina, A., & Schmitt, G. M. (1993). Die Theorie der persönlichen Konstrukte [The theory of personal constructs]. In J. W. Scheer & A. Catina (Eds.), *Einführung in die Repertory Grid-Technik: Band 1. Grundlagen und Methoden* (pp. 11–23). Bern: Hans Huber.

Chambers, R. (1991). Shortcut and participatory methods for gaining social information for projects. In M. M. Cernea (Ed.), *Putting people first: Sociological variables in rural development* (2nd ed., pp. 515–637). New York: Oxford University Press (for The World Bank).

Cresswell, T. (1996). Participatory approaches in the UK urban health sector: Keeping faith with perceived needs. *Development in Practice, 6*, 16–24.

Crick, M. R. (1982). Anthropology of knowledge. *Annual Review of Anthropology, 11*, 287–313.

D'Andrade, R. G., & Strauss, C. (Eds.). (1992). *Human motives and cultural models*. Cambridge, UK: Cambridge University Press.

DeWalt, B. R. (1994). Using indigenous knowledge to improve agriculture and natural resource management. *Human Organization, 53*, 123–131.

Ellen, R. F. (1993). *Nuaulu ethnozoology: A systematic inventory*. Canterbury, UK: University of Kent at Canterbury, Center for Social Anthropology and Computing in Cooperation with the Center of South-East Asian Studies, Canterbury.

Ellen, R. F. (1996). Putting plants in their place: Anthropological approaches to understanding the ethnobotanical knowledge of rainforest populations. In D. S. Edwards, W. E. Booth, & S. C. Choy (Eds.), *Tropical rainforest research—Current issues* (pp. 457–466). Dordrecht: Kluwer.

Ellen, R. F., & Harris, H. (2000). Introduction. In R. F. Ellen, P. Parkes, & A. Bicker (Eds.), *Indigenous environmental knowledge and its transformations: Critical anthropological perspectives* (pp. 1–33). Amsterdam: Harwood Academic.

Evers, H.-D. (2010). Local and global knowledge: Social science research on Southeast Asia. In T. Menkhoff, H. D. Evers, & C. Y. Wah (Eds.), *Governing and managing knowledge in Asia* (Innovation and knowledge management 2nd ed., Vol. 9, pp. 79–92). Singapore: World Scientific Publishing.

Farnell, B. (1999). Moving bodies, acting selves. *Annual Review of Anthropology, 28*, 341–373.

Fiske, S. T., & Taylor, S. E. (2013). *Social cognition: From brains to culture* (2nd ed.). London: Sage.

Fjellman, S. (1976). Natural and unnatural decision-making: A critique of decision theory. *Ethos, 4*, 73–94.

Flick, U. (1995). *Psychologie des Sozialen: Repräsentationen in Wissen und Sprache* [Psychology of the social: Representations in knowledge and language]. Reinbek: Rowohlt Taschenbuch Verlag.

Fransella, F. (2009). *George Kelly*. Key figures in counselling and psychotherap (W. Dryden, Series Ed.). London: Sage. (Original work published 1995).

Fransella, F., Bell, R., & Bannister, D. (2004). *A manual for repertory grid technique* (2nd ed.). Chichester: Wiley.

Friedberg, C. (1999). Diversity, order, unity: Different levels in folk knowledge about the living. *Social Anthropology, 7*, 1–16.

Fromm, M., & Paschelke, S. (2010). *GridPractice: Anleitung zur Durchführung und Auswertung von Grid-Interviews* [GridPractice: Guide to conducting and analyzing grid interviews]. Norderstedt: Books on Demand.

Geertz, C. J. (1983). *Local knowledge: Further essays on interpretative anthropology.* New York: Basic Books.

Gladwin, C. H. (1989). *Ethnographic decision tree modeling* (Qualitative research methods series, Vol. 19). Newbury Park: Sage.

Greaves, T. (Ed.). (1994). *Intellectual rights for indigenous peoples: A source book.* Oklahoma City: Society for Applied Anthropology.

Harrison, S. (1995). Anthropological perspectives on the management of knowledge. *Anthropology Today, 11,* 10–14.

Hauck, G., & Kößler, R. (2004). Universalität der Vernunft und lokales Wissen—Nicht nur epistemologische Überlegungen [Universality of reason and local knowledge—Not only epistemological considerations]. In N. Schareika, & T. Bierschenk (Eds.), *Lokales Wissen—Sozialwissenschaftliche Perspektiven* (Mainzer Beiträge zur Afrika-Forschung, Vol. 11, pp. 41–58). Münster: Lit Verlag.

Hobart, M. (1993). Introduction: The growth of ignorance. In M. Hobart (Ed.), *An anthropological critique of development: The growth of ignorance* (pp. 1–30). London: Routledge.

Honerla, S., & Schröder, P. (Eds.). (1995). *Lokales Wissen und Entwicklung* [Local knowledge and development]. Saarbrücken: Verlag für Entwicklungspolitik.

Hornidge, A.-K., & Antweiler, C. (Eds.). (2012). *Environmental uncertainty and local knowledge: Southeast Asia as a laboratory of global change* (Global studies). Bielefeld: Transcript.

Hunn, E. (1982). The utilitarian factor in folk biological classification. *American Anthropologist, New Series, 84,* 830–847.

Hutchins, E. (1980). *Culture and inference: A Trobriand case study.* Cambridge, MA: Harvard University Press.

Hutchins, E. (1996). *Cognition in the wild.* Cambridge, MA: MIT Press.

Irwin, A. (1995). *Citizen science: A study of people, expertise and sustainable development.* London: Routledge.

Jankowicz, D. (2004). *The easy guide to repertory grids.* Chichester: Wiley.

Jones, D. (1999). Evolutionary psychology. *Annual Reviews of Anthropology, 28,* 553–575.

Kalland, A. (2003). Environmentalism and images of the other. In H. Selin (Ed.), *Nature across cultures: Views of nature and the environment in non-Western cultures* (Science across cultures, Vol. 4, pp. 1–17). History and practice. Dordrecht: Kluwer.

Karamoy, A., & Dias, G. (Eds.). (1982). *Participatory urban services in Indonesia: People participation and the impact of government social services programmes on the kampung communities: A case study in Jakarta and Ujung Pandang.* Jakarta: Lembaga Penelitian, Pendidikan dan Penerangan Ekonomi dan Sosial (LPIIIES).

Kelly, G. A. (1955). *The psychology of personal constructs* (2 vols.). New York: W. W. Norton.

Knorr-Siedow, T., & Gandelsonas, C. (2004). Lokales Wissen in der Stadtpolitik und Quartiersentwicklung [Local knowledge in urban politics and quarter development]. In U. Matthiesen (Ed.), *Stadtregion und Wissen: Analysen und Plädoyers für eine wissensbasierte Stadtpolitik* (pp. 293–307). Wiesbaden: VS Verlag für Sozialwissenschaften.

Kronenfeld, D. B. (1996). *Plastic glasses and church fathers: Semantic extension from the ethnoscience tradition* (Oxford studies in anthropological linguistics, Vol. 3). New York: Oxford University Press.

Lambek, M. (1993). *Knowledge and practice in Mayotte: Local discourses on Islam, sorcery and spirit possession.* Anthropological horizons. Toronto: University of Toronto Press.

Lave, J. (1993). The practice of learning. In J. Lave & S. Chaiklin (Eds.), *Understanding practice: Perspectives on activity and context* (pp. 3–32). Cambridge, UK: Cambridge University Press.

Lave, J. (1996). The savagery of the domestic mind. In L. Nader (Ed.), *Naked science: Anthropological inquiry into boundaries, power, and knowledge* (pp. 87–100). New York: Routledge.

Lindblom, C. E. (1959). The science of "muddling through". *Public Administration Review, 19,* 79–88.

Lölke, U. (2002). *Zur Lokalität von Wissen: Die Kritik der local knowledge-Debatte in Anthropologie und internationaler Zusammenarbeit* [On locality of knowledge: The critique of the local-knowledge debate in anthropology and international cooperation] (Focus Afrika, IAK-Diskussionsbeiträge, Vol. 21). Hamburg: Institut für Afrika-Kunde.

Lyotard, J. -F. (1984). *The postmodern condition: A report on knowledge* (G. Bennington & B. Massumi, Trans.). Theory and history of literature, Vol. 10. Manchester: Manchester University Press. (Original work published 1979 as La condition postmoderne: rapport sur le savoir).

Malinowski, B. (1948). *Magic, science and religion and other essays*. Garden City: Doubleday Anchor.

Marsden, D. (1994). Indigenous management and the management of indigenous knowledge. In S. Wright (Ed.), *Anthropology of organizations* (pp. 39–53). New York: Routledge.

McCarthy, D. (1996). *Knowledge as culture: The new sociology of knowledge*. London: Routledge.

Menkhoff, T., Evers, H.-D., & Wah, C. Y. (Eds.). (2010). *Governing and managing knowledge in Asia* (Innovation and knowledge management 2nd ed., Vol. 9). Singapore: World Scientific Publishing.

Mersmann, C. (1993). *Umweltwissen und Landnutzung in einem afrikanischen Dorf: Zur Frage des bäuerlichen Engagements in der Gestaltung der Kulturlandschaft der Usambara-Berge Tansanias* [Environmental knowledge and land use in an African village: On the question of peasant engagement in shaping the cultural landscape of the Usambara Mountains of Tanzania] (Hamburger Beiträge zur Afrika-Kunde, Vol. 43). Hamburg: Institut für Afrika-Kunde.

Metzger, D. G., & Williams, G. E. (1966). Some procedures and results in the study of native categories: Tzeltal "firewood". *American Anthropologist, New Series, 68*, 389–407.

Meusburger, P. (2013). Relations between knowledge and economic development: Some methodological considerations. In P. Meusburger, J. Glückler, & M. El Meskioui (Eds.), *Knowledge and the economy* (Knowledge and space, Vol. 5, pp. 15–42). Dordrecht: Springer.

Miehlau, S., & Wickl, F. (2007). *Lokales Wissen und Entwicklung* [Local knowledge and development]. Bad Honnef: Horlemann.

Murdoch, J., & Clark, J. (1994). Sustainable knowledge. *Geoforum, 25*, 115–132.

Nader, L. (Ed.). (1996). *Naked science: Anthropological inquiry into boundaries, power, and knowledge*. New York: Routledge.

Nas, P. J. M. (Ed.). (1995). *Issues in urban development: Case studies from Indonesia*. Leiden: Research School of Asian, African, and Amerindian Studies CNWS.

Nasar, J. L. (1998). *The evaluative image of the city*. Thousand Oaks: Sage.

Nazarea, V. D. (1999). Introduction: A view from a point: Ethnoecology as situated knowledge. In V. D. Nazarea (Ed.), *Ethnoecology: Situated knowledge/located lives* (pp. 3–20). Tuscon: University of Arizona Press.

Nygren, A. (1999). Local knowledge in the environment–development discourse: From dichotomies to situated knowledges. *Critique of Anthropology, 19*, 267–288.

Pasquale, S., Schröder, P., & Schulze, U. (Eds.). (1998). *Lokales Wissen für nachhaltige Entwicklung: Ein Praxisführer* [Local knowledge for sustainable development: A practical guide]. Saarbrücken: Verlag für Entwicklungspolitik.

Pennington, D. C. (2000). *Social cognition*. London: Routledge.

Posey, D. A., & Dutfield, G. (1996). *Beyond intellectual property: Toward traditional resource rights for indigenous peoples and local communities*. Ottawa: International Development Research Centre.

Pottier, J. (1993). Harvesting words? Thoughts on agricultural extension and knowledge ownership, with reference to Rwanda. *Entwicklungsethnologie, 2*, 28–38.

Pottier, J., Bicker, A., & Sillitoe, P. (Eds.). (2003). *Negotiating local knowledge: Power and identity in development* (Series: Anthropology, culture and society). London: Pluto Press.

Prattis, J. I. (1973). Strategizing man. *Man, 8*, 45–58.

Preston, V. A., & Taylor, S. M. (1981). Personal construct theory and residential choice. *Annals of the Association of American Geographers, 71*, 437–451.

Quarles van Ufford, P. (1993). Knowledge and ignorance in the practices of development policy. In M. Hobart (Ed.), *An anthropological critique of development: The growth of ignorance* (pp. 135–160). London: Routledge.

Quinn, N. (Ed.). (2005). *Finding culture in talk: A collection of methods* (Culture, mind, and society series). New York: Palgrave Macmillan.

Quinn, N., & Holland, D. (1987). Culture and cognition. In D. Holland & N. Quinn (Eds.), *Cultural models in language and thought* (pp. 3–40). Cambridge, UK: Cambridge University Press.

Raethel, A. (1993). Auswertungsmethoden für Repertory Grids [Methods of analysis for repertory grids]. In J. W. Scheer & A. Catina (Eds.), *Einführung in die Repertory Grid-Technik: Band 1. Grundlagen und Methoden* (pp. 41–67). Bern: Hans Huber.

Richards, P. (1980). Community environmental knowledge in African rural development. In D. W. Brokensha, D. M. Warren, & O. Werner (Eds.), *Indigenous knowledge systems and development* (pp. 181–194). Lanham: University Press of America.

Rogoff, B., & Lave, J. (Eds.). (1984). *Everyday cognition: Its development in social context.* Cambridge, MA: Harvard University Press.

Saegert, S., & Winkel, G. H. (1990). Environmental psychology. *Annual Review of Psychology, 41,* 441–477.

Salmond, A. (1982). Theoretical landscapes: On cross-cultural conceptions of knowledge. In D. Parkin (Ed.), *Semantic anthropology* (ASA monographs, Vol. 22, pp. 65–87). London: Academic Press.

Sankoff, G. (1971). Quantitative analysis of sharing and variability in a cognitive model. *Ethnology, 10,* 389–408.

Schank, R. C., & Abelson, R. P. (1977). *Scripts, plans, goals, and understanding: An inquiry into human knowledge structures.* Hillsdale: Erlbaum.

Schareika, N. (2004). Lokales Wissen: Ethnologische Perspektiven [Local knowledge: Anthropological perspectives]. In N. Schareika & T. Bierschenk (Eds.), *Lokales Wissen— Sozialwissenschaftliche Perspektiven* (Mainzer Beiträge zur Afrika-Forschung, Vol. 11, pp. 9–39). Münster: Lit Verlag.

Schareika, N., & Bierschenk, T. (Eds.). (2004). *Lokales Wissen—Sozialwissenschaftliche Perspektiven* (Mainzer Beiträge zur Afrika-Forschung, Vol. 11). Münster: Lit Verlag.

Scheer, J. W. (1993). Planung und Durchführung von Repertory Grid-Untersuchungen [Planning and implementation of repertory grid studies]. In J. W. Scheer & A. Catina (Eds.), *Einführung in die Repertory Grid-Technik: Band 1. Grundlagen und Methoden* (pp. 24–40). Bern: Hans Huber.

Scheer, J. W., & Catina, A. (Eds.). (1993). *Einführung in die Repertory Grid-Technik: Band 1. Grundlagen und Methoden* (Introduction to repertory grid technique: Vol. 1. Basics and methods). Bern: Hans Huber.

Schönhuth, M., & Kievelitz, U. (1995). *Participatory learning approaches: Rapid rural appraisal/ participatory appraisal: An introductory guide.* Roßdorf: GTZ.

Scott, J. C. (1998). *Seeing like a state: How certain schemes to improve the human condition have failed* (Yale Agrarian studies series). New Haven: Yale University Press.

Seur, H. (1992). The engagement of researcher and local actors in the construction of case studies and research themes: Exploring methods of restudy. In N. Long & A. Long (Eds.), *Battlefields of knowledge: The interlocking of theory and practice in social research and development* (pp. 115–143). London/New York: Routledge.

Sillitoe, P. (1998a). The development of indigenous knowledge: A new applied anthropology. *Current Anthropology, 39,* 223–252. doi:10.1086/204722.

Sillitoe, P. (1998b). Knowing the land: Soil and land resource evaluation and indigenous knowledge. *Soil Use and Management, 14,* 188–193.

Sillitoe, P. (1998c). What know natives? Local knowledge in development. *Social Anthropology, 6,* 203–220.

Sillitoe, P. (Ed.). (2009). *Local science vs. global science: Approaches to indigenous knowledge in international development* (Studies in environmental anthropology and ethnobiology, Vol. 4). London: Berghahn.

Sillitoe, P. (2012). From participant-observation to participant-collaboration: Some observations on participatory-cum-collaborative approaches. In R. Fardon, O. Harris, T. H. J. Marchand,

M. Nuttall, C. Shore, V. Strang, & R. A. Wilson (Eds.), *The sage handbook of social anthropology:* 2 vols (Vol. 2, pp. 183–200). Los Angeles: Sage.

Sillitoe, P., Dixon, P., & Barr, J. (2005). *Indigenous knowledge inquiries: A methodologies manual for development programmes and projects.* London: Intermediate Technology Publications.

Siverts, H. (1991). Technology and knowledge among the Jivaro of Peru. In R. Gronhaug, G. Haaland, & G. Henriksen (Eds.), *The ecology of choice and symbol: Essays in honour of Fredrik Barth* (pp. 297–311). Bergen: Alma Mater.

Sperber, D. (1982). *Le savoir des anthropologues* [The knowledge of the anthropologists]. Collection Savoir. Paris: Hermann.

Spradley, J. P. (1979). *The ethnographic interview.* New York: Holt, Rinehart & Winston.

Stokols, D., & Altman, I. (Eds.). (1987). *Handbook of environmental psychology* (2 vols.). New York: John Wiley & Sons.

Strauss, A. L. (1987). *Qualitative analysis for social scientists.* Cambridge, UK: Cambridge University Press.

Strauss, C., & Quinn, N. (1997). *A cognitive theory of cultural meaning* (Publications of the Society for Psychological Anthropology, Vol. 9). Cambridge, UK: Cambridge University Press.

Tambiah, S. J. (1990). *Magic, science, religion, the scope of rationality.* Cambridge, UK: Cambridge University Press.

Tanner, C., & Foppa, K. (1996). Umweltwahrnehmung, Umweltbewußtsein und Umweltverhalten [Environmental perception, environmental consciousness, and environmental behavior]. *Kölner Zeitschrift für Soziologie und Sozialpsychologie, 36,* 245–271.

Thrift, N. J. (1996). *Spatial formations.* London: Sage.

Tognoli, J. (1987). Residential environments. In D. Stokols & I. Altman (Eds.), *Handbook of environmental psychology* (Vol. 2, pp. 655–690). New York: Wiley.

van der Ploeg, J. (1989). Knowledge systems, metaphor and interface: The case of potatoes in the Peruvian highlands. In N. Long & A. Long (Eds.), *Battlefields of knowledge: The interlocking of theory and practice in social research and development* (pp. 177–193). London: Routledge.

Vayda, A. P. (1983). Progressive contextualization: Methods for research in human ecology. *Human Ecology, 11,* 265–281.

von Cranach, M. (1995). Über das Wissen sozialer Systeme [On the knowledge of social systems]. In U. Flick (Ed.), *Psychologie des Sozialen: Repräsentationen in Wissen und Sprache* (pp. 22–53). Reinbek: Rowohlt Taschenbuch Verlag.

Warren, D. M., Slikkerveer, L. J., & Brokensha, D. W. (Eds.). (1995). *The cultural dimension of development: Indigenous knowledge systems* (IT studies in indigenous knowledge and development series). London: Intermediate Technology Publications.

Weller, S. C., & Romney, A. K. (1988). *Systematic data collection* (Qualitative research methods, Vol. 10). Newbury Park: Sage.

Werner, O., & Fenton, J. (1970). Method and theory in ethnoscience and ethnoepistemology. In R. Naroll & R. Cohen (Eds.), *Handbook of method in cultural anthropology* (pp. 537–578). New York: Natural History Press.

Werner, O., & Schoepfle, M. (1987). *Systematic fieldwork* (2 vols.). Newbury Park: Sage.

Worsley, P. (1997). *Knowledges: Culture, counterculture, subculture.* London: Profile Books.

Zeleny, M. (2010). Knowledge of enterprise: Knowledge management or knowledge technology? In T. Menkhoff, H. D. Evers, & C. Y. Wah (Eds.), *Governing and managing knowledge in Asia* (Innovation and knowledge management 2nd ed., Vol. 9, pp. 23–57). Singapore: World Scientific Publishing.

Chapter 9
Local Knowledge and Global Concerns: Artificial Glaciers as a Focus of Environmental Knowledge and Development Interventions

Marcus Nüsser and Ravi Baghel

Natural resource management systems in the Himalayas are prominent research topics of integrated human-environmental research. A number of such studies focus on issues of land-use and land-cover change (e.g., Crook and Osmaston 1994; Dame and Mankelow 2010; Nüsser 2000, 2001), often as a result of increasing utilization pressure and socioeconomic change. Other work treats characteristic questions of political ecology, such as power asymmetries, governance, territoriality, and uneven access to natural resources (e.g., Blaikie and Muldavin 2004; Forsyth 1996). A third direction of inquiry deals with various dimensions of local knowledge and institutional arrangements in natural resource management to deepen the understanding of coping and mitigating strategies (e.g., Aase et al. 2013; Aase and Vetaas 2007). Regardless of the specific point of departure, research aims, theoretical foundations, and methodological designs of case studies, it is widely accepted that the livelihoods of most of the Himalayan population are under stress and adversely affected by continuous degradation or depletion of the natural resource base. However, it is obvious that many findings from individual case studies cannot be extrapolated easily on a Himalayan scale (e.g., Ives 2004; Ives and Messerli 1989). Given the negative outcomes posed by projected climate change in the Himalayan region (e.g., Barnett et al. 2005; Nyaupane and Chhetri 2009; Orlove 2009; Raina 2009; Xu et al. 2009), one must assume that the status and availability of natural resources there will become even more problematic in the future.

M. Nüsser (✉)
South Asia Institute, Department of Geography, Heidelberg University,
Im Neuenheimer Feld 330, 69120 Heidelberg, Germany
e-mail: marcus.nuesser@uni-heidelberg.de

R. Baghel
Cluster of Excellence: Asia & Europe, Heidelberg University, Voßstraße 2,
69117 Heidelberg, Germany

Department of Geography, South Asia Institute, Heidelberg University,
Im Neuenheimer Feld 330, 69120 Heidelberg, Germany
e-mail: baghel@uni-heidelberg.de

© Springer International Publishing Switzerland 2016
P. Meusburger et al. (eds.), *Ethnic and Cultural Dimensions of Knowledge*,
Knowledge and Space 8, DOI 10.1007/978-3-319-21900-4_9

The dynamics of transformation (or persistence) of land-use practices and patterns in peripheral high mountain regions strongly depend on the interplay of local environmental knowledge and external development interventions. Beyond ethnic and cultural particularities, local communities in the Himalayan region help secure their livelihoods by pursuing altitudinal diversification across agro-ecological zones and by decreasing the threat that natural hazards pose to any of these environmental resources (e.g., Bisht et al. 2007; Nüsser 1998; Saxena et al. 2005; Semwal et al. 2004). Examples from the Himalayan region include the transformation of pastoral livelihood strategies (e.g., Bergmann et al. 2011, 2012; Kreutzmann 2012; Nüsser 2003, 2006; Nüsser and Clemens 1996), modifications in social organization, and changes in spatial patterns of irrigated mountain farming (e.g., Kreutzmann 1998, 2011; Nüsser et al. 2012).

In this context land-use practices such as irrigated crop cultivation are based on sophisticated forms of local environmental knowledge, which are commonly conceived of by scientists and development practitioners as a customary basis of knowledge and information about natural resources and their dynamics, functions, and services. More specifically, local knowledge refers to the idea that certain types of knowledge are geographically bounded and, even more important, engendered in specific localities that affect both its form and its content (Henderson and Gregory 2009). Although there is a tendency among scholars and development practitioners to romanticize local knowledge systems and position them as antithetical to modern (and predominantly western) scientific knowledge (e.g., Norberg-Hodge 1991; Shiva 1997), we depart from the assumption that local knowledge is inherently heterogeneous, dynamic, and contested (e.g., Agrawal 2005). Nevertheless, we maintain that local knowledge is site specific, that it emerges through people's practical engagement with their environmental setting, including its material resource base (e.g., Ingold and Kurttila 2000). By putting particular emphasis on the place-based character of knowledge generation, we intentionally deviate from the alternative concept of indigenous knowledge (e.g., Sillitoe 2010; Sillitoe and Bicker 2004; see also Sillitoe in this volume) that foregrounds stakeholders' ethnicity and tradition. In localities such as the Trans-Himalayan region of Ladakh in northern India, human-environmental interactions have historically been shaped by a diverse range of ethnic groups, communities, and external actors. For this reason it is of little practical value to search for a truly indigenous knowledge, especially given the fact that Ladakh is a hotspot for national and international development institutions, which contribute to the heterogeneity of *local* knowledge systems.

When we speak of the site-specific, place-based character of local knowledge, the most relevant features of place are nature as experienced in daily human life, that is, a particular set of environmental conditions and human interactions with them, including the process of ascribing meaning to their location. These features contribute to the specificity of local knowledge. However, the dynamism of place further implies that local knowledge cannot be timeless or static. Any such site-specific knowledge is itself part of the ever-changing constellation of human and environmental factors across boundaries of time and space.

Against the background of the Himalayan glacier debate, which is one of the most prominent demonstrations of the concern about global climate change (Le Masson and Nair 2012; Nüsser and Baghel 2014), this contribution explores developments in meltwater-dependent irrigated agriculture in Ladakh since the early 1990s. In particular, we focus on that region's so-called artificial glaciers, structures built to store frozen water in order to cope with seasonal water scarcity.

Most of the analyses of secondary sources and empirical research on the area were carried out in previous and ongoing projects on food security (see Dame 2015; Dame and Nüsser 2008, 2011) and glacier dynamics (see Schmidt and Nüsser 2012). The latest work was conducted by the authors in February 2014 and included field visits to selected artificial glaciers and interviews with stakeholders from local communities and development organizations. We use this case study of Ladakh to identify and analyze the modifications of local environmental knowledge that results from external interventions and environmental politics. At the same time, we aim to illustrate the ways in which global concerns about climate change inform strategies of local resource management. Himalayan glaciers have been cited as illustrations of pristine nature, of dangerous wilderness, and in recent times they have been seen as endangered by human activities and climate change (Nüsser and Baghel 2014). Our study thus offers an excellent setting to analyze the relationship between local knowledge and development interventions that introduce new knowledge and practices in the local environment.

The Evolution of the Socio-hydrological System in Ladakh

We use the term *socio-hydrological system* to conceptualize the complex interactions of specific hydrological conditions and dynamics, local water distribution technologies, socioeconomic development, institutional arrangements, and external development interventions (Nüsser et al. 2012, p. 52). This term aptly reflects the idea that the water cycle is not an inert natural resource contested by humans but rather something with which humans have an inevitably social relation (Robbins 2004; Swyngedouw 2007, 2009). Moreover, the very centrality of water to human existence arguably means that changes in its use and management always co-evolve with social changes; changes in the one are systematically reflected in the other (see Norgaard 1994). Even ideas of scarcity are anything but objectively quantifiable, for they can also be interpreted as social and political constructs (Mehta 2001).

Our focus on the artificial glaciers of Ladakh arises from the fact that irrigation is a human practice that exists at the intersection of social cooperation, livelihood strategies, political organization, and local environmental knowledge (Baghel 2014). The implication is that both human knowledge of glaciers and their role in irrigation and other practices are always set within a social context rather than being detached or objective knowledge. For this reason we explore the creation of artificial glaciers not as an invention but as a practice that arises from broad changes within the dynamics of geophysical processes and the sociopolitical framework.

The Trans-Himalayan region of Ladakh is especially useful as a case study of human interaction with Himalayan glaciers. First, the terrain is interpreted as an

Fig. 9.1 The idealized cultural landscape of Ladakh. This painting, hung in a fancy coffee shop in Leh, presents the region as a model of "sustainability" derived from traditional local knowledge (Photograph by M. Nüsser, August 25, 2013)

idealized cultural landscape[1] (Fig. 9.1), a place of ancient harmony of local people with the natural environment (e.g., Norberg-Hodge 1991). This image of a last Shangri La coincides with modern notions of sustainability and sustainable development (Goeury 2010). Second, Ladakh is a geopolitically sensitive border region (Aggarwal 2004; Dame and Nüsser 2008). Both these associations of an idyllic and sensitive setting are largely determined by Ladakh's peculiar geographical location, which once made it relatively difficult to access and which later gained strategic importance in an area with disputed and heavily militarized international borders.

Ladakh is a high mountain region in the Indian State of Jammu and Kashmir, sandwiched between the Great Himalayan Range to the south and the Karakoram Range to the north. It forms the westernmost part of the Tibetan plateau with an average altitude of over 3,500 m (roughly 11,500 ft). Its western and central parts are dominated by mountain ranges of altitudes well above 5,500 m (about 18,000 ft) (Fig. 9.2).

Because of its location in the rain shadow of the main Himalayan range, Ladakh is characterized by semiarid conditions with an average annual precipitation of less

---

[1] "Cultural landscapes are cultural properties and represent the 'combined works of nature and of man'…They are illustrative of the evolution of human society and settlement over time, under the influence of the physical constraints and/or opportunities presented by their natural environment and of successive social, economic and cultural forces, both external and internal". (UNESCO 2012, p. 14 [paragraph 47 of Chap. IIA, Definition of World Heritage]).

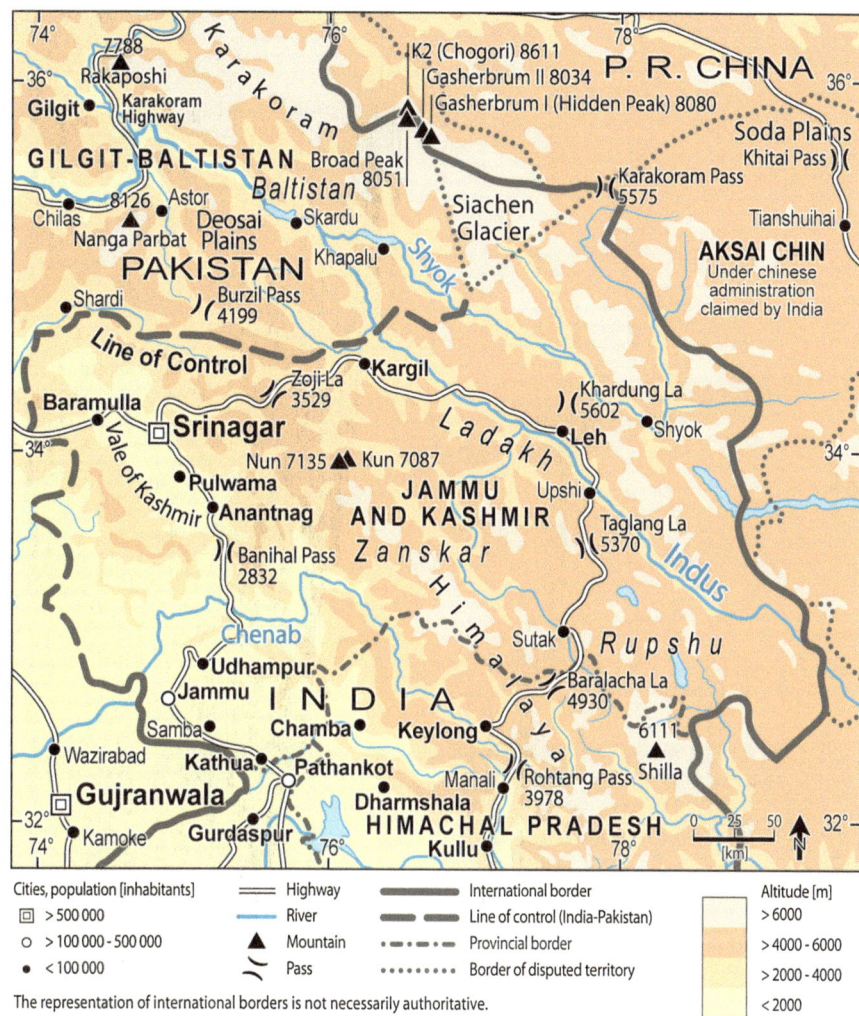

Fig. 9.2 The strategic location of Ladakh in the borderland region of India, Pakistan, and China (Modified after Dame and Nüsser 2011)

than 100 mm (about 4 in.) the upper Indus valley at Leh. Mean monthly tempera-tures values show a high seasonal variation of roughly 30 °C (86 °F) between the coldest and hottest months. These climatic conditions contribute to the small size and the high altitude of glaciers, almost all of which are located above 5,200 m (roughly 17,000 ft) (Schmidt and Nüsser 2012; see Fig. 9.3). Accordingly, scattered settlements and cultivated areas are only found in irrigated oases between 2,600 m and 4,500 m (about 8,500 to some 14,750 ft) situated in side valleys, on alluvial fans, or along the main rivers (Fig. 9.4).

Fig. 9.3 The arid high mountain landscape of central Ladakh. The photograph, taken from the upper slopes of Kang Yatze, shows that only the highest peaks and ridges are covered by glaciers and snowfields (Photograph by M. Nüsser, September 8, 2009)

Fig. 9.4 Irrigated cultivation of barley in central Ladakh. Crop cultivation is based completely on meltwater from glaciers and snowfields. The photograph shows the village of Hankar (3,950 m, or about 13,000 ft) in the Markha valley, with the Kang Yatze peak (6,401 m, or 21,000 ft) in the background (Photograph by M. Nüsser, September 5, 2009)

Agricultural production is based entirely on irrigation with a limited cropping period. Generally, channels divert meltwater from glaciers and snowfields or, where the topography allows, from the main rivers to the settlement and land-use oases. The Ladakh region is sparsely populated, with a total of 274,289 inhabitants (Census of India 2011). However, these figures do not include the number of army personnel and seasonal migrants to the region. Official census data show that the registered population expanded by 1.5 % annually between 2001 and 2011.

The aforegoing description of the region belies its varied and tumultuous geopolitical role over the centuries. Ladakh, especially the town of Leh, was historically a vital stop on the important trade and transit routes from the Indian subcontinent to Central Asia (Rizvi 1999). After the partition of India and the creation of Pakistan in 1947, however, military confrontations over the disputed region of Kashmir changed Ladakh's position from a trade hub to an international borderland. The border routes from China were closed in 1949. After the Sino–Indian war of 1962, the border of Ladakh once again became a site of military hostilities. In 1984 India occupied the ridges surrounding the Siachen glacier, which further raised the military importance of Ladakh as a logistic base for supplying the Indian armed forces (Baghel and Nüsser 2015). In 1999 western parts of the region became the site of a full-blown war between India and Pakistan, focused on Kargil, the second largest town of Ladakh, and its surrounding area. In Northern and Eastern Ladakh lies the Aksai Chin plateau, which is claimed by both India and China. Consequently, confrontations have escalated in recent years, leading to further military presence and infrastructural development.

The value of Ladakh as a strategic logistical base has continually appreciated with a subsequent buildup of infrastructure and a redrawing of transport routes. These developments have altered the economic base of the region, with direct employment in the armed forces and indirect employment in jobs for porters, road construction workers, and other paid activities becoming important sources of nonagricultural income (e.g., Dame and Nüsser 2008, 2011). The socioeconomic shifts over the last decades, together with population growth and migration, have produced new patterns of consumption and changing forms of agriculture, such as the cultivation of vegetables (Dame 2015). These sociopolitical transformations, changing uses of water, and the influx of alternative management strategies form the backdrop to the construction of artificial glaciers and offer interesting insights into the coevolution of the hydrosocial cycle conceptualized according to Swyngedouw (2007, 2009).

Development Interventions and Local Knowledge

The use of local knowledge as a concept has been influential, perhaps because of the multitude of possible interpretations. It therefore figures in all sorts of arguments. For this chapter it is especially relevant that local knowledge has often been valorized as essentially superior to other forms of knowledge, especially by critics with a postdevelopment or alternative development perspective. Shiva (1997), for example, has used the term local knowledge to describe a traditional and harmonious way of knowing nature—practiced particularly by women—that is threatened by western science. As part of a general criticism of postdevelopment, Kiely (1999) attacks this kind of thinking, asserting that it crudely identifies the local as good and the foreign as bad without any examination of the accuracy or applicability of such a characterization. In a similar vein, Nygren (1999) sees a need to cast local knowledge in more nuanced terms and to avoid fixing it in dichotomies such as rational/magical, universal/particular, theoretical/practical, and modern/traditional. These binaries, though well intentioned, are ultimately damaging because they do nothing to change the terms of the debate; they merely cast local knowledge as the threatened underdog in a good versus evil script (see also Agrawal 1995). This weakness lends some truth to Kiely's (1999) uncharitable characterization of such critique as being "the last refuge of the noble savage" (p. 30). A second, more balanced framing of local knowledge has a focus on practical *knowing*, as opposed to more theoretical, codified, and formal agglomerates of *knowledge* (Ibert 2007, p. 103).

Although the word *development* is generally a contentious term with a variety of meanings (for further discussion on the postdevelopment debate, see Blaikie 2000; Cowen and Shenton 1996; Escobar 1995; Peet and Watts 1996; Simon 2006; Watts 2005, among others), it is used in this chapter primarily in the sense of an intentional attempt to improve the conditions of a human population, namely, the conditions governing the livelihoods of the heterogeneous population of Ladakh. Development interventions can involve many agents and practices, including financial aid, technological assistance, and educational interventions. We specifically focus on interventions that involve the introduction of new knowledge into the socio-hydrological system. Our intention is thereby to sidestep many of the most polemic and often deadlocked debates on the ideology or language of development by focusing on concrete precedents that may open a new conversation on more empowering forms of development.

As commonly understood, development interventions often involve an introduction of external knowledge into the local space, whereas local knowledge is taken to be practice based and situated within a long tradition of interactions with the local environment. These formulations, though useful starting points, can lead practice and research into a dead-end if used carelessly. An uncritical use of these terms usually generates an extremely romanticized view of local knowledge and makes developmental interventions seem purely dominant and oppressive. We address the controversial topic of development by seeking to avoid this dichotomy and to offer a more nuanced position that balances critical rigor with a commitment to finding directions for positive engagement with development, especially with reference to local environmental knowledge. For this purpose we turn to the artificial glaciers of Ladakh.

Artificial Glaciers in Ladakh

Although livelihood strategies have diversified, agrarian land use remains the economic mainstay of the local population (Dame and Nüsser 2008). Adequate crop production therefore assumes a central role in the lives of these people, making the

Fig. 9.5 Artificial glacier at 3,900 m (about 12,800 ft), located above the village of Nang, Ladakh. The cascade is composed of a series of loose masonry walls ranging in height from 2 to 3 m (6½ to 10 ft), which help freeze water for storage (Photograph by M. Nüsser, February 20, 2014)

availability of irrigation water indispensable. The chief components of that resource in Ladakh are meltwater from glaciers and seasonal snow cover. However, seasonal melting of glaciers occurs quite late because of the region's high equilibrium line altitude (climatic snow line), which delays the availability of glacial meltwater until June. This seasonal water scarcity makes it essential to provide supplementary irrigation in order to take advantage of the complete growing season. The construction of artificial glaciers on south-facing slopes is regarded not only as a way to cope with water scarcity but also as an adaptation to climate change (Norphel 2009). These structures are built as a cascading series of rock walls in the river beds to reduce runoff velocity and guide meltwater into shadowed areas (Figs. 9.5 and 9.6). The resulting shallow pools begin to freeze as temperatures drop in winter, and ice accumulates. It is mainly an appropriation of the geomorphological phenomenon called icing, which results in the formation of superimposed ice. These artificial glaciers constructed as close to the villages as possible. Being at a lower altitude than the natural glaciers (see Fig. 9.7), artificial glaciers begin melting in April, providing irrigation just in time for the start of the agricultural season (see Fig. 9.8). Chewang Norphel, a well-known engineer of the Leh Nutrition Project (LNP), introduced this innovation of local technology to Ladakh in the 1980s and 1990s (Vince 2009).

Fig. 9.6 Artificial glacier at 4,420 m (14,500 ft), located above the village of Igu, Ladakh. The masonry wall in this construction is completely submerged by ice (Photograph by M. Nüsser, February 25, 2014)

This effort to create artificial glaciers can be related to traditional water harvesting technologies like the *zing*, which are small tanks where meltwater is collected through the use of an extensive and intricate network of channels. As in many other parts of India, Ladakh also has a long history of water harvesting and community management of water resources (Labbal 2000; Vohra 2000), evidenced in the institution of the *Chudpon*—a community official who ensures equitable distribution of water (Angchok et al. 2008). The nearby Spiti valley, also situated in the Trans-Himalayan region, has a local technology similar to the *zing*, called a *kul*, which relies on long rock-lined canals, some of them centuries old, that connect the glacier head to the village (Agarwal and Narain 1997). The mountain oases of the Hindu Kush and Karakoram ranges have similar irrigation networks (e.g., Kreutzmann 2000; Nüsser 2000). In the Gilgit-Baltistan region of northern Pakistan, one finds the local tradition of "birthing" glaciers by carrying pieces of "male" and "female" glaciers and placing them in caves at high altitudes. Glaciers containing rocks and debris are considered male, and glaciers containing blank ice and pure snow cover are considered female. This practice is used in places where water is scarce or where glaciers are in retreat (Douglas 2008; Tveiten 2007).

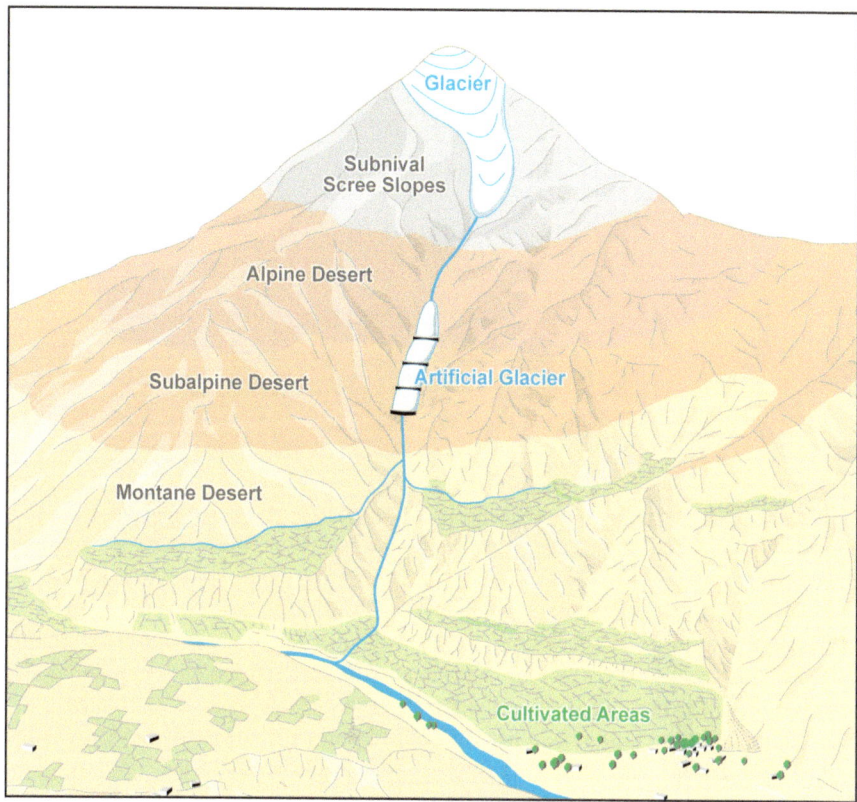

Fig. 9.7 Schematic overview of the position of artificial glaciers. These constructions are located at altitudes between the natural glaciers and the irrigation networks in the cultivated areas (Draft by M. Nüsser)

Discussion

Contemporary knowledge of the environment is reflected in a major survey in the region (Behera and Vaswan 2008), which found widespread perceptions that climate change is having local impact. They included "observations of rise in temperature and heavy snowmelt, less snowfall, heavy and untimely rainfall and biodiversity loss" (p. 1). Even though the scientific evidence on glacier retreat in the area as a response to climate change reveals a complex picture (Schmidt and Nüsser 2012), the local people have formed their perceptions on the basis of observed changes and have placed them in the context of the circulating knowledge of climate change. Accordingly, artificial glaciers appear to be a response to the globally circulating knowledge about climate change, but they are still produced through the dynamics of local knowledge and are not a disconnected invention.

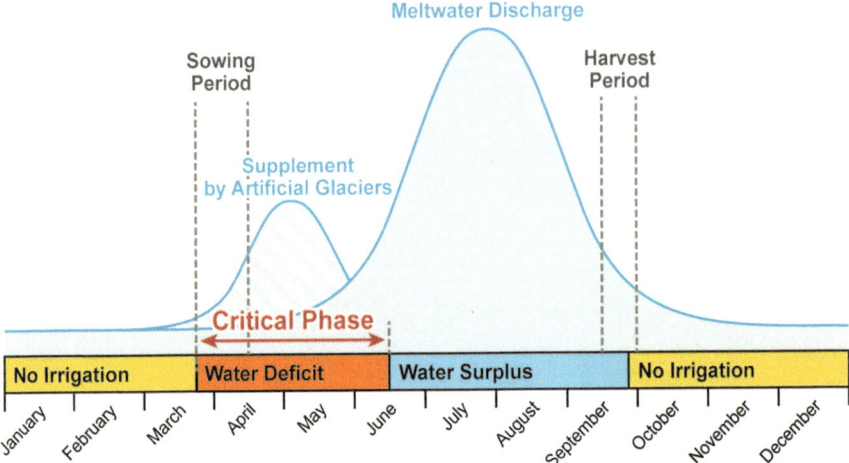

Fig. 9.8 Seasonal variation in the availability of irrigation water. The graph highlights the crucial role of artificial glaciers in bridging the phase of water scarcity in spring (Draft by M. Nüsser, building upon Kreutzmann 1998)

There also appears to be a change in the way knowledge is presented externally. Whereas this technology was long associated with rainwater harvesting and water scarcity, it is now coming across as adaptation to climate change. In this case the local can be understood as a site where many different kinds of circulating knowledge (cultural, scientific, and practical) interact and produce new practices and ideas. Consider, for instance, the widespread activities of nongovernmental organizations (NGOs) in Ladakh. In general, the tradition of regional development has been quite distinct and has figured in active endeavors to counter ideas of modernization or westernization as forms of development (see Fig. 9.1). Examples are the Ladakh Project launched in 1978 and the associated Ladakh Ecological Development Group, created in 1983 (Norberg-Hodge 1991). The members of these NGOs have a long tradition of presenting local practices of interacting with the environment as symbols of the strength and wisdom of an imagined sustainable land use in Ladakh (Page 1995). Ignoring the long history of trade and external interaction in Ladakh, they have operated with the simplistic assumption of an isolated Ladakh suddenly opened up to the outside world. However, these very assumptions have enabled them to increase confidence in the local practices of human-environmental interaction and give it a resilience that it might have otherwise lacked. The striving to build pride in an imagined homogeneous Ladakhi culture has created a close identification of local environmental knowledge with that culture while positioning it as part of the population's ethnic heritage (Fig. 9.1).

The creation of artificial glaciers can thereby be understood as a way of responding to the circulating knowledge of climate change by evolving appropriate strategies to deal with it, by drawing on knowledge perceived to be unique and traditional. This knowledge does not remain on the local scale; there are calls to replicate it not just in the nearby Spiti valley but in places further afield as well (Norphel 2009). Comparable intervention to create artificial glaciers has been reported from Western Tibet and Xinjiang (Field 1961). The aforementioned magical and ritualistic practices in northern Pakistan, sometimes also called "artificial glacier grafting" (Faizi 2007, p. 6), derive from similar intentions of generating ice bodies for the benefit of the local population. This pattern points to the emergent nature of local knowledge in that similar conditions can give rise to differing strategies by which to achieve similar goals in widely separated locations. Artificial glaciers are now also being evaluated in terms of development and their potential to improve living conditions of the local population. This circulation of local knowledge and its influence on development practice suggests that one does well to avoid regarding the exchanges between development and local knowledge as unidirectional or fixed within a static power asymmetry. Local knowledge is inherently heterogeneous and contested.

There is good reason that glaciers have this special place in the understanding of the environment, notably their visibility and their susceptibility to cultural framing (Orlove et al. 2008) and their role as repositories of scientific data on the environment (Carey 2007, 2010). Knowledge about glaciers is a negotiated outcome of various forms of circulation, such as scientific claims, media images, personal memory, cultural memory, witnessing, and exploration. Additionally, the crucial hydrological importance of glaciers to the world's freshwater supply has earned them the label *water towers* (Viviroli and Weingartner 2008) and has made the possibility of their loss especially threatening. Because of the multiple dimensions associated with glaciers and global warming, different ways of knowing about these dynamics coexist. This simultaneity gives *knowing* a different sense: Not only do people arrive at nonexpert or noncodified knowledge through daily interaction, but by knowing, they also build a *common knowledge* through various circulating media. Himalayan glaciers are known not only to Ladakhi farmers because of their relation with their livelihood but also to a London taxi driver who has seen a film on climate change and to a homeless man in New York who has collected a Greenpeace poster with which to decorate his shopping cart. In other words, local knowledge, in the form of artificial glaciers in Ladakh, finds an audience far beyond the scale on which it is produced (e.g., Douglas 2008). The local knowledge is not generated independently of what other actors on other scales and in other regions *know* about Himalayan glaciers. These epistemological interrelationships completely contradict any parochial understandings of what is local and demonstrate that the local tends to be a node rather than an origin of the knowledge produced as the artificial glaciers of Ladakh.

Conclusion

Changes in the local environment have widely differing impacts on local knowledge and on agency, depending on the magnitude of the changes. The view that local knowledge is modified slowly, if at all, can lead one to conclude that situations of rapid environmental or social change portend the inevitable introduction of external knowledge as development interventions (Blaikie et al. 1997, pp. 233–234). Our study of artificial glaciers does not bear out that reasoning. Martens and Rotmans (2005) use the concept of *transitions* to describe the phenomenon of an entire socio-ecological system shifting to a different state as a result of the combined effects of different events and processes. Calling attention to the human ability to adapt to, learn from, and anticipate new situations, they emphasize the role of human agency in moderating the effects of such potentially disruptive changes. In this sense artificial glaciers can be seen as a *transition* in local knowledge that arises in response to potentially disruptive changes represented by water scarcity, seasonal variations, glacier retreat, and climate change.

At the same time we emphasize that local knowledge is not just an analytical construct but a political one as well. The concept of local knowledge is a tactic for both addressing the asymmetries in the credence given to some forms of knowledge and for devaluing others. The focus on the shifts in asymmetries of power or knowledge does much to bring out the nuanced and fluid relations between development interventions and local knowledge, two concepts that have often been regarded as highly incompatible and sometimes even contradictory. The case study presented in this chapter demonstrates that local knowledge and external interventions are sometimes difficult to differentiate and that they tend to be in a dynamic and fluid relation that changes the local setting through local observation, external concerns, and strategies to sustain local livelihoods. This perspective shows the urgent need for innovative approaches to integrate scientific and local knowledge on the status, dynamics, functions, and use of natural resources.

Scientific approaches impart to knowledge a systemic view and an analytical differentiation between factors and parameters of environmental systems. By contrast, local actors have highly integrated, site-specific, and site-differentiating knowledge that reflects the locally relevant multifunctionality of sites and may represent systemic aspects at a more symbolic and phenomenological level than scientific approaches do. That is, combining the two knowledge systems means bringing together highly integrated site-specific knowledge with analytically differentiating knowledge on environmental systems and related single-resource components. It is necessary to avoid romanticizing local knowledge as part of timeless, harmonious tradition and to stress the production of local knowledge through ways of knowing that are intimately tied to the local environment yet placed within a much broader context than that. By acknowledging the dialectic relation between development interventions and local knowledge, we hope to point the way to forms of development that are potentially more empowering than those stemming from a dichotomous or hierarchical link.

Despite the problematic nature of assertions about the superiority of local knowledge, they have helped raise its profile. The emphasis on local knowledge as a central pillar of development is one of the most significant changes to come out of the turn from top-down to bottom-up approaches and has the ostensible objective of integrating local perspectives and perceptions. However, the politicization of local knowledge has made romantic views of local knowledge commonplace.

Their triteness can vary from the romanticization of the local as discussed above to the assumption that local knowledge is an almost unaltering property of the local environment. Taking artificial glaciers as objects of environmental knowledge has enabled us to exemplify the complex interactions and processes through which local knowledge is produced and practiced, especially when livelihood security is at stake.

Acknowledgements We gratefully acknowledge the support of the Cluster of Excellence *Asia and Europe in a Global Context* at Heidelberg University, Germany, for funding our project, *Himalayan Glaciers: Endangered and Dangerous Cryoscapes of Knowledge*. This chapter benefitted from various discussions and collaborations with Juliane Dame and Susanne Schmidt. Nils Harm helped with the cartography and graphics. The chapter also profited from the critical comments of an anonymous reviewer.

References

Aase, T. H., & Vetaas, O. (2007). Risk management by communal decision in Trans-Himalayan farming: Manang valley in central Nepal. *Human Ecology, 35*, 453–460. doi:10.1007/s10745-006-9057-6.

Aase, T. H., Chapagain, P. S., & Tiwari, P. C. (2013). Innovation as an expression of adaptive capacity to change in Himalayan farming. *Mountain Research and Development, 33*, 4–10. doi:10.1659/MRD-JOURNAL-D-12-00025.1.

Agarwal, A., & Narain, S. (1997). *Dying wisdom: Rise, fall and potential of India's traditional water harvesting systems*. New Delhi: Centre for Science and Environment.

Aggarwal, R. (2004). *Beyond lines of control: Performance and politics on the disputed borders of Ladakh, India*. Durham: Duke University Press.

Agrawal, A. (1995). Dismantling the divide between indigenous and scientific knowledge. *Development and Change, 26*, 413–439. doi:10.1111/j.1467-7660.1995.tb00560.x.

Agrawal, A. (2005). *Environmentality: Technologies of government and the making of subjects*. Durham: Duke University Press.

Angchok, D., Stobdan, T., & Singh, S. B. (2008, July). *Community-based irrigation water management in Ladakh: A high altitude cold arid region*. Paper presented at the Twelfth Biennial Conference of the International Association for the Study of Commons, Governing shared resources: Connecting local experience to global challenges, Cheltenham. Retrieved 13 July 2014, from http://dlc.dlib.indiana.edu/dlc /bitstream/handle/10535/1467/Angchok_125001.pdf?sequence=1

Baghel, R. (2014). *River Control in India: Spatial, Governmental and Subjective Dimensions*. Dordrecht/Heidelberg/New York/London: Springer.

Baghel, R., & Nüsser, M. (2015). Securing the heights: The vertical dimension of the Siachen conflict between India and Pakistan in the Eastern Karakoram. *Political Geography, 48*(September), 24–36. doi:10.1016/j.polgeo.2015.05.001.

Barnett, T. P., Adam, J. C., & Lettenmaier, D. P. (2005). Potential impacts of a warming climate on water availability in snow-dominated regions. *Nature, 438*, 303–309. doi:10.1038/nature04141.

Behera, B., & Vaswan, R. T. (2008, July). *Understanding the relationship between global and local commons: A study of household perceptions of climate change in Leh, India*. Paper presented at the Twelfth Biennial Conference of the International Association for the Study of Commons, Governing shared resources: Connecting local experience to global challenges, Cheltenham. Retrieved 13 July 2014, from http://dlc.dlib.indiana.edu/dlc/bitstream/handle/10535/404/Behera_120201.pdf?sequence=1

Bergmann, C., Gerwin, M., Sax, W. S., & Nüsser, M. (2011). Politics of scale in a high mountain border region: Being mobile among the Bhotiyas of the Kumaon Himalaya, India. *Nomadic Peoples, 15*, 104–129. doi:10.3167/np.2011.150207.

Bergmann, C., Gerwin, M., Nüsser, M., & Sax, W. S. (2012). State policy and local performance: Pasture use and pastoral practices in the Kumaon Himalaya. In H. Kreutzmann (Ed.), *Pastoral practices in High Asia: Agency of 'development' effected by modernisation, resettlement and transformation* (pp. 175–194). Dordrecht: Springer. doi:10.1007/978-94-007-3846-1_10.

Bisht, I. S., Mehta, P. S., & Bhandari, D. C. (2007). Traditional crop diversity and its conservation on-farm for sustainable agricultural production in Kumaon Himalaya of Uttaranchal State: A case study. *Genetic Resources and Crop Evolution, 54*, 345–357. doi:10.1007/s10722-005-5562-5.

Blaikie, P. (2000). Development, post-, anti-, and populist: A critical review. *Environment and Planning A, 32*, 1033–1050. doi:10.1068/a3251.

Blaikie, P. M., & Muldavin, J. S. S. (2004). Upstream, downstream, China, India: The politics of environment in the Himalayan region. *Annals of the Association of American Geographers, 94*, 520–548. doi:10.1111/j.1467-8306.2004.00412.x.

Blaikie, P., Brown, K., Stocking, M., Tang, L., Dixon, P., & Sillitoe, P. (1997). Knowledge in action: Local knowledge as a development resource and barriers to its incorporation in natural resource research and development. *Agricultural Systems, 55*, 217–237. doi:10.1016/S0308-521X(97)00008-5.

Carey, M. (2007). The history of ice: How glaciers became an endangered species. *Environmental History, 12*, 497–527. doi:10.1093/envhis/12.3.497.

Carey, M. (2010). *In the shadow of melting glaciers: Climate change and Andean society*. Oxford: Oxford University Press. doi:10.1093/acprof:oso/9780195396065.001.0001.

Census of India. (2011). Government of India. Ministry of Home Affairs. Retrieved 13 April 2014, from http://www.censusindia.gov.in/pca/default.aspx

Cowen, M. P., & Shenton, R. W. (1996). *Doctrines of development*. London: Routledge.

Crook, J., & Osmaston, H. (1994). *Himalayan Buddhist villages: Environment, resources, society and religious life in Zangskar, Ladakh*. Delhi: Motilal Banarsidass.

Dame, J. (2015). *Ernährungssicherung im Hochgebirge: Akteure und ihr Handeln im Kontext des sozioökonomischen Wandels in Ladakh, Indien* [Food security in a high mountain region: Actors and their strategies in the context of socio-economic change in Ladakh, India] (Erdkundliches Wissen, Vol. 156). Stuttgart: Franz Steiner Verlag.

Dame, J., & Mankelow, J. S. (2010). Stongde revisited: Land use change in central Zangskar. *Erdkunde, 64*, 355–370. doi:10.3112/erdkunde.2010.04.05.

Dame, J., & Nüsser, M. (2008). Development perspectives in Ladakh, India. *Geographische Rundschau: International Edition, 4*(4), 20–27 and supplement.

Dame, J., & Nüsser, M. (2011). Food security in high mountain regions: Agricultural production and the impact of food subsidies in Ladakh, northern India. *Food Security, 3*, 179–194. doi:10.1007/s12571-011-0127-2.

Douglas, E. (2008). How to grow a glacier. *New Scientist, 197*(2641), 37–39. doi:10.1016/S0262-4079(08)60298-5.

Escobar, A. (1995). *Encountering development: The making and unmaking of the third world*. Princeton: Princeton University Press.

Faizi, I. (2007). Artificial glacier grafting: Indigenous knowledge of the mountain people of Chitral. *Asia Pacific Mountain Network Bulletin, 8*(1), 6–7. Retrieved 10 July 2014, from http://www.mtnforum.org/sites/default/files/publication/files/icimod-apmn_bulletins2007.pdf

Field, A. R. (1961). Strategic development in Sinkiang. *Foreign Affairs, 39*, 312–318. Retrieved 24 January 2014, from http://www.foreignaffairs.com/articles/71613/a-r-field/strategic-development-in-sinkiang

Forsyth, T. (1996). Science, myth and knowledge: Testing Himalayan environmental degradation in Thailand. *Geoforum, 27*, 375–392. doi:10.1016/S0016-7185(96)00020-6.

Goeury, D. (2010). Ladakh, kingdom of sustainable development? Protecting the natural environment to protect identity. *Revue de Géographie Alpine/Journal of Alpine Research, 98*, 109–121. doi:10.4000/rga.1147.

Henderson, G., & Gregory, D. (2009). Local knowledge. In D. Gregory, R. Johnston, G. Pratt, M. J. Watts, & S. Whatmore (Eds.), *The dictionary of human geography* (5th ed., pp. 422–423). Malden: Blackwell.

Ibert, O. (2007). Towards a geography of knowledge creation: The ambivalences between 'knowledge as an object' and 'knowing in practice'. *Regional Studies, 41*, 103–114. doi:10.1080/00343400601120346.

Ingold, T., & Kurttila, T. (2000). Perceiving the environment in Finnish Lapland. *Body & Society, 6*, 183–196. doi:10.1177/1357034X00006003010.

Ives, J. D. (2004). *Himalayan perceptions: Environmental change and the well-being of mountain peoples*. London: Routledge.

Ives, J. D., & Messerli, B. (1989). *The Himalayan dilemma: Reconciling development and conservation*. London: Routledge.

Kiely, R. (1999). The last refuge of the noble savage? A critical assessment of post-development theory. *European Journal of Development Research, 11*, 30–55. doi:10.1080/09578819908426726.

Kreutzmann, H. (1998). From water towers of mankind to livelihood strategies of mountain dwellers: Approaches and perspectives for high mountain research. *Erdkunde, 52*, 185–200. doi:10.3112/erdkunde.1998.03.01.

Kreutzmann, H. (2000). Water management in mountain oases of the Karakoram. In H. Kreutzmann (Ed.), *Sharing water: Irrigation and water management in the Hindukush—Karakoram—Himalaya* (pp. 90–115). Karachi: Oxford University Press.

Kreutzmann, H. (2011). Scarcity within opulence: Water management in the Karakoram mountains revisited. *Journal of Mountain Science, 8*, 525–534. doi:10.1007/s11629-011-2213-5.

Kreutzmann, H. (2012). Pastoral practices in transition: Animal husbandry in High Asian contexts. In H. Kreutzmann (Ed.), *Pastoral practices in High Asia: Agency of 'development' effected by modernisation, resettlement and transformation* (pp. 2–29). Dordrecht: Springer. doi:10.1007/978-94-007-3846-1_1.

Labbal, V. (2000). Traditional oases of Ladakh: A case study of equity in water management. In H. Kreutzmann (Ed.), *Sharing water: Irrigation and water management in the Hindukush—Karakoram—Himalaya* (pp. 163–183). Karachi: Oxford University Press.

Le Masson, V., & Nair, K. (2012). Does climate modeling help when studying adaptation to environmental changes? The case of Ladakh, India. In A. Lamadrid & I. Kelman (Eds.), *Climate change modeling for local adaptation in the Hindu Kush-Himalayan region* (Community, environment and disaster risk management, Vol. 11, pp. 75–94). Bingley: Emerald Group. doi:10.1108/S2040-7262(2012)0000011011.

Martens, P., & Rotmans, J. (2005). Transitions in a globalising world. *Futures, 37*, 1133–1144. doi:10.1016/j.futures.2005.02.010.

Mehta, L. (2001). The manufacture of popular perceptions of scarcity: Dams and water-related narratives in Gujarat, India. *World Development, 29*, 2025–2041. doi:10.1016/S0305-750X(01)00087-0.

Norberg-Hodge, H. (1991). *Ancient futures: Learning from Ladakh*. New Delhi: Oxford University Press.

Norgaard, R. B. (1994). *Development betrayed: The end of progress and a coevolutionary revisioning of the future*. London: Routledge.

Norphel, C. (2009). Artificial glaciers: A high altitude cold desert water conservation technique. In E. Pedersen & D. Campana (Eds.), *Energy and climate change in cold regions of Asia: Proceedings of the Seminar, Ladakh, India, April 21–24, 2009* (pp. 62–64). Leh: Geres. Retrieved 24 Jan 2014, from http://www.geres.eu/images/publications/seminar_proceedings.pdf

Nüsser, M. (1998). *Nanga Parbat (NW-Himalaya): Naturräumliche Ressourcenausstattung und humanökologische Gefügemuster der Landnutzung* [Nanga Parbat (NW-Himalaya): Natural resources and human-ecological pattern of land use] (Bonner Geographische Abhandlungen, Vol. 97). Bonn: Dümmler.

Nüsser, M. (2000). Change and persistence: Contemporary landscape transformation in the Nanga Parbat area, northern Pakistan. *Mountain Research and Development, 20*, 348–355. Retrieved 24 Jan 2014, from http://www.jstor.org/stable/3674056.

Nüsser, M. (2001). Understanding cultural landscape transformation: A re-photographic survey in Chitral, eastern Hindukush, Pakistan. *Landscape and Urban Planning, 57*, 241–255. doi:10.1016/S0169-2046(01)00207-9.

Nüsser, M. (2003). Ressourcennutzung und Umweltveränderung: Mensch-Umwelt-Beziehungen in peripheren Gebirgsräumen [Resource management and environmental change: Human-envirnoment relationships in peripheral mountain areas]. In P. Meusburger & T. Schwan (Eds.), *Humanökologie: Ansätze zur Überwindung der Natur-Kultur-Dichotomie* (pp. 327–342) (Erdkundliches Wissen, Vol. 135). Stuttgart: Franz Steiner.

Nüsser, M. (2006). Ressourcennutzung und nachhaltige Entwicklung im Kumaon-Himalaya (Indien) [Resource management and sustainable development in the Kumaon Himalaya, India]. *Geographische Rundschau, 58*(10), 14–22.

Nüsser, M., & Baghel, R. (2014). The emergence of the cryoscape: Contested narratives of Himalayan glacier dynamics and climate change. In B. Schuler (Ed.), *Environmental and climate change in South and Southeast Asia: How are local cultures coping?* (pp. 138–156). Leiden: Brill.

Nüsser, M., & Clemens, J. (1996). Impacts on mixed mountain agriculture in the Rupal Valley, Nanga Parbat, northern Pakistan. *Mountain Research and Development, 16*, 117–133. Retrieved 24 Jan 2014, from http://www.jstor.org/stable/3674006.

Nüsser, M., Schmidt, S., & Dame, J. (2012). Irrigation and development in the upper Indus Basin: Characteristics and recent changes of a socio-hydrological system in central Ladakh, India. *Mountain Research and Development, 32*, 51–61. doi:10.1659/MRD-JOURNAL-D-11-00091.1.

Nyaupane, G. P., & Chhetri, N. (2009). Vulnerability to climate change of nature-based tourism in the Nepalese Himalayas. *Tourism Geographies, 11*, 95–119. doi:10.1080/14616680802643359.

Nygren, A. (1999). Local knowledge in the environment–development discourse: From dichotomies to situated knowledges. *Critique of Anthropology, 19*, 267–288. doi:10.1177/03082 75X9901900304.

Orlove, B. (2009). Glacier retreat: Reviewing the limits of human adaptation to climate change. *Environment: Science and Policy for Sustainable Development, 51*(3), 22–34. doi:10.3200/ENVT.51.3.22-34.

Orlove, B., Wiegandt, E., & Luckman, B. H. (2008). The place of glaciers in natural and cultural landscapes. In B. Orlove, E. Wiegandt, & B. H. Luckman (Eds.), *Darkening peaks: Glacier retreat, science, and society* (pp. 3–19). Berkeley: University of California Press.

Page, J. (1995). The Ladakh project: Active steps towards a sustainable future. In H. Norberg-Hodge, P. Goering, & S. Gorelick (Eds.), *The future of progress: Reflections on environment and development* (pp. 208–225). Dartington: Green Books.

Peet, R., & Watts, M. (1996). Liberation ecology: Development, sustainability, and environment in an age of market triumphalism. In R. Peet & M. Watts (Eds.), *Liberation ecologies: Environment, development, social movements* (pp. 1–45). London: Routledge.

Raina, V. K. (2009). *Himalayan glaciers: A state-of-art review of glacial studies, glacial retreat and climate change.* New Delhi: Ministry of Environment and Forests, Government of India. Retrieved 24 Jan 2014, from http://www.moef.nic.in/downloads/public-information/MoEF%20 Discussion%20Paper%20_him.pdf

Rizvi, J. (1999). *Trans-Himalayan caravans: Merchant princes and peasant traders in Ladakh.* New Delhi: Oxford University Press.

Robbins, P. (2004). *Political ecology: A critical introduction.* Malden: Blackwell.

Saxena, K. G., Maikhuri, R. K., & Rao, K. S. (2005). Changes in agricultural biodiversity: Implications for sustainable livelihood in the Himalaya. *Journal of Mountain Science, 2*, 23–31. doi:10.1007/s11629-005-0023-3.

Schmidt, S., & Nüsser, M. (2012). Changes of high altitude glaciers from 1969 to 2010 in the Trans-Himalayan Kang Yatze Massif, Ladakh, Northwest India. *Arctic, Antarctic, and Alpine Research, 44*, 107–121. doi:10.1657/1938-4246-44.1.107.

Semwal, R. L., Nautiyal, S., Sen, K. K., Rana, U., Maikhuri, R. K., Rao, K. S., & Saxena, K. G. (2004). Patterns and ecological implications of agricultural land-use changes: A case study from central Himalaya, India. *Agriculture Ecosystems & Environment, 102*, 81–92. doi:10.1016/S0167-8809(03)00228-7.

Shiva, V. (1997). Western science and its destruction of local knowledge. In M. Rahnema & V. Bawtree (Eds.), *The post-development reader* (pp. 161–167). London: Zed Books.

Sillitoe, P. (2010). Trust in development: Some implications of knowing in indigenous knowledge. *Journal of the Royal Anthropological Institute, 16*, 12–30. doi:10.1111/j.1467-9655.2009.01594.x.

Sillitoe, P., & Bicker, A. (2004). Introduction: Hunting for theory, gathering ideology. In A. Bicker, P. Sillitoe, & J. Pottier (Eds.), *Development and local knowledge: New approaches to issues in natural resources management, conservation and agriculture* (pp. 1–18). London: Routledge.

Simon, D. (2006). Separated by common ground? Bringing (post) development and (post) colonialism together. *The Geographical Journal, 172*, 10–21. doi:10.1111/j.1475-4959.2006.00179.x.

Swyngedouw, E. (2007). Technonatural revolutions: The scalar politics of Franco's hydro-social dream for Spain, 1939–1975. *Transactions of the Institute of British Geographers, 32*, 9–28. doi:10.1111/j.1475-5661.2007.00233.x.

Swyngedouw, E. (2009). The political economy and political ecology of the hydro-social cycle. *Journal of Contemporary Water Research & Education, 142*, 56–60. doi:10.1111/j.1936-704X.2009.00054.x.

Tveiten, I. N. (2007). *Glacier growing: A local response to water scarcity in Baltistan and Gilgit, Pakistan*. Master's thesis, Norwegian University of Life Sciences, Department of International Environment and Development Studies. Ås, Norway. Retrieved 24 Jan 2014, from http://www.umb.no/statisk/noragric/publications/master/2007_ingvar_tveiten.pdf

UNESCO. (2012). *Operational guidelines for the implementation of the world heritage convention*. Paris: UNESCO World Heritage Centre. Retrieved 2 July 2014, from http://whc.unesco.org/archive/opguide12-en.pdf

Vince, G. (2009). Glacier man. *Science, 326*, 659–661. doi:10.1126/science.326_659.

Viviroli, D., & Weingartner, R. (2008). "Water towers": A global view of the hydrological importance of mountains. In E. Wiegandt (Ed.), *Mountains: Sources of water, sources of knowledge* (pp. 15–20). Dordrecht: Springer.

Vohra, R. (2000). Notes on irrigation and the legal system in Ladakh: The Buddhist 'Brog-Pa'. In H. Kreutzmann (Ed.), *Sharing water: Irrigation and water management in the Hindukush—Karakoram—Himalaya* (pp. 146–162). Karachi: Oxford University Press.

Watts, M. (2005). Postmodernism and postdevelopment. In T. Forsyth (Ed.), *Encyclopedia of International Development* (pp. 546–550). London: Routledge.

Xu, J., Grumbine, R. E., Shrestha, A., Eriksson, M., Yang, X., Wang, Y., & Wilkes, A. (2009). The melting Himalayas: Cascading effects of climate change on water, biodiversity, and livelihoods. *Conservation Biology, 23*, 520–530. doi:10.1111/j.1523-1739.2009.01237.x.

Chapter 10
Political Economy, Power, and the Erasure of Pastoralist Indigenous Knowledge in the Maghreb and Afghanistan

Diana K. Davis

This chapter argues that the sophisticated indigenous veterinary knowledge of nomads in the Maghreb, along with their related knowledge of herding and range management, has been and continues to be eroded, often to the detriment of the well-being of both the nomads and the environment. I show that changes made to the regional political economy during the French colonial period, the imposition of new laws, and the further exercise of colonial power in the form of environmental knowledge production have all had a negative impact on nomadic pastoralist knowledge systems and practice. I additionally suggest that the erosion of pastoralist indigenous knowledge persisted under the postcolonial government via retention of many colonial laws and policies as well as mainstream development projects that privilege "expert" knowledge. At the other end of the Middle East, in Afghanistan, I suggest that a similar eradication of indigenous veterinary knowledge, especially the knowledge of women, is taking place among the Koochi nomads. The combination of biased views of Muslim women by western development "experts" in conjunction with the extremely conservative religious government of the country is curtailing women's work with livestock in ways that endanger their knowledge, practice, and well-being. I employ the term *indigenous knowledge* to refer to knowledge that the nomads and other pastoralists have garnered, often over centuries, from working with animals in an arid, stochastic environment. Indigenous knowledge is very dynamic and the phrase is a contested term, but in this chapter it means knowledge and practice executed at the local level as opposed to a variety of expert knowledges such as those of international agencies and national agricultural extension agents. It is used interchangeably with *local knowledge*.[1] The intent is not to

---

[1] For an excellent discussion of the debates over "indigenous knowledge" and the slightly less problematic term "local knowledge," see the chapter 8 in this volume by Christoph Antweiler. He

D.K. Davis (✉)
Department of History, The University of California,
1 Shields Avenue, Davis, CA 95616, USA
e-mail: geovet@ucdavis.edu

© Springer International Publishing Switzerland 2016
P. Meusburger et al. (eds.), *Ethnic and Cultural Dimensions of Knowledge*,
Knowledge and Space 8, DOI 10.1007/978-3-319-21900-4_10

211

highlight the differences in indigenous/local knowledge and western/expert knowledge, for this kind of approach has facilitated the marginalization and oppression of indigenous peoples (Agrawal 1995, 2005; Davis 2005a). I am more interested in exploring the ways that local knowledge becomes suppressed and expert knowledge becomes privileged and how that process facilitates certain power relations with far-reaching consequences. The power to label knowledge as expert or indigenous/local is very meaningful, especially in development contexts, and this chapter highlights the importance of knowledge construction and valuation in assessing and implementing environmental and developmental interventions.

The Precolonial Maghreb and Pastoralist Indigenous Knowledge

When the French invaded Algeria in 1830, only 5 % of the local population was urban and 60–65 % of the population was nomadic (Boukhobza 1992).[2] The rest of the rural population was settled and agricultural, with most of the inhabitants raising livestock of various kinds, including small ruminants, cattle, camels, and poultry. Pastoralism, the raising of animals of different kinds both intensively and extensively, was widely practiced by a large majority of the population (Bedrani 1991).[3] Fire was frequently used to prepare agricultural land (clearing weeds and pests and providing a fertilizing ash) and to regenerate pastures for livestock-grazing, even in forested areas (Nouschi 1959, 1961). Given the Mediterranean ecology of the region, which is highly adapted to fire and grazing, this land use was sustainable at the contemporary population level of about three million people.[4] Indigenous knowledge of farming was quite sophisticated and included not only the appropriate and successful use of fire and grazing management but also legume rotation, irrigation, and soil conservation methods (Pfeifer 1986).

In the more arid areas of the High Plateaus and the desert south, extensive pasture lands were used by nomadic and seminomadic groups to raise livestock, including sheep, goats, and camels (see Fig. 10.1). This arid to hyperarid region was by far the largest part of the Algerian territory, and the nomads and seminomads living there accounted for about half of the Algerian population. Rainfall in this region was erratic and sparse, and the primary livelihood tool was the timely movement of livestock and people in response to rainfall patterns. During the dry summers,

---

also highlights the fact that all knowledge is "situated" knowledge and must be considered within multiple contexts (e.g., historical, social, economic, and political). See also Davis (2005a) for a discussion of indigenous knowledge and expert knowledge in contemporary Morocco.

[2] The first part of this chapter concentrates on Algeria because it was colonized much earlier by the French than either Tunisia or Morocco and because it was in Algeria that the foundation was laid for so much of what was later transferred to these two French protectorates.

[3] Pastoralism ranges from the fully sedentary pastoralism (often called agropastoralism) that is practiced with field agriculture to fully nomadic, long-range pastoralism. There exist many different variants of pastoral nomadism (or nomadic pastoralism) that include various kinds of seminomadism or transhumance (moving from low- to high-altitude pastures).

[4] For a discussion of the region's ecology, see Davis (2007, pp. 177–186).

Fig. 10.1 Camel herd of a tribe of Touareg of the Hoggar, Algeria (From *Les Productions Algériennes* (p. 75), by Jean Blottière, 1930, Paris: Publications du Comité national métropolitain du centenaire de l'Algérie. Copyright holder unknown)

animals were moved to more northerly pastures or to areas of higher elevation in the High Plateaus. During the winter months, when the weather cooled and became wetter, livestock were moved back to the desert or lowland areas.

The success of the Algerians' knowledge and skills at managing this difficult environment is evident in the large numbers of livestock estimated at the time of occupation. In 1830 sheep alone were calculated at approximately 8 million head, a population nearly three times larger than that of all humans in the region, with goats almost certainly being even more numerous (Julien 1964, p. 8). The bountiful live-stock production was matched by similarly productive agricultural harvests. Before French occupation, Algeria was fully self-sufficient in agricultural products, and the surpluses were exported, especially grain. Early in the colonial period, however, local knowledge began to be constrained by a series of related laws, policies, and changes to the political economy that had a disastrous effect on the Algerians' way of life and eventually on their body of local knowledge. What amounted to an erasure of indigenous knowledge was perhaps most significant among pastoral populations.

French Colonialism and the Erasure of Pastoralist Indigenous Knowledge

French antipathy for pastoralists, especially for nomads and their way of life, was becoming apparent in Algeria as early as 1834. In an official report to the Council of Ministers in that year, the leader of a cavalry troop in Algeria wrote, "during all the time that you [nomads] have occupied the soil, it has been destroyed because you are nomads. [Therefore,] we can take your lands if we wish" (Nouschi 1961, p. 161).[5] The officer, M. Poinçot, concluded that it would be easy to obtain land needed by the French for colonization in this way at low cost. In 1838, the year the Algerian Forest Service was created, two laws were passed that significantly restricted the traditional ways of life of most Algerians. The first forbade "the burning, for any cause, of woods, copse, brush, hedges…[, and] standing grasses and plants" and imposed a heavy penalty on violators (Marc 1916, p. 208). The second law specified a penalty of 50–100 francs for any burning of brush for the purpose of fertilizing the soil (Ageron 1968, p. 107). By 1846 the fine had been modified to include a period in jail ranging from 6 days to 6 months or more (p. 107). This law effectively criminalized the common technique of using fire to improve agricultural and grazing lands, which was sustainable at the low population levels of the period.

Nomads were restricted in other ways as well early in the colonial period. In 1845 a village had been built by the Arab Bureau as a model of how to sedentarize the nomads in order to control them and their allegedly destructive way of life.[6] In 1847 a new law introduced *cantonnement*, or delimitation, of the nomadic populations. Through two new land laws, those of 1844 and 1846, the policy of *cantonnement* allowed land to be taken from nomadic groups on the premise that the nomads controlled more land than they could possibly need.[7] They were left only what little land the French government deemed adequate for their needs. Other land laws were passed with the Senatus Consultus in 1863 and the Warnier laws of 1873 and 1887. The end result of these laws was the destruction of indigenous communal land tenure, the institution of private property rights (along with a booming property market), and the loss of more nomad land.

Changes enacted in the name of forest protection had an especially deleterious effect on rural Algerians' way of life. In the year of occupation, 1830, France claimed all the forests of Algeria as state property, later recognizing small areas of "communal" forests for the local tribes. These changes were codified into law in 1851 (Ageron 1968, p. 107). As a result, Algeria's forests were automatically placed under the French Forest Code of 1827, which checked many traditional activities such as the gathering of food and medicines and the Algerians' use of fire and grazing in forested areas. Infractions continued, and forest fires, soon as much a form of protest as a livelihood tool, caused the French nearly insurmountable problems despite fines and individual imprisonments. By 1856 the colonizers had considered the draconian measure of punishing an entire tribe for any given fire (Boudy 1948, pp. 233, 235). Modifications of the forest code as applied to Algeria in 1874 did institute collective punishment, further restricted the use of fire, and stiffened other punishments for infractions. In 1885 the code added land expropriation for

---

[5] Translations from French sources cited in this chapter are my own.

[6] Center for Overseas Archives, Aix-en-Provence, France: CAOM, ALG, GGA, 8H/10, Letter from Capitaine Lapasset of the Arab Bureaux to the duc d'Aumale, Governor General of Algeria, 12 January 1848.

[7] For details on *cantonnement*, see Lacoste et al. (1960, pp. 363–364).

reforestation nearly everywhere in the colony. These changes culminated in the new 1903 Algerian Forest Code, which was more punitive than the French Forest Code with regard to fires, retained collective punishment of entire tribes, and had a strong focus on reforestation.

The reach of the Forest Service was very broad. Plants well outside anything that could be thought of as forest were placed under its jurisdiction and subjected to the forest code. One of the most important examples was the hardy perennial grass known as Alfa grass (*Stipa tenacissima*), widely used by pastoralists for grazing in hard times (see Fig. 10.2). Classified as a forest product, it could not be burned (often done to foster the new, tender growth palatable to livestock), and grazing was drastically reduced (Trabut 1889). Even sand dunes, important water sources for nomads, were claimed by the state or by private French citizens in the name of reforestation. By the last quarter of the nineteenth century, almost every traditional use the Algerians had made of the forest and other parts of their environment had been severely limited or criminalized. Thus, it was nearly impossible to collect firewood, use fire for agricultural or grazing management, collect food or medicinal products, graze livestock, or live in or near an area classified as forest. Even bare soil could be classified as forest if it was claimed that forest could potentially grow in that area.

Policies outside the environmental realm of forests and land also had adverse impacts on pastoralists' way of life during the colonial period. Taxes, for example,

Fig. 10.2 A sea of Alfa grass in the south of Oran, Algeria (From *Notes sur les forêts de l'Algérie, Collection du centenaire de l'Algérie* (p. 593), by H. Marc, 1930, Paris: Librairie Larose. Copyright holder unknown)

first instituted as the *impôts Arabes*, were declared payable only in cash by 1845. These taxes enmeshed the Algerians in a money economy, forcing many to sell belongings, land, and livestock in order to pay. It forced others into wage labor in the cities or in the growing capitalist agricultural ventures, both of which took pastoralists away from tending animals at home. By 1914 32 % of rural Algerians were working as sharecroppers, and wage labor "had become one of the most characteristic features of colonial agricultural production" (Ruedy 1992, p. 98).

By the end of the colonial period, as a result of colonial laws and policies, most of the best agricultural and grazing lands in Algeria were in the hands of the French and other European settlers. The settler population, only 3.5 % of the agricultural population, owned nearly 40 % of the prime agricultural land, including 75 % of the irrigated crop land (Saint-Germes 1955a, p. 158, 1955b, p. 5). The Algerians, comprising 90 % of the total population, were left with only 62 % of the arable land and about 12 % of the forested land (Boudy 1955, p. 369; Saint-Germes 1955a, p. 158). Pastoralists had lost access to nearly 3 million hectares (11,583 square miles) of forest pasture, approximately 26,000 hectares (100.4 square miles) of forest and nonforest land locked up in national parks, and 777,000 hectares (3,000 square miles) of alfa pastures categorized as forest (Saint-Germes 1955a, p. 68).

Equally important, mobility—which in this context is the ability to move people and livestock over long distances and which is essential for extensive, nomadic pastoralism—had been drastically curtailed by many policies, especially *cantonnement*, as well as by public health regulations. These policies forced many pastoralists to settle in villages and/or sell their herds, devastating both nomadic pastoralism and settled agropastoralism and in many areas ending them completely (Ageron 1968; Boukhobza 1982; Lacoste et al. 1960; Trautmann 1989). Livestock owned by Algerians had declined dramatically. Sheep, for instance, decreased from about 10.5 million in 1887 to only 3.8 million in 1955 (Sari 1978, p. 106). Nomads, once accounting for about 60 % of the Algerian population, constituted only 5 % by independence (Boukhobza 1992, p. 26). An unknown number of seminomads and agropastoralists no longer raised animals. As two French scholars have put it, the pastoralist way of life "was degraded into a miserable sedentarization" (Poncet and Raymond 1971, p. 42). Closely related policies and changes were enacted in Tunisia and Morocco with results that were all too similar.

Historians and others have documented the long and sophisticated history of Arab medicine in the Middle East and North Africa, including the treatment of animals. Although colonial archival documentation is slim, traditional veterinary medicine was practiced in the Maghreb when the French occupied Algeria. North African pastoralists continued to do so with a good degree of success, judging by the size of the herds, until colonial laws and policies began to wreak havoc on traditional animal production in the ways described above. With this precipitous drop in the number of people raising animals went the knowledge of how to raise livestock, how to manage the arid and semiarid environment, and how to maintain the animals' health and treat their diseases. The loss of indigenous veterinary knowledge was intensified with a series of laws that circumscribed and criminalized the practice of traditional veterinary medicine.

One of the earliest laws pertaining to veterinary medicine was that which created the sanitary animal police in 1887 (Geoffroy Saint-Hilaire 1919, p. 431). These new veterinary police were responsible for implementing a series of French government decrees regulating infectious animal diseases. The 1879 decree by the Ministry of War, for example, was intended to prevent the spread of dourine, the trypanosomiasis of horses that was the bane of the French military. Similar legislation establishing veterinary police was applied to Tunisia in 1885 and to Morocco in 1914 (Geoffroy Saint-Hilaire 1919). Although these measures may seem benign at first, their complicated execution and their complex attendant rules and regulations had sweeping effects that further disempowered the pastoralist populations and drove their sedentarization and dependence in the name of rationalized animal agriculture and public health.

In Tunisia and Morocco these veterinary laws became particularly complex and far-reaching. Beginning in 1885 in Tunisia and 1914 in Morocco, all contagious diseases of domestic animals—and all diseases resembling those on the official list of contagious diseases—were placed under the jurisdiction of the veterinary police (Geoffroy Saint-Hilaire 1919, pp. 432–451). Identification and treatment of these infectious diseases were declared mandatory and could be carried out only by licensed veterinarians holding the appropriate French diploma or by their official assistants. These contagious diseases were defined by specific name and by "any parasitic, microbial and nonmicrobial diseases that may be transmitted from sick to healthy animals" (p. 445). Such a definition includes practically all animal diseases and states of ill health and makes even ubiquitous and relatively harmless infections such as mange a contagious disease treatable only by a licensed veterinarian.

Moreover, the treatment of animals for any contagious disease (basically any animal disease) by nonlicensed veterinarians was made strictly illegal and punishable by steep fines and imprisonment (Geoffroy Saint-Hilaire 1919). This legislation resulted in the criminalization of traditional animal healers and the practice of indigenous veterinary medicine. It also mandated treatment of contagious diseases and therefore forced the local people to seek treatment from licensed (French) veterinarians. In the case of large, commercial transactions in livestock, such as importing or exporting animals, an inspection by the *vétérinaires sanitaires* was mandatory and had to be paid for by the owner or the importer of the animals.

As onerous as this legislation may have been for such groups, however, its insidious impact was more profound among the poor smallholders of livestock and the pastoral nomads, who could ill afford the expense of such services. Like the imposition of taxes, it forced them into a money economy that fundamentally altered their way of life, in many ways making it more precarious than it had been. A different approach was taken in the remote areas of the French empire in North Africa, where the enforcement and regulation of these laws was relatively difficult. Similar in many ways to programs promoting human health and hygiene in these areas, free veterinary clinics were set up to habituate the herders to "modern" veterinary care and to encourage its use. A more important goal, though, was to demonstrate the superiority and benevolence of the French veterinary system and the ultimate good of the French protectorate. The stationing of French veterinarians, usually *vétérinaires militaires*, in these remote areas also facilitated the implementation of

Fig. 10.3 Mobile veterinary group, providing free veterinary services in southern Morocco near Marrakesh (From *Le service vétérinaire et le service de la remonte aux colonies, les armées françaises d'Outre-Mer* (p. 22), by L'inspection du service vétérinaire de l'armée, 1930, Paris: Imprimerie Nationale. Copyright holder unkown)

the new laws and the surveillance and punishment of illegal veterinary activities by nonlicensed, traditional, practitioners (Davis 2006b). Many of these military veterinarians were mobile and traveled in groups with their assistants (see Fig. 10.3).

By the end of the colonial period, then, the numbers of pastoralists across the Maghreb had plummeted, certainly by 50 % or more, along with the numbers of livestock, especially small ruminants and camels, the animals mostly preferred by pastoralists in arid environments. Nearly all traditional ways of life and practices of land management of the North Africans had been greatly restricted and/or criminalized, and many people had been forced into a wage economy that did not involve the raising of animals. These changes alone diminished indigenous knowledge of range management and veterinary care. The subsequent introduction of the colonial veterinary laws further exacerbated the erasure of this valuable knowledge.

The Postcolonial Maghreb and the Politics of Indigenous Knowledge

Since the onset of the postcolonial period, many of the laws, policies, and attitudes of the governing powers have changed little from those of the colonial period with respect to pastoralism and the environment. In Morocco, for example, the colonvial forest code of 1917 has remained in effect with few amendments. Traditional land management techniques such as fire and grazing are still either banned or tightly controlled, and migration of pastoralists and livestock is strictly monitored and limited. Migration with livestock is not allowed across departmental or regional borders without a permit (Davis 2001).[8] It is still illegal to practice veterinary medicine without a Moroccan license. The following discussion focuses on Morocco because it survived the colonial period with the highest proportion of nomads and pastoralists in North Africa and because it has followed the western free-market model more closely than Algeria has. Tunisia, for all intents and purposes, has no significant extensive pastoralism today; nomads there were sedentarized by the end of the colonial period.

Despite the hardships facing pastoralists in general, and nomads in particular, an estimated 17 % of the rural population in Morocco was nomadic in the mid-1970s, the highest proportion in the Maghreb (El-Ghonemy 1993, p. 32). Although official statistics do not provide a category called pastoralist or nomad, a large proportion of rural Moroccans still raise a variety of livestock. Most of these activities, though, are small, family affairs with a single cow or a few goats subsisting on graze next to roadways and on what supplementation the family members can buy or gather. Families with more animals will frequently rent fields of stubble after harvest. Few Moroccan pastoralists have the opportunity for extensive pastoralism where they can practice range-management skills. One of the few surveys of the practice and utilization of traditional medicine, including veterinary medicine, in Morocco showed a very small percentage of traditional medicines or practices being applied to animals. Of 2,800 interview responses in which the various uses of traditional medicines and techniques were mentioned, only 1.6 % listed veterinary medicine as an application of this knowledge (Bellakhdar 1997, p. 99).[9] This finding indicates an erosion of indigenous veterinary knowledge, attrition likely due to the changes outlined above.

The trend since independence, and especially since about 1980, has been the capitalization of sheep and cattle-raising. Because of government incentives, wealthy Moroccans across the country have steadily bought out smallholders and created large commercial herds that rely on modern inputs and veterinary care and that are raised mostly on the ranch model.[10] This restructuring has increased the quantity of red meat and milk produced in Morocco, but it has also disenfranchised many traditional pastoralists. In addition to the fact that livestock are being bought up by these commercial interests, tribal collective lands have been appropriated by the state or privatized and sold to parties such as the commercial ranching companies at an alarming rate. In the Middle Atlas region, for example,

---

[8] I conducted 2 years of field work with nomads in Morocco in the late 1990s.

[9] A further 3.5 % listed pastoral and agricultural uses, a large category that is hard to breakdown for potential animal- or range-management usage.

[10] Western aid agencies such as the United States Agency for International Development (USAID) have been primarily responsible for fostering this ranch-style model of raising livestock and for promoting the privatization of range land.

collective lands, which occupied nine tenths of the zone at the beginning of the century, have been reduced to less than one quarter of the area, while more than half of it has become state land and almost one quarter has been privatized. The restriction of collective rights is assisted by the monopolization of numerous sectors...by large holders. (El Khyari 1987, p. 383; transl. by Davis)

Much of this lost collective land has been put into cereal production despite the fact that most of the areas do not receive enough annual average rainfall for reliable grain production (Davis 2001).

Two main motivations the Moroccan state has for restricting pastoralists, especially nomads, and for actively reducing collective tribal properties are its long-standing distrust of nomads and its focus on maintaining social order. Since the late 1970s, a third main motivation has been the government's embrace, to a large extent, of neoliberalism. Although the outcomes of reforms under neoliberalism in Morocco have been complex, the results in the dryland and pastoral areas have been primarily negative for traditional livestock-raisers (Davis 2006a). The state has adopted the colonial story that pastoralists, especially nomads, ruin the environment (e.g., overgrazing) and that collective land must therefore be privatized and ideally put to what it regards as proper productive use, such as raising cereals.[11] New laws have been passed that facilitate the privatization of collective land by tribal members simply if they prepare it for cultivation. This constellation of forces has brought a large amount of former grazing land under the plow, and cereals have expanded onto increasingly marginal land, with results that are often disappointing (Davis 2006a). The cereal harvests have not been adequate, and the disruption of native vegetation by the clearing and plowing of the fields has ruined these native grazing grounds for years to come.

Despite all the changes detailed above, indigenous pastoralist knowledge still exists in Morocco, as I found during field work with the Aarib camel nomads in the southern part of the country, south of Zagora, just north of the Algerian border (see Fig. 10.4). This community of an estimated 3,000 people is about 60 % sedentarized, with most members living in and around the small village of M'hamid (Davis 2005a). The Aarib saw their traditional ways of life first mildly bounded by the French (whom they fondly remember today for having treated Aarib camels for trypanosomiasis, a major problem in the region) and are now heavily restricted by the independent Moroccan government. Formerly migrating as far as Tomboctou, the Aarib today must obtain government permits to move their animals even a few kilometers and can no longer migrate south across the militarized Algerian border to their former customary grazing grounds. Importantly, they chose to sell off about three quarters of their camels rather than watch them starve to death on the reduced grazing lands left to them during the closure of the Moroccan–Algerian border from 1975 to 1980 and Morocco's occupation of the Western Sahara (Davis 2005a).

This decision points to the Aarib's sophisticated body of knowledge of range management and veterinary care. Probably because these people exist on the fringes

---

[11] For details on the false French colonial environmental history of North Africa, including the claim that nomads ruined the environment, see Davis (2007).

Fig. 10.4 Nomad brother and sister, southern Morocco (Photo by Janice S. Davis, 1998)

of the Moroccan kingdom and figure as marginal except as a potential threat to the government, the state has largely left them alone.[12] It has granted them little to no veterinary care or range management advice, and not until the late 1990s did they start to receive basic services such as electricity and water in the areas where they have sedentarized (e.g., M'hamid). I lived with these nomads for over 6 months and found that they have sophisticated knowledge of range plant ecology, climate, weather, and animal health and disease (Davis 2001, 2005a). Indeed, their understanding of range ecology and range management resembles cutting-edge research in these fields more closely than does the knowledge of most Moroccan government range experts. The Aarib also use a broad variety of traditional plants and other treatments to care for their animals as best they can. Despite the laws on the books, they did not appear to be reprimanded for practicing traditional veterinary medicine.

I have argued elsewhere that the Moroccan government, rather than continuing to suppress this indigenous knowledge and practice of extensive pastoralism, should consider building on it and radically reform range-management policies in the country (Davis 2005b). I suspect that other extensive pastoralists, such as the Ait Atta in southern Morocco, retain a similar knowledge base despite being mostly sedentarized. However, it may largely vanish in another generation or two if current

---

[12] Armed Moroccan military guards are stationed at the only entrance to M'hamid, the main village where many Aarib have sedentarized. Aarib movements are strictly monitored and confined by the government (for details see Davis 2001).

trends persist. At the other end of the Middle East, in Afghanistan, pastoralist indigenous knowledge is also under threat by related, but more complicated, forces.

Postwar Afghanistan and the Case of Nomad Women's Indigenous Knowledge

Unlike the Maghreb, Afghanistan was never successfully colonized, despite several attempts by the British. Archival materials for Afghanistan are thus very sparse, and reliable estimates of pastoralist or nomadic populations are not available other than to say that it is almost certain that a majority of Afghans raised livestock in some form in the nineteenth and early twentieth centuries (Dupree 1980). Based on census material beginning in the 1970s, current estimates are that up to 25 % of the population may have been nomadic in the late nineteenth and early twentieth centuries (Nyrop 1986). By the 1970s about 15 % were estimated to be nomadic; as of 2003, about 10 %, or 2,500,000 people (UNOCHA 2001, 2003).

Conducting ethnoveterinary research with Afghan Koochi nomads, I found that they have a rich array of knowledge of animal diseases and their treatments.[13] Though not perfect, it serves many of them well in caring for their herds. Given that literature on nomad women rarely mentions them doing any labor with animals other than milking, it is important and surprising to have found that Koochi women work with livestock frequently and that they know a great deal about livestock disease and about care for ill and injured animals.[14] Milking is indeed a woman's job most of the time, and in many cases women sell any surplus milk in order to buy what they need for their household (Davis 1995). In a sense the milk is their property, although I use the term loosely in this context. Milking and the production of dairy products are two of their most important duties in Koochi society.

In addition, Koochi women regularly care for sick and newborn animals (which stay at the tent); help with difficult births; and clean, feed, and water the animals when they are kept near the tent (see Fig. 10.5). In fact, caring for sick animals is one of the women's main tasks (although men do so, too). About the only thing women do not do is take the livestock to graze far from the tent, something that normally ranks as men's work in the Koochi division of labor. Women are responsible for cleaning and preparing the carcass after men slaughter an animal for food. Koochi women therefore have the opportunity to correlate certain disease states

---

[13] I conducted research with Afghan nomads for 3 months in the early 1990s. A story told to me at that time illustrates why it is so important not to romanticize indigenous knowledge or conceive of it as some kind of panacea for development problems. One Afghan nomad I interviewed was skilled at using a local vaccine technique (ear-slit vaccine) for preventing contagious caprine pleuropneumonia. When he tried to use it for anthrax, however, he killed most of his herd.

[14] The significant role of women in pastoral societies is beginning to be recognized by some researchers and development agencies. For an instructive overview and a discussion of past oversights, see Flintan (2008).

Fig. 10.5 Koochi woman
and two of her children
near Quetta, Pakistan
(Photo by Diana K. Davis,
1994)

with symptoms and treatment, as is the case with gastrointestinal parasites and mastitis (infection of the udder). This scope of experience helps explain why Koochi women are skilled at understanding and treating many livestock diseases. In several cases, women had a more accurate understanding of livestock diseases than men did. For instance, 90 % of the women interviewed could name at least one disease that causes mastitis, whereas 70 % of men could not, and only women reported that mastitis can be infectious—which it usually is (Davis 1995).

Koochi women are proud of their work with animals and their knowledge of how to care for livestock. In fact, they would like to expand their knowledge of livestock health and other things (e.g., reading and religion) and would even like to work as providers of animal health care. In research I conducted in the context of a Basic Veterinary Worker (BVW) program that was being implemented for male Afghan herders by an NGO with funding from USAID, I found that Koochi women and men agreed that women would be capable providers of animal health care, could receive training in basic livestock care, and could then provide that care for a fee because very few women observed purdah inside the tribe (Davis 2005a).[15]

The NGO and USAID staff, however, had assumed that women do no work with animals other than milk them, that they had little or no knowledge of animal health and disease, and that their mobility was highly circumscribed by traditions of purdah. These notions of women are inherent in a wider set of essentialized

---

[15] Purdah is the practice of veiling women and keeping them secluded from public view. In practice, the levels of purdah observed vary a great deal from country to county and within countries.

assumptions about Muslim societies held by many western organizations and individuals (Davis 2005b). Their development initiative, the BVW program, trained only men to treat livestock diseases that had traditionally been dealt with by women (e.g., mastitis).[16] By training men to treat udder infections with antibiotic infusions, the staff members of the BVW program not only create the possibility for men to appropriate the treatment of udder infections but also enable men to appropriate milk, the distribution (sale) of milk, and milking (Davis 2005b). By making an economic investment in the antibiotic udder infusion, men may begin to see the milk as their property.

Given livestock development that encourages the commercialization of milk production and/or the conversion to commercial meat production, the disenfranchisement of Koochi women from one of their most significant productive spaces becomes even more likely than it would be without this change. If milk production is commercialized, the results will probably be very similar to the disenfranchisement of pastoral women in other societies when subsistence dairy production is commercialized (Ensminger 1984; Waters-Bayer 1994). Not only will these women lose work that they value, they may lose the power to control milk distribution in the family. The women's and children's health status may thus also suffer from loss of milk and income. Women's overall status in the community declines in these cases, further disadvantaging them in their daily lives.

This situation raises an important point because women account for 60–70 % of the country's population (Khanna 2002) as a result of so many Afghan men having been killed over the last several decades of war. These women have lived with and are now again confronting the influence of a resurgent Taliban whose notions of Islam, the role of women, and gender relations are particularly conservative, even by the norms of other conservative Islamic states around the world. One expert on women in the Middle East, Valentine Moghadam (1999), has pointed out that, under the Taliban, "women's reproductive roles [are] fetishized in the context of a kinship-ordered patriarchal structure… [and] it [is] difficult to see women in other roles, such as students, citizens, or income-earning employees" (p. 173).

Not surprisingly, Afghanistan is widely regarded as an "extreme case" of classical patriarchy, the institutions of which dominate social, economic, and political life to varying degrees throughout the Muslim world today (Barr 2013; Moghadam 1999; Suad and Slyomovics 2001). When the proper roles of women are not understood to be that of citizen or income earner, when their mobility is so hemmed that it is nearly impossible for them to go shopping or to the doctor, it all but rules out ways to support the country's many families now headed by women. Moreover, these women now face a male-dominated, western development industry whose conventional plans and projects for development bring to postwar Afghanistan many stereotypes and false notions about women in general and about Muslim women in particular.

---

[16] The BVW program taught only "western" veterinary treatments to the men and aimed to commercialize veterinary medicine in Afghanistan with a supply chain for veterinary treatments and pharmaceuticals. See Davis (2005b) for details.

Are Koochi women's productive spaces, their sophisticated knowledge, their extensive work, and their future spaces of development being eroded in several ways by patriarchal, western-led development operating within a sternly conservative religious ideological climate? The BVW project does in fact suggest that likelihood. I believe the current geopolitically defined development situation in Afghanistan indicates, unfortunately, that contemporary reconstruction activities may compound the disenfranchisement of these nomad women and undermine their knowledge and skills well into the future.

Conclusion

These contemporary trends in political-economic restructuring, including development initiatives, mirror earlier ones in the nineteenth and twentieth centuries with respect to negative effects on pastoral communities in the Maghreb and Afghanistan. By defining local knowledge and practice as less correct than western approaches, and often as wrong and damaging to livestock and the environment, the colonial state identified—just as the postcolonial state and international agencies have identified—what they take to be a problem that needs to be solved or improved by experts with specialized knowledge and tools. Once established, this problematization has been rendered technical (Li 2007) and thus governable, by a variety of experts and related development plans in veterinary medicine, range management, ecology, agriculture, forestry and other sectors. This conceptualization facilitates the exercise of state power in the form of new laws, policies, and development programs that, by and large, suppress local knowledge and practice while privileging expert knowledge and tending to benefit elites.

In the case of the colonial Maghreb, changes in property laws, rules governing natural resource use (especially forestry and grazing laws), the imposition of western veterinary medicine, and the spread of capitalist social relations all had profound, mostly adverse impacts on pastoralists' ways of life and knowledge systems. The fact that many of these colonial changes are often more pronounced today than they were then and that they continue having inimical effects on pastoralists and their knowledge underscores the importance of seriously considering history and political economy when studying any dimension of indigenous knowledge. In Afghanistan, nomad women's knowledge of and work with livestock has been rendered invisible to the Afghan government, which is strongly influenced by the Taliban at present, and to western development experts—in both cases because of their prejudices and conservative ideological beliefs about Muslim women. This bias is resulting in development projects that reduce women's work with livestock and marginalize them within limited social roles and simultaneously commercialize livestock production in general in ways that privilege western expert veterinary knowledge. The result will almost certainly be a further decrease in local pastoral knowledge.

Conceivable alternative approaches could begin with respectful acknowledgement of the many strengths that the local knowledge and practice of pastoralists have. Although not a panacea, building on them with hybrid approaches, as suggested in ideas of "high reliability pastoralism" (Roe et al. 1998), might help preserve local knowledge and empower pastoralists.[17] Recent research shows that increasing the mobility of pastoralists will also reduce their vulnerability to drought, increase their resilience to predicted climate change, and improve their management of pasture resources (Freier et al. 2014). All too often, the erasure of pastoralists' indigenous/local knowledge is a sign of their marginalization in a political economy geared to benefit others. Empowering pastoralists by granting them secure land tenure, guaranteeing their access to adequate pastures and migration routes, and restoring their authority in making decisions about range management and veterinary treatment would likely be highly effective at dynamically preserving indigenous/local knowledge in ways more appropriate and successful within pastoral societies and their environment than current development trends are.

References

Ageron, C.-R. (1968). *Les Algériens musulmans et la France (1871–1919) [The Algerian Muslims and France (1871–1919)]*. Paris: Presses Universitaires de France.

Agrawal, A. (1995). Indigenous and scientific knowledge: Some critical comments. *Indigenous Knowledge and Development Monitor, 3*(3), 3–5.

Agrawal, A. (2005). *Environmentality: Technologies of government and the making of subjects*. Durham: Duke University Press.

Barr, R. (2013). 'Dark future' for women's rights in Afghanistan. *Human Rights Watch*, May 22, 2013. Retrieved from http://www.hrw.org/news/2013/05/22/dark-future-womens-rights-afghanistan

Bedrani, S. (1991). Legislation for livestock on public lands in Algeria. *Nature & Resources, 27*, 24–30.

Bellakhdar, J. (1997). *La pharmacopée Marocaine traditionnelle: Médicine Arabe ancienne et savoirs populaires [Traditional moroccan pharmacopoeia: Ancient Arabic medicine and common knowledge]*. Paris: Ibis Press.

Boudy, P. (1948). *Économie forestière nord-africaine: Milieu physique et milieu humain [North African forest economy: Physical and human environment]* (Vol. 1). Paris: Éditions Larose.

Boudy, P. (1955). *Économie forestière nord-africaine: Description forestière de l'Algérie et de la Tunisie. [North African forest economy: Description of forests in Algeria and Tunisia]* (Vol. 4). Paris: Editions Larose.

Boukhobza, M. (1982). *L'agro-pastoralisme traditionnel en Algérie: De l'ordre tribal au désordre colonial [Traditional agropastoralism in Algeria: From tribal order to colonial disorder]*. Algiers: Office des Publications Universitaires.

Boukhobza, M. H. (1992). *Monde rural: Contraintes et mutations [Rural world: Constraints and mutations]*. Algiers: Office des Publications Universitaires.

Davis, D. K. (1995). Gender-based differences in the ethnoveterinary knowledge of Afghan nomadic pastoralists. *Indigenous Knowledge and Development Monitor, 3*(1), 3–5.

---

[17] For other ideas on securing pastoralist livelihoods in ways that benefit people and the environment, see Flintan (2012).

Davis, D. K. (2001). *Overgrazing the range? A political ecology of pastoralism in southern Morocco.* Unpublished Ph.D. dissertation, University of California, Berkeley.

Davis, D. K. (2005a). Indigenous knowledge and the desertification debate: Problematizing expert knowledge in North Africa. *Geoforum, 36,* 509–524.

Davis, D. K. (2005b). A space of her own: Women, work, and desire in an Afghan nomad community. In G.-W. Falah & C. Nagel (Eds.), *Geographies of Muslim women* (pp. 68–90). New York: Guilford.

Davis, D. K. (2006a). Neoliberalism, environmentalism, and agricultural restructuring Morocco. *The Geographical Journal, 172,* 88–105.

Davis, D. K. (2006b). Prescribing progress: French veterinary medicine in the service of empire. *Veterinary Heritage, 29,* 1–7.

Davis, D. K. (2007). *Resurrecting the granary of Rome: Environmental history and French colonial expansion in North Africa.* Athens: Ohio University Press.

Dupree, L. (1980). *Afghanistan.* Princeton: Princeton University Press.

El Khyari, T. (1987). *Agriculture au Maroc [Agriculture in Morocco].* Mohammedia: Editions OKAD.

El-Ghonemy, M. R. (1993). *Land, food, and rural development in North Africa.* Boulder: Westview Press.

Ensminger, J. (1984). Theoretical perspectives on pastoral women: Feminist critique. *Nomadic Peoples, 16,* 59–71.

Flintan, F. (2008). *Women's empowerment in pastoral societies.* Nairobi: IUCN-WISP.

Flintan, F. (2012). *Making rangelands secure: Past experience and future options.* Rome: The International Land Coalition.

Freier, K. P., Finckh, M., & Schneider, U. A. (2014). Adaptation to new climate by an old strategy? Modeling sedentary and mobile pastoralism in semi-arid Morocco. *Land, 3,* 917–940. doi:10.3390/land3030917.

Geoffroy Saint-Hilaire, H. (1919). *L'Élevage dans l'Afrique du Nord [The livestock in North Africa].* Paris: Augustin Challamel.

Julien, C.-A. (1964). *Histoire de l'Algérie contemporaine: La conquete et les débuts da la colonisation (1827–1871) [The history of modern Algeria: The conquest and early colonization (1827–1871)].* Paris: Presses Universitaires de France.

Khanna, R. (2002). Taking a stand for Afghanistan: Women and the left. *Signs, 28,* 464–465.

Lacoste, Y., Nouschi, A., & Prenant, A. (1960). *L'Algérie: Passé et present [Algeria: Past and present].* Paris: Éditions Sociales.

Li, T. (2007). *The will to improve: Governmentality, development, and the practice of politics.* Durham: Duke University Press.

Marc, H. (1916). *Notes sur les forêts de l'Algérie [Notes on the forests in Algeria].* Algiers: Typographie Adolphe Jourdan.

Moghadam, V. M. (1999). Revolution, religion, and gender politics: Iran and Afghanistan compared. *Journal of Women's History, 10*(4), 172–185.

Nouschi, A. (1959). Notes sur la vie traditionelle des populations forestières algériennes [Notes on the traditional lifestyle of Algerian forest dwellers]. *Annales de Géographie, 68,* 525–535.

Nouschi, A. (1961). *Enquête sur le niveau de vie des populations rurales constantinois de la conquête jusqu'en 1919 [Survey on the living standard of the rural population from the conquest of Constantine to 1919].* Paris: Presses Universitaires de France.

Nyrop, R. (Ed.). (1986). *Afghanistan: A country study.* Washington, DC: The American University.

Pfeifer, K. (1986). The development of commercial agriculture in Algeria, 1830–1970. *Research in Economic History, 10,* 277–308.

Poncet, J., & Raymond, A. (1971). *La Tunisie [Tunisia].* Paris: Presses Universitaires de France.

Roe, E., Huntsinger, L., & Labnow, K. (1998). High reliability pastoralism. *Journal of Arid Environments, 39,* 39–55.

Ruedy, J. (1992). *Modern Algeria: The origins and development of a nation.* Bloomington: Indiana University Press.

Saint-Germes, J. (1955a). *Économie algérienne [Algerian economy]*. Algiers: La Maison des Livres.

Saint-Germes, J. (1955b). *La réforme agraire algérienne [Algerian agrarian reform]*. Algiers: La Maison des Livres.

Sari, D. (1978). *La dépossession des fellahs [The dispossession of the fellahin]*. Algiers: Société Nationale d'Édition et de Diffusion.

Suad, J., & Slyomovics, S. (Eds.). (2001). *Women and power in the Middle East*. Philadelphia: University of Pennsylvania Press.

Trabut, L. (1889). *L'Halfa [Halfa]*. Algiers: Giralt, Imprimeur de Gouvernement Général.

Trautmann, W. (1989). The nomads of Algeria under French rule: A study of social and economic change. *Journal of Historical Geography, 15*, 126–138.

UNOCHA. (2001). *Afghanistan: FAO launches appeal for agricultural rehabilitation* (Press Release). Islamabad: UN Office for the Coordination of Humanitarian Affairs. Retrieved from http://www.irinnews.org/report/13869/afghanistan-fao-launches-appeal-for-agricultural-rehabilitation

UNOCHA. (2003). *Afghanistan: Heavy toll on civilians in years of war* (Press release). Kabul: UN Office for the Coordination of Humanitarian Affairs. Retrieved from http://www.irinnews.org/indepthmain.aspx?InDepthId=31&ReportId=70546

Waters-Bayer, A. (1994). Studying pastoral women's knowledge in milk processing and marketing—for whose empowerment? *Agriculture and Human Values, 11*(2–3), 85–95.

Chapter 11
"*Masawa—bogeokwa si tuta!*": Cultural and Cognitive Implications of the Trobriand Islanders' Gradual Loss of Their Knowledge of How to Make a *Masawa* Canoe

Gunter Senft

> *Kwatuyavesa waga,* Turn round the sail of the canoe,
> *rakeda milaveta!* its course is to the open sea!
>
> (*Oruvekoya song cycle, first stanza*)

A few days after I had set foot on the Trobriand Islands for the first time in 1982,[1] I spied a *masawa* canoe sailing close-hauled toward Kiriwina, the main island of the Trobriands. Although I had seen Malinowski's (1922/1978) photographs of these impressive canoes in his monograph *Argonauts of the Western Pacific*, I was overwhelmed by the grace and beauty of this sight. Three weeks later I had the opportunity to sail on such a canoe from Kaibola, the northernmost village on Kiriwina Island, back to Tauwema, the village on Kaile'una Island that has been my place of residence during my field research on the Trobriands for more than 25 years now. My sailing experience with a *masawa* canoe is one of my dearest memories so far.

In Malinowski's times Kaile'una Island as well as Vakuta Island, Kitava Island, and the village of Sinaketa on Kiriwina Island were known as the best centers for canoe-building and as the places where most expert canoe-builders and carvers lived (Malinowski 1922/1978, pp. 121–145). In 1982 Tauwema had eight *masawa*,

---

[1] This chapter is based on more than 40 months of field research on the Trobriand Islands. I thank the National and Provincial Governments in Papua New Guinea; the Institute for PNG Studies, especially Don Niles; and the National Research Institute, especially James Robins, for their permission for and assistance with my research projects. I express my great gratitude to the people of the Trobriand Islands and, above all, to the inhabitants of Tauwema and my consultants for their hospitality, friendship, and patient cooperation over all these years. Without their help, none of my work on the language of the Trobriand Islanders (Kilivila) and on the Trobriand culture would have been possible.

G. Senft (✉)
Max Planck Institute for Psycholinguistics, 310, 6500 AH Nijmegen, The Netherlands
e-mail: gunter.senft@mpi.nl

© Springer International Publishing Switzerland 2016
P. Meusburger et al. (eds.), *Ethnic and Cultural Dimensions of Knowledge,*
Knowledge and Space 8, DOI 10.1007/978-3-319-21900-4_11

each of which had a proper name and were proudly owned by men living in the village.[2] In 1992, when I returned to Tauwema for my third period of field research, only one of these canoes remained (Nusai's *Seguvagava*), and ever since 1996 there have been no *masawa* in Tauwema anymore. My fascination with these canoes led me to collect descriptions of how to make a *masawa* from Kilagola, at that time both the chief of Tauwema and a *toliwaga* (an owner of a *masawa*), and from Tosulala, another *toliwaga* in Tauwema, during my first 15 months of field research in 1982 and 1983. Mokeilobu, an expert on sails, told me how to make one, and Kilagola described how canoe prows are carved. He also informed me about canoe magic (as did Yoya, another expert magician in 1989). The men of Tauwema greatly appreciated my interest in their big canoes. I was thus quite shocked to observe the gradual loss of the *masawa* in Tauwema over the years. In 1996, when I asked former owners of such canoes why nobody in Tauwema had taken the initiative to build a new *masawa*, their answer was simply, *"Masawa—bogeokwa si tuta!"* (*Masawa* canoes—their time is over!).

In what follows I draw on the collected descriptions[3] to describe how the Trobriand Islanders of Papua New Guinea used to construct their large seagoing *masawa* canoes and how they used to make their sails, what different forms of knowledge and expertise they needed in order to do so during various stages of the construction processes, how this knowledge was socially distributed, and what the social implications were of all the joint communal activities necessary before a new canoe could be launched. I then try to answer the question of why the complex distributed knowledge of how to make a *masawa* has been gradually getting lost in most of the village communities on the Trobriand Islands. Lastly, I outline and discuss the implications of this loss for the culture of the Trobriand Islanders, their social construction of reality, and their indigenous cognitive capacities.

The Art of Making a *Masawa* Canoe

The Trobriand Islanders differentiate between six types of canoe[4]:

1. *kekwaboda* the smallest outrigger canoe (Malinowski 1922/1978, p. 112) (*kewo'u*)

---

[2] The names of the canoes and, in parentheses, the names of their owners were *Seguvagava* (Nusai), *Tovivila* (Tosulala), *Mogerai* (Bulasa), *Meraga* (Topiesi), *Dedayasi* (Tosobu), *Vaneyaba* (Moligogu), *Genare'u* (Taidyeli), and *Topasi* (Tomtava). All these canoes used traditional sails made of pandanus leaves. The orthography of Kilivila is based on Senft (1986).

[3] I also heavily rely on the descriptions of the *masawa* canoes and their construction as recorded in Malinowski (1922/1978, pp. 105–146).

[4] Gerrits (1974) also mentions, documents, and describes burial canoes of the Trobriands (on Kitava) and the Marshall Bennet Islands (on Iwa). These objects, called *nalio'ema*, were small coffins carved like miniature canoes. However, even his informants did not "remember anybody to have been buried in this way" (Gerrits 1974, p. 229). I have never heard anything about these canoes on the Trobriands.

Fig. 11.1 Children playing with a *kemolu* canoe on the reef of Tauwema, with a *masawa* canoe passing in front of the reef (all photos by author)

2. *kemolu* a small outrigger canoe that can be sailed for fishing and relatively short trips (not mentioned in Malinowski 1922/1978; see Fig. 11.1)
3. *ligataya* a canoe larger than a *kemolu* and used for relatively long trips but not as carefully constructed and decorated as a *masawa* canoe (see Fig. 11.2). The *ligataya* resembles the *kalipoulo* type of canoe described by Malinowski (1922/1978, p. 112) but has no prow boards.
4. *masawa* large *kula* canoe[5] (see Fig. 11.3)
5. *mesolaki* very large *kula* canoe (not mentioned in Malinowski 1922/1978)
6. *nagega* largest *kula* canoe; neither used nor built by the Trobriand Islanders but rather by other Islanders who participate in the *kula* (Malinowski 1922/1978, pp. 144–145)

The generic term for all these dugout canoes (and other vehicles except airplanes) is *waga*. The expression *waga parai* refers to any kind of sailing vessel. This chapter concentrates solely on the building of a *masawa* canoe. However, before explaining this complex and time-consuming activity, I briefly outline what Malinowski (1922/1978) called "the canoe's sociology" (p. 113) and describe essentials of a *masawa*'s construction.

---

[5] Gerrits (1974, p. 230) reports that the *masawa* canoe came originally from Dobu Island. *Kula* is the name of a ritualized trade in which shell valuables are exchanged. The *kula* covers a wide area of Melanesia (see Malinowski 1922/1978).

Fig. 11.2 A *ligataya* canoe with a sail made of plastic

Fig. 11.3 Nusai's *masawa* (*Seguvagava*) under full sail (1989)

The *Masawa* Canoe's Sociology

According to Malinowski (1922/1978, p. 114), only chiefs or headmen of a village sector could own *masawa* canoes. A closer look at the canoe owners in Tauwema reveals that this general rule did not hold in that village (nor, as far as I know, for the villages on the islands of Kaile'una, Kuiava, and Simsim). None of the eight canoe owners in Tauwema was a chief, and none of them, at least in 1996, was the head-man of a village sector. Only Nusai, Tosulala, and Taidyeli were members of the highest ranking *Malasi* clan; Topiesi (chief Kilagola's eldest son). Tosobu, Moligogu, and Tomtava were members of the third-ranking *Lukwasisiga* clan, and Bulasa was even a member of the lowest ranking *Lukulabuta* clan (but was one of the first local missionaries—*misinari*—of Tauwema). However, in Tauwema and on the other islands mentioned above, the *toliwaga* were as solely responsible for the building of their *masawa* as were the *toliwaga* on Kiriwina, Vakuta, and Kitava, the other islands of the Trobriand chain. They had to initiate the building of their canoe, acted as the spokesmen in everything related to constructing and sailing it, and, last but certainly not least, had to accumulate enough food and other items of wealth such as stone blades (*beku*), belts (*kaloma*), and small necklaces (*kuwa*) to pay experts and other workers involved in making the *masawa*. Besides this strong eco-nomic position, the *toliwaga* had to be respected persons who had gained a degree of influence and authority within their village community. The *toliwaga* was the one who—

- made all kinds of decisions during the process of building the canoe;
- selected the indispensable experts (craftsmen and magicians);
- decided whether to rent the canoe to others (for which service he would receive payment, traditionally in yams, betelnuts, fish, and so forth);
- selected the crew of his canoe; and
- saw to it that all magical rites and duties inextricably intertwined with the *masawa* were properly performed from the very outset of its construction to ensure the canoe's performance and safety at sea.

This list of a *toliwaga*'s main duties shows that making and using a *masawa* were a communal matter. Along with the experts that were needed for building the canoe, the *toliwaga* also needed the help of a smaller group of workers, usually his relatives and friends, who helped him at certain stages of the construction. Moreover, whenever communal work was necessary during this building process, all villagers supported him and took part in the work. If the *toliwaga* put an expert canoe builder (a *tota'ila waga*) in charge of making the canoe, which used to happen quite often, the *tota'ila waga* was entitled to make all the decisions during the construction process, including the selection of other experts. However, the final responsibility for the canoe and the obligation to reimburse the experts and helpers for their efforts and to sponsor the communal meals remained with the *toliwaga*.

The crew of a *toliwaga* consisted of sailing experts (again normally kinsmen and friends) who specialized (or would eventually specialize) in performing a particular task (e.g., a steersman would always be responsible for the rudder (*kuliga*)). The *toliwaga* usually acted as captain, provided he was a skilled sailor. If he was not, he could take over some other assignment and leave the command to the best seaman of his crew.

During the building and use of the canoe, the *toliwaga* had to rely on experts specialized in specific aspects of craftsmanship. An especially important individual was the one who performed the magic to guarantee the *toliwaga*'s success at achieving his aim of owning a beautiful, admirable, swift, and elegantly sailing canoe that was protected against all the harms it could encounter at sea or on the beach. Malinowski's (1922/1978) description of the *toliwaga*'s role (p. 120) implies that the latter person knew all the various forms of magic crucial to the realization of this ambitious goal. I know of nobody on the Trobriands these days who claims to have the paramount competence in canoe and sailing magic. Even chief Kilagola, a canoe-builder of renown, did not dare claim this ability. Thus, the *toliwaga* depended on an expert crew but also on experts in the manifold forms of canoe and sailing magic.

Essentials of the *Masawa* Construction[6]

The dugout, ranging in length from 6 to 10 m (some 20–33 ft) and featuring massive pointed ends, is watertight, immersible, and able to carry even heavy loads, but it has no lateral stability. Any load put on the hollowed-out log will cause it to capsize. To solve this problem, the Trobriand Islanders equip their canoes with outriggers (*lamina*). That is, they attach a second, slightly smaller, solid log to the dugout. However, the greater stability afforded by this float is not symmetrical. Stability is guaranteed only when downward pressure is exerted on the opposite side of the canoe's body. This force lifts the outrigger out of the water and produces "momentum . . . proportional to the displacement, and the rest of the canoe will come to equilibrium" (Malinowski 1922/1978, pp. 109–110). This asymmetrical stability has important consequences for sailing an outrigger canoe: The float must always remain windward! Another factor that increases the stability of the canoe is the volume and depth of the dugout. Both factors can be easily increased by adding longitudinal planks to them. These planks are closed in at the ends of the canoe by transversal prow boards that break the waves. Sagittal prow boards are mounted on the front of the transversal ones to reinforce them. This construction lends the canoe "a good deal of freeboard to prevent water from breaking in" (Malinowski 1922/1978, p. 110). Yet another stability-increasing factor is the distance between dugout and float. With the *masawa* this distance is ordinarily about a quarter of the length of the dugout and is covered with a platform.

The Making of a *Masawa*

In late July 1983, Tosulala demonstrated to my wife and me how he burned coral to make lime out of it. He then gave a detailed and quite sophisticated verbal description of what he had done. I thereupon asked him whether he would like to tell me how one builds a *masawa*. On the evening of July 30, he came to my house and agreed to do so. Most of what follows is based on Tosulala's tape-recorded account,

---

[6]This section relies on Malinowski (1922/1978, pp. 109–111), Haddon and Hornell (1991, Vol. II, pp. 266–271), and Helfrich (1984, pp. 47–48).

with a few additions from the descriptions by Kilagola and Mokeilobu mentioned above, my field notes, and other relevant information from Haddon and Hornell (1936–1938/1991), Helfrich (1984), and Malinowski (1922/1978).[7]

The first thing the *toliwaga* has to do is to go into the bush and find a tree that meets the criteria for making a good dugout.[8] Helfrich (1984, p. 47) assumes that this tree may belong to the species *Calophyllum inophyllum*. The tree has to have a particular length and width, and it should be slightly bent so that it will not be too much work for the expert to give the dugout the sickle-like shape that improves its ability to cut the waves. It should not be unduly cumbersome to transport the tree from its place of origin to the village. When the *toliwaga* has found such a tree, he publicly announces his intention to build a *masawa*. For this purpose a communal meal called a *katuyuvisa* is prepared by his wife and members of his own clan. On the next day the toliwaga and a few other men take their axes and go to the bush where the tree is growing. After they clear the bush around the tree, the *toliwaga* or another expert magician performs the first magical rite. The magician makes a small incision in the trunk of the tree and puts a piece of food or a betelnut into it. He then recites a magical formula that addresses the *tokway*, harmful dwarf-like beings in whose existence the Trobriand Islanders believe. They have long beards and sleek long hair, play unpleasant tricks on the Islanders, steal yams, cause sicknesses,[9] and can make themselves invisible. The incantation makes the *tokway* leave the tree. To give them time to do so, the party leaves the spot in the bush and has a second breakfast at another place. Tosulala emphasized that this meal includes eating fish. I have no idea whether this choice of food has any connection with the magical rite itself, but if the rite is not performed, the *tokway* of this tree will become angry and sabotage the construction of the canoe in one way or another. The log may turn out to be full of knots or may split just before the process of scooping it out is finished. In those cases, the canoe will not perform properly at sea or will quickly rot. After a while the men return to the tree and cut it down with their axes. A special form of magic is whispered over the axe(s) that will fell the tree. After the tree is felled, it is measured. Tosulala drew attention to the indigenous Trobriand way of measuring long objects: The length of the log should be four or five times the span of two extended arms.[10] After measuring the tree, the men trim it, lop off the branches, and

---

[7] For similar descriptions see Pule (1983) and Whakataka-Brightwell (1994). See also Finney (2006), Nayak (2008), and Thompson and Taylor (1980).

[8] Munn (1977, p. 42) mentions a canoe origin myth on Gawa, on the northern tip of Elcho Island in the Arnhem Land of Northern Territory, Australia.

[9] The *tokway* can cause sicknesses by inserting small sharp and pointed objects into persons' bodies. There are magicians who know extraction magic and produce the inserted objects at the end of their magical healing session (see Schiefenhövel 1986). *Tokway* can also have names. The Trobriand Islanders' inventory of string figure games (*ninikula*) contains a figure that refers to a *tokway* called Tokemtuya (see Senft and Senft 1986, pp. 149–150). See also Munn (1977, p. 41).

[10] The *masawa* that was exhibited in Berlin in 1985, the *Meraga*, was 5.93 m (19 ft 6 in.) long, 1.08 m (3 ft 6 in.) high, and 1.45 m (4 ft 9 in.) wide. It was built in Tauwema, owned by chief Kilagola, and acquired for the Ethnographic Museum in Berlin–Dahlem by Wulf Schiefenhövel in 1983.

cut the log to both the desired length and rough shape of the canoe. The tree is left in the bush, and the men return to their village.

On the following day the *toliwaga* tells the people in his village that he and his friends are going into the bush to the felled tree to start hollowing it out. All the villagers accompany him, and they join in when he commences with the work. The account by Malinowski (1922/1978) differs from Tosulala's description at this point. In Tauwema all villagers, not just the *toliwaga*'s kinsmen and friends, help cut the log to the desired length and shape. They even begin hollowing it out. In the afternoon they all go back to their village, where the *toliwaga* has organized another communal meal, for which he even provides a pig besides the typical food and betelnuts.

A few days later the *toliwaga*, a magician, and many other men return to the log to bring it into the village. The undertaking is not easy, but the magician knows special magic, which Malinowski (1922/1978, p. 129) calls the "*Kaymomwa'u* Spell." It makes the log lighter. He whips the log with a rope. Malinowski also mentions both a dry banana leaf that is lain on the log and a bunch of dry lalang grass that serves as a kind of whip with which to reduce the weight of the log. After the magician has whipped the canoe, he discards this heavy grass and takes another bunch of grass, the light bunch, and hits the log again to make it lighter. Then the men pull it to the village. In most cases on Kaile'una Island the men try to find a path from the bush down to the sea so that they can float the log at least part of the distance. Even so, they have to drag it out of the bush by using small round timber as rollers. During this heavy work, the *toliwaga* again distributes food, betelnuts, and pieces of cooked pig to the magician and other men who work very hard. These gifts, the *puvaya*, are announced like prizes awarded during a harvest competition. Finally, the men arrive with the log in their village, where it is left bottom up either on the village square, near the house of the *toliwaga*, or on a spot close to the beach.

Malinowski (1922/1978, p. 130) mentions that on the following day the *toliwaga* ceremonially cuts off the rope with which the log had been hauled to the village. Tosulala and Kilagola did not mention this detail. Tosulala pointed out that the *toliwaga* will cut and hollow out the log with his adze (*ligogu*). However, this work can also be done by another expert, the builder of the canoe (*tota'ila waga*). It takes a long time, during which the kinsmen and friends of the *toliwaga* usually construct a shed for the canoe. Tosulala mentioned that it is an excellent omen if in this phase the *tota'ila waga* wakes up at night, goes outside his house, and sees shooting stars in the south, the direction of what the Trobriand Islanders regard as the mountains— the D'Entrecasteaux Islands and Dobu Island in Papua New Guinea—where the Kula partners of the Trobriand Islanders live. Tosulala also emphasized that the *tota'ila waga* knows a specific kind of magic for his adze. Malinowski (1922/1978, pp. 130–131) refers to this formula as the "*Kapitunena Duku* Spell." It takes a long time to chant it over the *ligogu*, but this magic is performed repeatedly while the canoe-builder works on the outside of the log. When he has finished this part of his job and the outside of the dugout has received its final polish, the *tota'ila waga* turns the log over so that it is bottom-side down. He takes a different kind of adze, the *lali* or *kavilali*, which is inserted in a handle with a movable part that "allows cutting to be done at varying angles to the plane of striking" (Malinowski 1922/1978, p. 132). He uses this adze to hollow out and cut the inside of the log. It goes without saying that another specific kind of magic, the "*Ligogu* Spell" (p. 132), goes with this tool.[11] Sometimes the *tota'ila waga* also uses a pickaxe (*kabolu*) to do such work. Scooping out the log is difficult; it takes great skill, especially during the final phase when the walls of the dugout have to be made sufficiently thin. This stage, too, takes a relatively long time. When it is finished, an expert carver gives the canoe its specific decoration pattern, the *pusa*, at both ends of the dugout and another decoration, the *talapwapwa*, on the upper part of the dugout. To execute these ornamentations, the master must know a particular form of carving magic.

After the *tota'ila waga* has finished this task,[12] he sets about making the other parts that are necessary for the construction of the canoe. Tosulala began this part of his account by mentioning the *gelu*, which are L-shaped ribs that will hold the side planks and stabilize the connection between the body of the dugout and the boards at its sides. Depending on the length of the canoe, 12–20 pairs of these ribs have to be made and joined into a U-shape within the canoe. The four prow boards are carved, first the initial set of two *lagim* and then the two *tabuya*. All four boards differ in their carved designs. Many of these boards are marvelous and admirable pieces of art (see Campbell 2002; Munn 1977, pp. 47–50; Senft 1993, 2005; Scoditti 1990) by expert and renowned master carvers (*tokabitam*), whom the *toliwaga* pays handsomely for their work if he is not a carver himself (see Fig. 11.4). The carving of the prow boards, too, calls for specific forms of magical formulae and rites. Kinsmen and friends of the *toliwaga* assist him again when it comes to making four broad and long planks that will form the gunwale. At the same time other men also prepare long poles (*kesuya*) critical for the longitudinal strengthening of the ribs and for the outrigger platform (*pitapatila*). They cut short poles (*kaitota*) that must serve as transversals for this platform and as main supports of the outrigger. They also fashion the small sticks that are to connect the float with the transversals. (Tosulala did not mention these sticks in his description but did refer to them when they were needed for constructing the canoe.) After these kinsmen and friends have finished their work, the *toliwaga* organizes another communal meal for them as payment for their efforts.

Then the outrigger has to be made. Another tree, Tosulala referred to a black palm, is felled (after the *tokway* magic has been performed, of course). The float is cut in the bush, either by the *toliwaga* or his friends and kinsmen, and is brought into the village. At the same time, the *liu* sticks are cut. These cross-girths connect the canoe with the outrigger and constitute the basis of the outrigger platform, which is on the vessel's windward side (*kelamila*). In the meantime the *tota'ila waga* inserts the *gelu* ribs into the canoe, assembling each pair of them into their characteristic U-shape. The men cut the gunwale's lower plank (*budaka*)—or "strake" (Haddon and Hornell 1936–1938/1991, Vol. II, p. 270, Fig. 157)—and

---

[11] This logic was obvious to Tosulala, which is probably why he did not mention the spell in his description.

[12] Malinowski (1922/1978, p. 133) notes that "parallel with the process of hollowing out, the other parts of the canoe are made ready." This observation does not seem to hold for what is done in Tauwema.

Fig. 11.4 The prow of a
masawa canoe used by
visitors from Vakuta Island
on the beach of Tauwema
(1983)

upper plank (*sipa*). Thereafter, the *tota'ila waga* takes a traditional drill (*kegigiu*)
and creates holes (*kogwa*) in the *budaka*, through which planks will be attached to
the outrigger side (*kelamina*). These holes will later receive the *liu* sticks. All parts
of the canoe are now ready.

Malinowski (1922/1978, p. 133) reports that on Kiriwina the magical rite named
kapitunela nanola waga is performed at this juncture. The term means "cutting the
canoe's mind" and implies the shaping of the canoe's determination. The *toliwaga*
or another expert magician recites a short spell over drops of coconut oil, which are
subsequently wrapped up in a bundle. The incantation is recited over the adze, and
the bundle is put inside the canoe and struck with the adze. This magic is intended
to make the canoe sail at high speed.[13]

---

[13] Neither Tosulala nor Kilagola mention this kind of magic. The incantation I received from Yoya
in 1989 belongs to the types of formulae that are used to make canoes sail swiftly. However, magicians usually cast this magic on the canoe before it starts to sail away. The following account by
Tosulala differs substantially from Malinowski's (1922/1978, pp. 134–144) description of the next

Now the canoe parts are ready to be put together. First, an expert magician performs the *katuliliva tabuyo* rite over the *lagim* and *tabuya* prow boards. These boards are thereupon inserted into their grooves at both ends of the canoe, and the side planks are put onto the edge (*tolena*) of the dugout to constitute the gunwale.[14] The side planks are preliminarily lashed to the *gelu* ribs and onto the dugout's upper rim (*kilababa*). The canoe-builders now take a 4-day rest.

To resume their work, the *toliwaga* and his helpers take ropes and *veyugwa* lashings, or lashing creepers made of lianas (*Lygodium circinnatum* [?]; Helfrich 1984, p. 48), and retie the side planks and the *gelu* ribs. Before the *veyugwa* are used to join the canoe parts, a magician recites the *Wayugo* Spell over them (Malinowski 1922/1978, pp. 137–139). The *liu* sticks are inserted through the holes in the *budaka* on the outrigger side. They rest upon both edges of the dugout and are tied to the *gelu* as well.

The construction of the outrigger platform follows. Everything that has to do with this phase of the canoe's construction is closely monitored either by the *toliwaga* himself or by an expert known as a *tolilamila*. Long poles called *kaitota* are set perpendicular over the *liu* sticks and tied to them, and the *vatota* sticks that connect the float with the outer top pole of the platform are sharpened. The next day the float is hardened and blackened over an open fire (see Fig. 11.5). Pairs of long undercrossed *vatotua* sticks are hammered into the float as well as tied to and interlocked with the platform's first vertical outer top pole (*kedudula*). In addition there are strengthening spars (*donaba*), in some cases only two, each of which is tied to the *vatotua* sticks at the float and made to slant upward to about the strake under the platform and over the *liu* sticks (Haddon and Hornell 1991, Vol. II, p. 270, Figure 157).[15] Below the *budaka*, there is another longitudinal pole, the *kesukwava*, which is tied to the body of the canoe. In the space between the *kesukwava* and the *liu* sticks are the *donaba*, which are tied to the *kesukwava* pole. This work is very difficult because the canoe-builder must ensure that the float is perfectly parallel to the dugout to optimize the maneuverability of the vessel when it is sailed. That is the manner in which the float is connected with the outrigger platform. The workers make sure that all parts of the canoe are tied together properly. Tosulala mentioned four different knots used for this purpose: *nabwasiko* (from Vakuta Island), *namkilavava* (from Tauwema), *mumyobikwa* (from Simsim Island), and *yumnukwausa* (from Kitava Island). Simultaneously, others lending the *toliwaga* a hand build a shed (*buneyova*) for the canoe.

---

stages of constructing a canoe. Malinowski mentions that the next stage commences with another magical rite (the *Katuliliva tabuya*). Thereafter, the prow boards are mounted on the dugout, which is ceremonially launched for the first time. After magicians have recited a number of different formulae over the dugout, it is ceremonially washed and heaved ashore, its parts tied together, and the assembled canoe caulked.

[14] Tosulala forgot to mention that the four prow boards (the *lagim* and *tabuya*) are first put in place.
[15] Neither Tosulala nor Malinowski (1922/1978) mentioned these sticks, but see Haddon and Hornell (1991, Vol. II, p. 270).

Fig. 11.5 Topiesi hardening a new outrigger for a *masawa* canoe in the fire on the beach of Tauwema (1983)

The next day the canoe is caulked. People cut out roots of a specific tree (Helfrich 1984, p. 48, assumes it could be the parinarium nut tree, *Maranthes corymbosa*, Bl.). They scrape off parts of the roots and mix them with water, and the *toliwaga* or another expert magician recites a charm over this substance (*kebasi*), with which that person caulks the canoe with the utmost care. This job is concluded by another communal meal that the *toliwaga* organizes for all the people who have aided him during these phases of the canoe's construction.

The day thereafter people burn chalk to make lime with which to paint the canoe prows, the canoe boards, and the tips of the dugout (*kabununa*). They also go and collect red soil, the ochre with which to paint the *lagim* and *tabuya* boards and the side planks of the canoe. Coconut husks are burned to produce charcoal with which to paint these planks and prows black. After the various canoe parts are painted, white cowrie shells (*Ovula ovum*) are tied to the canoe prows and the tips of the dugout to serve as a decoration termed *buna*. Sticks are bound together and also tied to the canoe. Lastly, pandanus streamers are tied below the outrigger platform as extra decoration.[16]

---

[16] Malinowski (1922/1978, pp. 139–140) mentions three more magical rites that have to be performed before the canoe is painted. They are "exorcisms against evil influences" (p. 139). In the *Vakasulu* the magician must prepare a veritable witch's cauldron containing "the wings of a bat, the nest of a very small bird,…some dried bracken leaves, a bit of cotton fluff, and some lalang grass," which he subsequently burns beneath the canoe, an act that has a cleansing and speed-giving influence. The *Vaguri* is an exorcism in which the magician strikes the body of the canoe with a wand, expelling evil witchery. In the *Kaytapena waga* rite the magician puts a spell on a coconut-leaf torch and fumigates the canoe with it. This rite, too, cleanses the canoe and increases its speed.

The next day one of the brothers of the *toliwaga* goes to the bush early in the morning to cut *yakwara* and other trees. He will bring them to the canoe shed and use them to build a slipway from the shed to the sea, work that takes all day. On the following night the *toliwaga*—Tosulala also referred to him as the *totatai* (the cutter of the canoe) sleeps in the canoe shed, but not on the canoe, for that act is tabooed. During the night he will hear whether the canoe was tied together properly. If the *liu* sticks or *vatotua* sticks shake, he will know that parts of the canoe must be retied in the morning. The following day the *toliwaga* or other expert recites magic over the canoe, the logs, the pandanus streamers, and the conch shell trumpet (*tauya*) that is put into the canoe just behind the *lagim* prow. These magical rites take all day.

The next morning an expert magician crawls under the outrigger platform of the canoe, sits there, and recites another magical formula. He strikes the canoe at several places with fibers on which he has cast a specific spell. He also shakes the vessel with his hands while reciting another charm over the canoe. These forms of magic will make the canoe light so that it will be easy to move from the shore into the sea or from the sea onto the shore. Before the toliwaga launches his new canoe for the first time, he names it (Munn 1977, p. 41). He takes the conch shell and holds it in front of the canoe as strong men pull it down from its shed, over the slipway, and into the sea. The magic put onto the conch shell will make the canoe lighter for the men who lug it into the sea. This kind of magic is done only before and after the canoe's first four trips.

When the canoe is afloat, the magician blows the conch shell. When the villagers hear the sound, they come to the beach and inspect and give exhaustive critique of the new canoe. The *toliwaga* or an expert magician first washes the canoe and puts a spell on it to make it swift. The *toliwaga* and his crew then board the canoe to paddle and test it on the reef at high tide. Eventually, they paddle it ashore again and pull it to its shed. Before it may be pushed into its shed, however, the villagers attack the canoe as it were. They throw stones, sticks, papayas, young coconuts, and rotten coconuts at the *lagim* prow until it breaks (the reason why the initial set of lagim boards is usually thinner and less artistically carved than the final set). A broken *lagim* after such an attack is taken by the *toliwaga* and the whole village community as a positive sign of the quality of the canoe, especially its swiftness and safety. The person who manages to break the *lagim* receives a pig from the *toliwaga* as a prize. The *toliwaga* treats all the villagers to a major feast during which he not only feeds them well but also distributes valuables like stone-axe blades, belts, small necklaces, pigs, and betelnuts. His guests sing the traditional *Kapoka* songs that celebrate new canoes and praise their owners (Malinowski 1922/1978, pp. 146–149). After this feast the broken *lagim* is replaced by the stronger and more elaborate real one.

At this stage the sail (*naya*) is made. The *toliwaga* and many other villagers go into the bush and cut pandanus leaves. The leaves are brought into the village, the thorns on the sides of the leaves are cut off, and the leaves are dried in the sun. They are thereafter rubbed with specific stones and dried in the sun again. The drier the leaves, the stronger and smoother the sail will be. When the leaves have become soft, they are rolled up. After this task the *toliwaga* makes a rope, a so-called *tasiu*. When the rope is finished, he or an expert sail-maker marks the shape of the canoe on the ground. He drives stakes into the ground to outline the triangular frame of the sail and connects the stakes with the *tasiu*. As Malinowski (1922/1978, p. 140) indicates, an old sail may sometimes serve as a pattern. Sails match the size of the canoe, and the better this match, the better the canoe will sail. Then the pandanus leaves are unrolled and laid out on the ground within the frame, with people immediately lying down on them for a while to keep them flat. This practice is probably the funniest part of the work, both for participants and observers. When the pandanus leaves finally remain flat on their own, the sail-makers sit on them and sew them together, starting at the apex of the triangle. This technique gives the sail an intricate pattern. The sail-makers formerly used bones of a flying fox as needles. Their threads are strips of specially toughened pandanus leaves. Two layers of leaves are sewn one on top of the other to produce a solid fabric, the typical triangular Oceanic lateen sail (*Velum latinum*; Höver 1957). When the sail is finished, the *toliwaga* organizes another communal meal for the people who have worked with him.

If the *toliwaga* also needs a new mast for the canoe and poles for the gaff, they have to be made, too. He needs the main mast (*vania*), the long pole on top of the sail (*kunaya*), the pole at the bottom of the sail (*kenaya*), the mast support (*kena'ila*), and a good deal of rope (both *soya* and *tapwai*) for setting the sail and handling it when the vessel is in use (see Fig. 11.6). When all these parts are ready, the *toliwaga* and his crew do a short-distance trial run with the new canoe, sailing it for a few hours. Tosulala referred to this trial run as *i-valakola-si tolilamila*.

Fig. 11.6 Nusai's *masawa*, *Seguvagava*, under full sail, with Mokeilobu leaning against its mast

When the members of the crew sail the canoe into the wind, they immediately know whether the frame that holds the dugout and the float together is as strong and flexible as it has to be for a seaworthy *masawa* canoe. All the villagers gather at the beach again when the new canoe is pushed into the sea—the moment at which the crew puts up the mast at the third or fourth *liu* stick and sets sail. And they are there when the canoe returns in the afternoon. Thus ended Tosulala's description of how to make a *masawa*.[17]

His account shows that the construction of such a canoe is accomplished in many stages that require organized and coordinated individual and communal labor and the assistance of many experts (see Table 11.1). These experts are paid by the *toliwaga*, and after phases of the building process that must involve communal labor, those who have contributed are honored and their work is acknowledged with communal meals organized and sponsored by him. The building of a canoe is always guided by magic. Only the traditionally appropriate coordination of craftsmanship and effective magic guarantees the successful outcome of this

Table 11.1 The traditional process of building a *Masawa* canoe on the trobriand Islands

Phases and duration	Protagonist(s)	Activities	Magical rites	Communal meals and prizes sponsored by *toliwaga*
1. Several days	*Toliwaga*	Searching for a tree for the dugout		
2. One day	*Toliwaga*	Publicly announcing intention to build a *masawa*		Meal (*katuyuvisa*) for all villagers
3. One day	*Toliwaga* and helpers (kinsmen, friends) Expert magician (if needed) *Toliwaga* *Toliwaga* and helpers	Clearing the bush around the tree Felling the tree and cutting the log to the desired length	*Tokway* magic Magic for axe(s)	Picnic for helpers and magician
4. One day	*Toliwaga* and fellow-villagers *Toliwaga*	Cutting the log to the rough shape of the dugout		Meal for all villagers

(continued)

[17] It is interesting to compare Tosulala's account with Malinowski's (1922/1978, pp. 105–150) description of how canoes are built on the Trobriands. Tosulala's description and my morpheme-interlinearized transcription of it, along with Mokeilobu's description of how to make a sail and my morpheme-interlinearized transcription of that explanation, can be read and heard at http://www.mpi.nl/people/senft-gunter/research

Table 11.1 (continued)

Phases and duration	Protagonist(s)	Activities	Magical rites	Communal meals and prizes sponsored by *toliwaga*
5. One day	Expert magician *Toliwaga* and helpers *Toliwaga*	Bringing log into the village	*Kaymomwa'u* Spell to lighten log	Food and prizes for helpers
6. Several months	*Toliwaga* Helpers Expert carver	Cutting off the rope with which the log was pulled; hollowing out the log Building a canoe shed Carving *pusa* and *talapwapwa* decoration	*Ligogu* Spell on adzes Carving magic	
7. Several months	*Toliwaga* and helpers Expert carver	Making other parts of the canoe Carving *lagim* and *tabuya*	Carving magic	
8. One day	*Toliwaga*			Communal meal for helpers
9. A week or two	*Toliwaga* and helpers	Making the float and the outrigger construction; ensuring that all parts of the canoe are ready to be bound together		
10. One day	Expert magician *Toliwaga* and helpers	Making the gunwale and lashing the planks together	*Kapitunela nanola waga* magic; *Katuliliva tabuyo* rite over *lagim* and *tabuya*	
11. Four days	All	Resting from work		
12. A few days	Expert magician *Toliwaga* and helpers	Tying the planks together with lashing creeper; constructing the connection between canoe and float; hardening the float; making the outrigger platform; connecting float to platform and windward side of canoe; tying all parts together; building a shed for the canoe	*Wayugo* Spell on lashing creeper	
13. A day or two	*Toliwaga* and helpers Expert magician *Toliwaga* or other expert	Making the *kebasi* substance for caulking the canoe Caulking the canoe	*Kebasi* Spell on the caulking substance	

(continued)

Table 11.1 (continued)

Phases and duration	Protagonist(s)	Activities	Magical rites	Communal meals and prizes sponsored by *toliwaga*
14. One day	*Toliwaga*			Communal meal for helpers
15. A day or two	Expert magician *Toliwaga* and helpers	Making colors and painting the canoe, its planks, prows, and boards; decorating the canoe	Magical exorcisms against evil influences	
16. One day and night	Brother of *toliwaga* *Toliwaga*	Cutting trees for slipway and constructing it from canoe shed to the shore Sleeping in shed and checking whether the canoe is properly tied together		
17. One day	*Toliwaga* and expert magician		Magical formulae over canoe, slipway, and conch shell to lighten canoe before first four or five voyages	
18. One day	Expert magician *Toliwaga* and crew Expert magician Villagers *Toliwaga* and expert magician *Toliwaga* and crew Villagers *Toliwaga* Villagers	Naming and launching the canoe Blowing the conch shell Inspecting new canoe and critiquing it Trying out the canoe, paddling on the reef at high tide Attacking the canoe when it approaches the beach and breaking its *lagim* with objects which are thrown at it Singing *kapoka* songs	Magic under outrigger platform to lighten canoe Magical washing of canoe to make it swift	Donation of a pig to the person who breaks the *lagim*; communal meal for all villagers Distribution of valuables to helpers
19. A few days	*Toliwaga* and helpers *Toliwaga*	Making the sail, the mast, its support and poles for the gaff		Communal meal for helpers
20. A day	*Toliwaga* and crew	Trial run, clause-hauled sailing of the canoe		
21. Half a day	Expert magician		Special magic after first four or five runs of the new canoe to make it swift and light	

enterprise. As Malinowski (1922/1978) puts it, the organisation of labour in canoe-building rests on the one hand on the division of functions, those of the owner, the expert and the helpers, and on the other on the co-operation between labour and magic (p. 116).

The Social Distribution of Knowledge Necessary for Making a *Masawa* and the Social Implications of the Joint Communal Activities during This Process

Table 11.1 reiterates that making a canoe directly engages not only a select number of people and the *toliwaga* but, on three occasions, the whole village community as well. The *toliwaga* relies on the aid of his kinsmen and his friends. Constructing the canoe properly demands expertise in more than ten forms of magical rites and the knowledge of the respective formulae. It is simply inconceivable to a Trobriand Islander that a *masawa* could be constructed without the appropriate magical rites being performed.[18] Indeed, at least two additional rites and formulae have to be performed by the master carver who decorates the *pusa* and *talapwapwa* ornaments and the *lagim* and *tabuya* prow boards. The only former *toliwaga* I knew to be versed in the *tokway*, *ligogu*, and *kapitunela nanola waga* forms of magic, magical exorcisms against evil influences, and the magic to make his canoe safe and fast was Kilagola, the chief of Tauwema, who died in 1991. He was also a master carver who knew all the formulae and rites for carving the *lagim* and *tabuya* prows. Nevertheless, even he did not claim to know all the different forms of magic essential for building a *masawa*. This knowledge is distributed among expert magicians within a village or, in some cases, within an island community.

Ordinarily, there are also expert craftsmen who have specialized either in making specific parts of the canoe and its equipment, for which they use 15 different kinds of wood,[19] or in assembling parts of the construction. Moreover, the *toliwaga* may ask other experts to seek out the appropriate tree for the dugout and caulk the canoe for him, or at least to do it together with him. Making the sails, too, normally entails a number of specialists. Mokeilobu, Moagava, Mogega, and Nusai are all expert sail-makers in Tauwema. Thus, all the knowledge that must go into making a *masawa* canoe and its sail is distributed within a village (or an island) community.

This distributed knowledge ought to be integrated, of course. Meeting that need is one of the central functions of the communal meals arranged by the *toliwaga*

---

[18] On the importance of magic to the Trobriand Islanders, see Malinowski (1922/1978, 1935/1965, Vol. II; 1974) and Senft (1985, 1997a, 2001).

[19] See the Appendix for the named parts of a *masawa*, the implements and materials needed to make such a canoe, and some of the equipment needed to sail it. Lists like it have a rather old tradition (e.g., Schnepper 1908). See also Breidbach (1988) and the literature cited there.

during or after specific phases of the building process. As noted in the previous section, the *toliwaga* treats his kinsmen and his other helpers to three full meals, even four if the sail has to be made. He organizes a picnic for them while they wait for the *tokway* to leave the tree selected for the dugout. And he distributes valuables among them when they bring the log into the village. But that is not all. The *toliwaga* puts on three full meals for all the inhabitants of his village, and at the final communal meal he again distributes valuables and awards a pig to the person who manages to break the *lagim* of the canoe when it is returned to the beach after its first trial run. All these communal meals serve at least the following two quite antagonistic functions (Eibl-Eibesfeldt 1989):

> First, as strategies of rank striving and self-presentation in the highly competitive Trobriand society, the communal meals and the distributions of valuables and gifts serve to increase the status of the already quite exposed *toliwaga*; they confirm publicly that he is an economically and politically influential person who can afford to build a *masawa*.
>
> Second, with their bonding function as rituals of reciprocal care and unification, these communal meals and the distributions of valuables and gifts maintain at the same time group harmony amongst the villagers; they integrate not only the experts and the other helpers of the *toliwaga*, but also all his other fellow-villagers into his canoe building enterprise. Moreover, as strategies of support, they also contribute to the resolution of possible conflict which could arise just because of the exposed position of the *toliwaga* as a person of high status. (pp. 520–521)

Thus, despite the fact that these communal activities sponsored by the *toliwaga* increase his personal status, they are, above all, a means of bringing together the expert canoe-builders, the magicians, and the village community as a whole. Group harmony is achieved. The village community and the group of expert craftsmen and magicians involved in making the canoe accept that one of them demonstrates his rank as an economically powerful person because he is willing and can afford to redistribute a relatively high amount of his accumulated wealth to them.

"Masawa—bogeokwa si tuta!" (*Masawa* Canoes—Their Time Is Over!)

This chapter has now indicated what a complex enterprise the making of a *masawa* canoe is, how the knowledge and expertise for doing so is socially distributed, and what kind of social implications the joint communal activities during this process has for the *toliwaga*, his experts, and the village community as a whole. In what follows I first try to answer the question of why most of the village communities on the Trobriand Islands have been gradually losing the knowledge of how to make a *masawa*. In the final section of this chapter I outline and discuss the implications of this loss for the culture of the Trobriand Islanders, their social construction of reality, and their cognitive capacities.

As outlined above, making a *masawa* canoe required the *toliwaga* to be someone who disposed over all the food resources and other items of wealth and value
that he needed in order to sponsor all the communal meals and other activities
inextricably intertwined with the canoe-making enterprise. In 1983 the National
and Provincial Governments of Papua New Guinea strongly promoted the policy
of transforming their national economic systems, most of which were still based
on barter, into a modern capitalist money economy. In those days the national currency, the Kina, was strong, so it was amazing how much money flooded the
market everywhere in the country. The Trobriand Islanders, at least those living
near the ocean and having fishing rights, earned well by selling fish to a trawler
that was hired and subsidized by the Milne Bay Provincial Government. Many of
the Trobriand Islanders also had relatives who worked either in the capital of
Milne Bay Province (Alotau) or in other, larger cities in the country (e.g., Madang
and Port Moresby). These Trobrianders had always felt obliged to send fairly
significant sums of money to their relatives back home on the Islands. When I
returned to the Trobriands in 1989, I immediately realized that the new national
and provincial politics had been highly successful within a few years. There was
a great deal of money on the Islands. Petrol and oil were relatively cheap at that
time, so it had simply proven to be less expensive to buy and use a dinghy powered by an outboard engine than to organize the making of a *masawa* canoe. In
1989 three men in Tauwema owned their own dinghy. In 1992 there were already
seven dinghies lying on the beach of the village. Within 10 years the dinghies had
replaced almost all the *masawa* canoes in Tauwema. Only one of these traditional
vessels had survived.

This development had implications for the experts who were needed to make a
masawa canoe. The skills of the magicians were no longer in demand during the
process of constructing a *masawa*. The dinghies could make do with nothing more
than magical exorcisms to protect them and their crews against evil influences. This
change also affected other expert craftsmen who had cooperated with the *toliwaga*
to make a canoe. Their skills, too, had become obsolete. Neither these expert craftsmen nor the vast majority of the expert magicians could continue finding appren-
tices to whom they could bequeath their knowledge. It had lost its value, and
members of the younger generation no longer saw sense in learning these craft skills
and magical formulae in lengthy instruction under their elder relatives. Most of
these expert craftsmen and magicians have since died, taking their knowledge with
them.

The *misinari* observed these changes with great interest and strongly favored
them. Ever since 1894, when the first missionaries commenced work in the Trobriand
Islands, they had been attempting to displace the magicians with respect to their
official status, rank, and power. In 1983 Christians on the Trobriand Islands lived in
an interesting form of syncretism in which traditional belief in magic and Trobriand
eschatology (Malinowski 1974) were combined with Christian ideas (Senft 1994,
1997b, p. 53). By 1992 these syncretic features of Trobriand Islands Christianity
had decreased dramatically. Belief in magic was not denounced directly by the

misinari as something heathen. The strategy they pursued to fight these pagan customs was much more subtle: The *misinari* argued that there were two ways to live one's life. One was the old, traditional way, which includes magic and the eschatological belief in the immortal spirits of the dead living in the underground paradise on Tuma Island (see Malinowski 1974; Senft 2011). The other way was the new Christian way of life, with its specific Christian beliefs and its own eschatological ideas. Both ways are mutually exclusive, or, as the local priests put it, "one can either walk the path of the ancestors or take the Christian way together with Jesu Keriso, the Lord Jesus Christ." The modern developments that led to the transformation of their indigenous economic barter system into a capitalist economy in which buying a dinghy with an outboard engine was cheaper than making a *masawa* catalyzed the missionaries' strategies in their fight against magic and magicians. The fact that the influential magicians who contributed decisively to making a *masawa* were no longer either needed or powerful signaled an important victory for the *misinari* in their long fight against these experts. It proved the *misinari* right in their calls to abandon the ways of the ancestors. I return below to the question of how the local *misinari* derived further advantage from this development.

Processes of globalization have reached the Trobriand Islands as well in recent years. In the meantime the soaring price of oil has made it quite expensive to use dinghies. The only Trobriand Islanders who can still afford to use them are those who had invested in small businesses like fisheries or public interisland transport from which they earned well. The vast majority of the former dinghy owners, however, can no longer afford the high price of zoom, the mix of gasoline and oil they use as fuel for outboard engines. Instead, they now use the *ligataya* to paddle and sail from one island to the other or the rather simple and relatively small *kemolu* for fishing and coastal traffic. These types of canoe types can be made with a minimum of ritual knowledge and do not entail knowledge of magical formulae at all. They have superseded both the splendid *masawa* canoe and, it seems, the common use of dinghies powered by outboard engines. Because the *ligataya* canoe is not really suitable for covering great distances—it is nowhere near as safe as the *masawa* canoe—most Trobriand Islanders use the airplane for such travel, or they use a ship that regularly connects Losuia, the district center on Kiriwina Island, with Alotau. Since 1982, cultural and social change initiated by a gradually growing capitalist economy and catalyzed first by missionaries and more recently by globalization have, in most village communities on the Trobriand Islands, led to the erosion or even complete loss of the complex distributed knowledge of how to make a *masawa* canoe and its sails.

Implications of This Loss

In this chapter I have pointed out that the construction of the impressive and beautiful *masawa* canoes involved from the outset a number of different experts and their knowledge of magic. It also called for strict adherence to the rules of various rituals in which magic played an important part. Although the construction of such a canoe was always initiated by an individual, namely, its future owner, the whole enterprise had important social implications. The process of building a canoe continuously tested and monitored the security and stability of a village community's social network. Initiating the construction of a canoe meant a communal effort that relied on the aid of expert magicians, carvers, and sail-makers. They all had to cooperate in good faith to ensure the success of the canoe under construction, and they all had to be paid in the form of adequate food-distribution ceremonies after certain stages in the construction process. Such ceremonies were among the highlights of the Trobriand Islanders' year and automatically involved experts as well as the entire village community (or the entire sector community in relatively large villages). The experts were more publicly honored than paid during these ceremonies.

In 1992 only five men in Tauwema still knew the correct rituals and ceremonies that accompany the making of a *masawa* canoe. However, they themselves were no longer able to initiate the construction of such a canoe, and they had no one to whom they could pass on their knowledge. The knowledge of how to make such a canoe has since been lost in Tauwema and many other villages on the Trobriand Islands.

However, technologies are not all that have been lost; social events that used to be intertwined with those technologies have disappeared, too. These social events had the important function of rituals as forms of social bonding. All such rituals and ceremonies were accompanied by speeches that were clearly defined by their appropriateness to the stage of the construction process. From what my informants told me, I infer that these speeches had their own pragmatics. The knowledge of these pragmatics is lost as well now. Social events that are intended today to take over the key social functions of the rituals and ceremonies accompanying the construction of the *masawa* canoes are activities organized by the missionaries. They include outdoor communal prayers and hymn-singing in the village center and meetings that involve the whole village community and that are conducted by the *misinari* for missionaries from neighboring villages or for Christian women's associations. These get-togethers are also accompanied by communal meals, and sometimes even a kind of food distribution ceremony precedes these meals (Senft 1997b, pp. 53–54). These activities are rooted in Christian beliefs. They are completely different in structure from the traditional events, especially with respect to speeches and other forms of verbal communication that once took place during the long process of making a *masawa* canoe.

The loss of knowledge about how to make a *masawa* canoe also implies the loss of the corresponding specialized vocabulary in the Kilivila lexicon (Senft 1992, p. 78) and the loss of the text category known as *canoe magic*. As emphasized above, all the experts who were needed to make a *masawa* were convinced that they could perform their work properly only by invoking the power of their magic. The making of a *masawa* required the use of a broad and rather complex variety of incantations. When the *masawa* were no longer needed, neither were these formulae, and they are now lost.

Their loss, however, is embedded in more general processes of cultural change on the Trobriands. Until recently, all Trobriand Islanders used magical formulae to achieve certain aims, firmly convinced that they could thereby influence and control nature, the course of their lives, and events therein (Senft 1997a). The Trobrianders differentiated between various forms of magic. They knew weather magic, black magic, healing magic, garden magic, fishing magic, dance magic, beauty magic, love magic, sailing and canoe magic, and magic against witches and sharks. There were specialists for certain kinds of magic, and all magic was regarded as personal property. When I first arrived on the Trobriand Islands in 1982, magic still played a dominant role and the power of magicians and their magical formulae clearly pervaded everyday life there. In 1983 chief Kilagola gave me parts of his canoe magic as a present when he adopted me as one of his sons. His brother Weyei bestowed me with a similar present: five formulae of his weather magic as a sign of his friendship (Senft 1985). And Vaka'ila, one of the oldest men of the village, presented me with several formulae of his garden magic because I reminded him of his late brother Keyalabwala. These three men were the only persons who offered me such personal and secret information, and I was rather proud of being honored by them in this way. In 1989, at about the same time that I initially realized there were only a few *masawa* canoes left in Tauwema, more than 12 women and men approached my wife and me and offered to sell magical formulae for money and tobacco. We felt as though we were in the middle of a fire sale on magic. The gesture was clear evidence of the fact that magical formulae had lost their importance for the majority of Trobriand Islanders. It was the obvious outcome of the old fight between traditional magicians and Christian missionaries. Because the belief of the Trobrianders in the magical power of words included the conviction that magic was a means of controlling nature and the incidents affecting their personal lives, the loss of that conviction resulted in a political and ritual power vacuum, which the *misinari* used for their own ends (Senft 1992, pp. 79–80). In 1992 the magician's ritual and political power in Trobriand society was finally superseded by the priest's.

Metalinguistically, the magical formulae constituted a nondiatopical variety of Kilivila, the *biga megwa*—the language of magic. This variety of Kilivila is now moribund (Senft 2010). The demise of *biga megwa* has had implications for the Trobriand Islanders' incredible capacity to memorize forms of indigenous knowledge, be it magical formulae or mythical stories and tales—their oral history. In this nonliterate society all these forms of knowledge had been transmitted verbally only. The members of the younger generation had had to learn and laboriously memorize this knowledge from the older generation with great motivation and dedication and had had to train themselves to retrieve it when it was needed. Given that more and more Trobriand Islanders are persuaded these days that the former belief in the power of magic was heretical and completely ungrounded, this transmission of the knowledge of magic has ceased. And people who no longer practice retrieving extensive chunks of memorized texts soon lose this capacity—a rather trivial insight of the psychology of memory and learning.[20] Thus, the loss of this capacity implies

---

[20] On memory in nonliterate societies, see, for example, Baddeley (1990, pp. 150–160; 1999, pp. 287–291), De Groot (1965), and Miller (1956). Miller (1962) points out that "many psychologists prefer to speak of memory as something a person does, rather than something he has" (p. 192).

the end of all forms of indigenous knowledge and oral history that have not been documented in some form of print. Obviously, this loss must have important consequences for the Trobriand Islanders: their individual, social, and cultural identity; their self-evaluation; and the complex construction of their social reality. Therefore, the fact that the Trobriand Islanders have lost, or are gradually losing, their knowledge of how to make a *masawa* canoe is only one minor facet of the massive processes driving the cultural change presently affecting the Trobriand Islanders.[21]

Appendix

Table A1 Bilingual glossary of parts of a *Masawa* Canoe

Kilivila	English	Construction material
Dugout canoe		
Waga	Dugout canoe	*Reyawa* (wood) (?
Lopola waga	Bilge ("belly of the canoe")	*Calophyllum inophyllum?*)
Sibula	Line ahead (of keel)	
Kilababa	Top rim of dugout	
Tolena	Edge of the gunwale (on the *kilababa*)	
Katala	Leeward side of dugout	
Kelamina	Outrigger side of dugout	
Kabununa	Tips of dugout	
Gunwale		
Budaka	Lower plank	*Kaga* (wood)
Sipa	Upper plank	
Gelu	L-shaped ribs	*Reyawa* (wood)
Compartments		
Liku	Compartments (from rib to rib)	
Poles and sticks		
Kesuya	Long poles that go through the holes in the *gelu*	*Kesesa* (wood)
Kaitota	Transversal poles on gunwales inside (optional)	Wood
Liu	Cross-girths going from inside the canoe through a hole in the gunwale's upper plank—windward side—basis for outrigger platform	*Yawoura* (wood)

(continued)

---

[21] In my long project of documenting the indigenous knowledge of the Trobriand Islanders, this chapter is the first work in which I deal with Trobriand canoes and the art of making a *masawa* canoe.

Kilivila	English	Construction material
Pitapatila	1. Long outrigger platform sticks 2. Outrigger platform	*Wagewa* (wood)
Kedudula	First horizontal pole on top of platform	*Kwetaola* (wood)
Kesukwava/kemsukwa	Lower longitudinal pole below the *budaka* plank at the *kelamina* side	*Yadadiga* (wood)
Vatotua	Vertical prop from outrigger (float) to platform	*Bokeyala* (wood)
Donaba	Slanting prop from outrigger (float) to upper plank and *liu* cross-girths	*Usari* (wood)
Float		
Lamina/lamila	Outrigger	*Riga* (wood)
Prow boards		
Lagim	Transverse board	Malea (wood)
Tabuya	Sagittal prow in front of *lagim* on tip of dugout	*Yuyuwi* (wood)
Holes		
Kogwa	Holes in upper rim of the gunwale's lower plank	
Sesuya	Hole in *gelu* rib	
Pwanina	General term used to refer to other holes	
Lashing		
Veyugwa	Lashing creeper	Liana (? *Lygodium circinnatum*?)
Kelugwesi	Lashing of the *kesuya* poles to the lower plank	
Ketawaga	Lashing of the *kesuya* poles to the dugout	
Caulking		
Kebasi	Caulking substance	Made of the scraped parts of specific roots mixed with water (? roots of the parinarium nut tree *Maranthes corymbosa*?)
Beluma	Caulking substance	Seaweed
Mast		
Vania	Mast	*Yowai* (wood)
Kununa	Forked top of mast	
Odinakwau	Forked lower part of mast which will be put onto the *liu*	
Kena'ila	Mast support sitting on the outrigger platform	*Rewaya* (wood)

(continued)

Kilivila	English	Construction material
Sail		
Naya	Sail	Yagawana/kebwibwi
Matala naya	Front of sail (the sail's eye)	(pandanus leaves)
Dabala naya	Back of sail (the sail's head)	
Gaff[a]		
Kunaya	Long pole at the top of the sail	Riga (wood)
Kenaya	Bottom pole at the bottom of the sail	
Unakeli	Carved part at the end of bottom pole	
Ropes		
Soya	Rope to set the sail at top of the gaff	Kind of bast fiber
Tapwai	Rope to set the sail in the middle of the gaff	
Other equipment		
Kuliga	Rudder, helm	Kakaya (wood)
Kaikaila/kaikela	Paddle	Kesesa (wood)
Sususta	Tip of paddle	
Kavala	Punting pole	Wood
Yasika	Piece of wood to sit on in the canoe	
Kweyapapa	Rescue wood	Light floating wood
Yatula	Scoop for bailing water out of the canoe	reyawa (wood)
Tauya	Conch-shell trumpet	
Buna	White cowrie shell decoration	
Vatila	Construction for transporting pots and other items	

[a]The stout, rounded, usually wooden or metal piece on which the head of a sail is extended

References

Baddeley, A. D. (1990). *Human memory: Theory and practice*. London: Lawrence Erlbaum.
Baddeley, A. D. (1999). *Essentials of human memory*. Hove: Psychology Press.
Breidbach, W. (1988). *Die Schiffsbezeichnungen des Alt-und Mittelhochdeutschen: Ein onomasiologisches und etymologisches Glossar* [Ship terms in Old and Middle High German: An onomasiological and etymological glossary]. Arbeitspapier Nr. 6 (Neue Folge). Cologne: Universität zu Köln, Institut für Sprachwissenschaft.
Campbell, S. F. (2002). *The art of kula*. Oxford: Berg.
De Groot, A. D. (1965). *Thought and choice in chess*. The Hague: Mouton.
Eibl-Eibesfeldt, I. (1989). *Human ethology*. New York: Aldine de Gruyter.
Finney, B. (2006). Ocean Sailing Canoes. In K. R. Howe (Ed.), *Vaka moana: Voyages of the ancestor: The discovery and settlement of the Pacific* (pp. 100–153). Honolulu: The University of Hawai'i press.

Gerrits, G. J. M. (1974). Burial canoes and canoe-burials in the Trobriand and Marshal Bennett Islands (Melanesia). *Anthropos, 69,* 224–231.

Helfrich, K. (1984). Boote aus Melanesien und Australien [Canoes of Melanesia and Australia]. In G. Koch (Ed.), *Boote aus aller Welt* (pp. 33–54). Berlin: Staatliche Museen Preußischer Kulturbesitz.

Hornell, J. (1991). *Canoes of Oceania: Three volumes combined into one.* Honolulu: Bishop Museum Press. (Original work published 1936–1938).

Höver, O. (1957). Das Lateinsegel—Velum latinum—Velum laterale [The Latin sail]. *Anthropos, 52,* 637–640.

Malinowski, B. (1965). *Coral gardens and their magic: The description of gardening* (2 Vols.). London: Allen & Unwin. (Original work published 1935).

Malinowski, B. (1974). *Magic, science and religion and other essays.* London: Souvenir Press. (Original work published 1948).

Malinowski, B. (1978). *Argonauts of the Western Pacific.* London: Routledge & Kegan Paul. (Original work published 1922).

Miller, G. A. (1956). The magical number seven, plus or minus two: Some limits on our capacity for processing information. *Psychological Review, 63,* 81–97.

Miller, G. A. (1962). *Psychology: The science of mental life.* Harmondsworth: Penguin Books.

Munn, N. D. (1977). The spatiotemporal transformations of Gawa canoes. *Journal de la Société des Océanistes, 33,* 39–53.

Nayak, G. (2008). The wisdom of 'tied logs': Traditional boats of India's Orissa coast during the colonial era. *International Institute for Asian Studies, Newsletter, 47,* 22.

Pule, R. T. (1983). *Binabina: The making of a Gela war canoe.* Suva: Solomon Islands Centre and The University of the South Pacific.

Schiefenhövel, W. (1986). Extraktionszauber: Domäne der Heilkundigen [Magic of extraction: Domains of healers]. In W. Schiefenhövel, J. Schuler, & R. Pöschl (Eds.), *Traditionelle Heilkundige—Ärztliche Persönlichkeiten im Vergleich der Kulturen und medizinischen Systeme* (pp. 353–372) [Sonderband]. Braunschweig: Friedrich Vieweg & Son.

Schnepper, H. (1908). *Die Namen der Schiffe und Schiffsteile im Altenglischen. Eine kulturgeschichtlich-etymologische Untersuchung* [An etymology of the names of ships and their parts in Old English]. Kiel: H. Fiencke.

Scoditti, G. M. G. (1990). *Kitawa: A linguistic and aesthetic analysis of visual art in Melanesia.* Berlin: Mouton de Gruyter.

Senft, G. (1985). Weyeis Wettermagie: Eine ethnolinguistische Untersuchung von fünf wettermagischen Formeln auf den Trobriand Inseln [Weyei's weather magic: An ethnolinguistic study of five formulae of weather magic on the Trobriand Islands]. *Zeitschrift für Ethnologie, 110*(1), 67–90 & *110*(2), Erratum (2–4).

Senft, G. (1986). *Kilivila: The language of the Trobriand Islanders.* Berlin: Mouton de Gruyter.

Senft, G. (1992). As time goes by ...: Changes observed in Trobriand Islanders' culture and language, Milne Bay Province, Papua New Guinea. In T. Dutton (Ed.), *Culture change, language change: Case studies from Melanesia* (pp. 67–89). Canberra: Pacific Linguistics.

Senft, G. (1993). Review of the book Kitava a linguistic and aesthetic analysis of visual art in Melanesia by Giancario M. G. Scoditti. *Journal of Pragmatics, 19,* 281–290. doi:10.1016/0378-2166(93)90033-L.

Senft, G. (1994). Darum gehet hin und lehret alle Völker: Mission, Kultur- und Sprachwandel am Beispiel der Trobriand-Insulaner von Papua-Neuguinea [Thus go and teach all peoples: Mission, culture, and change of language as exemplified by the Trobriand Islanders of Papua New Guinea]. In P. Stüben (Ed.), *Seelenfischer: Mission, Stammesvölker und Ökologie* (pp. 71–91). Gießen: Focus.

Senft, G. (1997a). Magical conversation on the Trobriand Islands. *Anthropos, 92,* 369–391.

Senft, G. (1997b). Magic, missionaries, and religion: Some observations from the Trobriand Islands. In T. Otto & A. Borsboom (Eds.), *Cultural dynamics of religious change in Oceania* (pp. 45–58). Leiden: KITLV Press.

Senft, G. (2001). 'Kevalikuliku': Earthquake magic from the Trobriand Islands (for unshakeables). In A. Pawley, M. Ross, & D. Tryon (Eds.), *The boy from Bundaberg: Studies in Melanesian linguistics in honour of Tom Dutton* (pp. 323–331). Canberra: Pacific Linguistics.

Senft, G. (2005). Review of the book *The Art of Kula* by Shirley F. Campbell. *Anthropos, 100,* 247–249.

Senft, G. (2010). Culture change—Language change: Missionaries and moribund varieties of Kilivila. In G. Senft (Ed.), *Endangered Austronesian and Australian Aboriginal languages: Essays on language documentation, archiving and revitalization* (pp. 69–95). Canberra: Pacific Linguistics.

Senft, G. (2011). *The Tuma underworld of love: Erotic and other narrative songs of the Trobriand Islanders and their spirits of the dead.* Amsterdam: John Benjamins.

Senft, B., & Senft, G. (1986). Ninikula—Fadenspiele auf den Trobriand Inseln, Papua Neuguinea: Untersuchungen zum Spiele-Repertoire unter besonderer Berücksichtigung der Spielbegleitenden Texte [*Ninikula*—String figures on the Trobriand Islands, Papua New Guinea: Studies on the game repertoire with special consideration of texts accompanying games]. *Baessler Archiv, Beiträge zur Völkerkunde NF, 34,* 93–235.

Thompson, J., & Taylor, A. (1980). *Polynesian canoes and navigation.* Laie: Brigham Young University, Hawaii Campus, Institute for Polynesian Studies.

Whakataka-Brightwell, M. (1994). *Waka.* Wellington: Ministry of Education.

Chapter 12
Beyond Merry-Making: Customs of Indigenous Peoples and the Normative Functions of Ceremonies in Precolonial Igbo Societies

Ikechi Mgbeoji

For centuries, Africanist scholars have marveled at the variety and gaiety of ceremonies in traditional African societies. In wondrous bemusement, the notion has taken hold in some anthropological circles that ceremonies in traditional African society were a manifestation of primitive merry-making, anarchic gyrations of inhumane societies. This pauperized construction of the nature and functions of ceremonies in traditional African societies has hindered a fuller and richer understanding of the normative, didactic, and constructivist dimensions of traditional ceremonies that existed in precolonial African societies.

By using the Igbo societies of Southeast Nigeria as a template, I argue in this chapter that ceremonies in the traditional knowledge framework were veritable instruments that reflected, transmitted, shared, and modified the society's sense of social justice, legitimacy of law, public participation in governance, integrity of the human person, and protection of the family.[1] By traditional knowledge, I mean the awareness and understanding of the perceived workings of the environment as imparted to peoples living in a setting informed by their ancestral customs and usages. That awareness and understanding often traversed various pursuits of life in response to how the local peoples conceived of and understood the vagaries and dynamics of their environment, including the universe. I posit that the normative significance of ceremonies among precolonial Igbo of Southeast Nigeria was of peculiar importance. Properly understood, such ceremonies were norm-bearing activities, even though they failed the Eurocentric test of "repugnancy" at the end of

---

This chapter is dedicated to my son, Nnazimizu, for taking time out to distract me from work.

[1] There are arguments in some circles that the notion of human rights is entirely a Western creation and unknown in other cultures (see Donnelly 1989, pp. 49–65). For a critique of this school of thought, see Ejidike (1999).

I. Mgbeoji (✉)
Osgoode Hall Law School, Ignat Kaneff Building, York University,
4700 Keele Street, Toronto, ON, M3J 1P3, Canada
e-mail: imgbeoji@osgoode.yorku.ca

© Springer International Publishing Switzerland 2016
P. Meusburger et al. (eds.), *Ethnic and Cultural Dimensions of Knowledge*,
Knowledge and Space 8, DOI 10.1007/978-3-319-21900-4_12

257

the twentieth century (High Court Law of Lagos State 1994, Section 26, CAP.H3, p. 11; for a critique of this doctrine, see Nwabueze 2002).

I assert in this chapter that ceremonies accomplished norm-bearing and norm-iterating functions within the framework of indigenous and traditional peoples' knowledge (Adler 1931, p. 91). Ceremonies and festivities among precolonial Igbo marked certain main events that affected social bonds and complex interrelationships (*Eshugbayi Eleko v. Government of Nigeria*, 1931). Further, ceremonies validated existing obligations from which members of society could not unilaterally deviate without sanction (Isichei 1976). At the hub of these activities lay a sense of justice, the Igbo sense of justice. For *Ndigbo* (as they call themselves), justice went beyond the Aristotelian idea of giving each person his or her due and extended to the concept of *ikwuba aka oto* 'keeping one's hands straight in society'.[2]

This chapter is divided into three sections. The first summarizes the limitations of the Austinian conception of law within the jurisprudential framework among the Igbo. The second briefly explores traditions of Igbo origin, culture, and jurisprudence (Afigbo 1981a[3]). The third section details and evaluates the key ceremonies and festivities among the Igbo. I deal particularly with some of the well-known ones—birth, rites of passage into adulthood, marriage, war treaties, and burial of the Igbo—which occur in the three eminent phases of Igbo life. Certain ceremonies were peculiar to a given pertinent cultural area. I emphasize, though, that some of the ceremonies analyzed in this chapter are no longer practiced, and those that survived European colonialism have undergone considerable adulteration, Westernization, or both (*Balogun v. Oshodi*, 1929). Given the distorting impact of western colonialism on Igbo life (Anya 1982), the analysis underscores the salience of ceremonies as norm-bearing activities among contemporary Igbos (Animalu 1990).

The Limitations of Austinian Positivism

Austinian positivism (Austin and Austin 1885) and its latter-day variants have long been assailed for the unsophisticated dismissal of uncodified and noncoercive directions or prescriptions of social interaction as nonlaw or mere morality. Noncoercive rules of human behavior have been recognized as variants of normativity (Cohen 1941, p. 737). Law includes both sovereign commands and myriad "law-like" (in the sense of formal and psychological forces) institutions (Fuller 1969, p. 1). Prescribed behavior in the precolonial Igbo regime, as transmitted and iterated in ceremonies (Emenanjo 1996, 2001), epitomized this holistic dimension of

---

[2] The straight hand is a metaphorical way of saying the truth without fear or favor, maintaining fair relationships with your neighbor, never tilting the balance in ontological relationships, and so on.

[3] In Igboland the *Ahiajoku* lecture, the body of literature to which this sources belongs, is the intellectual feast at which leading scholars of Igbo pedigree address a subject of interest in commemoration of the yam's importance in Igbo culture.

normativity (Hart 1961, p. 518). In effect, one cannot undertake a study of rule-governed actions in society by pretending that law is an arrogant "statements of facts" (Olivecrona 1971) or a set of mathematical equations (Frank 1932, p. 568) royally indifferent to and hermetically separate from the social milieu and context to which it is relevant.

On this broad canvass neither Holmes's postulations on the "bad man's" anxious inquiry about the law (Holmes 1897, p. 457), nor Llewellyn's (1960) pretentious "public officials" (p. 9), nor Augustinian rejection of the legality of "unjust law" (*Non videtur esse lex quae justa non fuerit*) capture the profound complexity of indigenous legal regimes. Normative regimes that transcend narrow Austinian postulations may no longer be relegated "as a kind of museum piece offering an object of serious study only to anthropologists curious about the ways of tribal peoples" (Fuller 1969, p. 11). These so-called ethnic bodies of jurisprudence are institutions concerned in very sophisticated ways with the duties and rights of members of their societies. I now turn to the Igbos and their traditional practices as they relate to traditional knowledge systems.

The Igbo: Origins, Culture, and Jurisprudence

Though the Igbo need no introduction to the world (Equiano 1794; Herskovits 1937), the meaning and etymology of the word *Igbo* is a subject of debate. According to Umeh (1999), the word *Igbo* derives from the longer phrase *Olu na Igbo* 'riparian and hinterlands', an ancient geographical description of area occupied by the Igbo.[4] *Igbo* also means a "community of people" (Onwuejeogwu 1972, p. 16). Regardless of the differences between these schools of thought, the word *Igbo* in modern times may mean (a) the speaker of the language, (b) the areas occupied by Igbo, (c) the language spoken by Igbo people, or (d) the ethnic group itself (Amadiume 1987a; Umezuruike 1992).[5] The Igbo language challenges both non-Igbo and native speakers of the language (Echeruo 1979). Despite common beliefs, there is also cultural diversity among the Igbos (Agbasiere 2000; Ohadike 1994; Onwuejeogwu 1972, 1981; Oriji 1998).

Although Igbo territories have always been densely populated (Horton 1969; Umeh 1999), the origin of the Igbo remains unresolved. Oral traditions of Igbo origins have ranged from the celestial (Umeh 1999) to the migratory (Nwoga 1984) and the autochthonous (Onwuejeogwu 1987). Extant Igbo child-naming ceremonies, religious practices, culture, and language (Greenberg 1966) substantiate the existence of links with some East and Central African groups, but the theory of authochthonous Igbo origin may suggest a common African origin (Amadiume

---

[4] In some languages of the same Kwa family, such as Yoruba, *Igbo* means bush or forest. Yoruba mythology posits that Oduduwa, the father of the Yorubas, encountered the dwellers in the forests called *Igbo* when he and his followers settled at Ile-Ife.

[5] The Igbo language broke away from the Kwa subfamily around 600 B.C.

1987b). Regardless of which theory of origin one subscribes to, however, incontrovertible evidence of the Igbos' long and continuous occupation of their present territory in modern-day Nigeria abounds (Afigbo 1981b). Pottery shards, stone tools, rock shelter, and other artifacts show that current Igbo territories were peacefully occupied from 12,000 to 15,000 B.C. Igbo civilization is distinct from the Nok, Ife, and Benin civilizations (Okigbo 1986). Most scholars of Igbo history agree that Igbos have engaged in waves of migration from their original base. These migrations have given rise to multiple cradles of Igbo origin and civilization spanning several thousand years (Hartle 1967). These cradles of Igbo civilization gave birth to 12 minicivilizations of the Igbo, which make up the pan-Igbo civilization (Dike and Ekejiuba 1990). Of the 12 civilizations, the Ngwa and Aniocha developed symbolic communication in written forms (MacGregor 1909, p. 215). These forms of writing are African in origin, with no Arabic or European root (Dayrell 1911).

Like other societies, that of the Igbo has been heavily influenced by its geography (Morel 1911). It seems that the relative infertility and dense population of Igboland influenced the Igbo penchant for trade, competitiveness, and itinerancy (Isichei 1973, p. 18). Igbos today relentlessly pursue achievements, money, and the esteem of their peers in society (Ndem 1961; Nwabueze 1985). Igbo culture prizes individual achievement and an unquenchable thirst for education (Ohuche 1991). However, this individualism has always been "rooted in group action" (Njoku 1990, p. 16). The nuclear and extended family is very important to the Igbo. Evening tales and storytelling under the moonlight constantly reinforce the bonds of family and drill into the minds of the young the normative values held by the community (Obiechina 1994).

Furthermore, Igbo society is essentially participatory.[6] Most anthropologists agree that the Igbos exhibit republican tendencies with an ingrained mistrust of institutionalized and concentrated political power (Henderson 2000). The Igbos may have had kings in ancient times (the Igbo language has numerous words pertaining to royalty), but over the millennia it seems that the exaggerated aphorism among the Igbos—*Igbo enwe eze* 'the Igbo have no kings' has come to symbolize (Animalu 1990) the relatively nonhierarchical type of political organization with which Igbos have been largely identified (Anyanwu and Aguwa 1993). In the Igbo worldview, the individual has relevance only in her or his function as a member of the group, family, and clan. Thus, the primary duty of the individual is to maintain the dynamic cosmological balance that, in turn, guarantees stable and successful life on earth and the promise of a worthy life after death. Individual responsibility is contingent on age, social status, gender, and mental capacity (Chijioke 1989).

---

[6] In the words of J. Umeh (1999), "You can't know the inside facts about *Mmanwu* [masquerade] unless you are admitted to and participate in operating *Mmanwu*. You cannot say much about the various Ozo title societies [Igbos have societies of titled men and women; the titles are in cadres] unless you are admitted to them and participate in their rituals and activities. You can't know the implications of various socialization rites, ceremonies, including rites of passage unless you participate in them" (p. 17).

Normative rules among Igbos (Finnis 1980) do not derive from any observations of human nature but rather from *Ala*—the earth goddess. In addition to *Ala*, there were other deities such as *Ahiajoku*, the god of yams. The genius of precolonial Igbo jurisprudence lay in the fact that it was inspired by the concept of dynamic duality and balance between opposites and the interactive roles of the entities and spiritual forces in both the cosmic realm and the temporal plane. The spirit world, an animate and inanimate place, is also the abode of both the creator known as *Chukwu-Okike* 'God the creator' and the ancestral spirits. *Chukwu* is a source of the *anyanwu* 'light and knowledge' and of *chi* 'procreation'.

The world is a marketplace for both the dead and the living. The dead are expected to come back to life to rejoin the lineage. The norms created by *Ala* are interpreted and pronounced by the *Amala* 'elders' and, if need be, are enforced by various executive organs, including the *otu ebiri* 'age grades', the *umuada* 'daughters', and *mmanwu* 'masquerades'. Violations of laws constitute a disturbance of the harmony between the spiritual and the temporal. The events that could upset the equilibrium include natural disasters, such as long continuous droughts, famine, and epidemics: sorcery and other antisocial forces; and frivolous litigation, homicide, violation of taboos, or similar antisocial acts that the Igbo define as *alu* or *nso* 'evil or sin' (Uchendu 1965).

Maintaining the social and cosmological equilibrium may take the form of several types of *ichu aja* 'sacrifices' (Arinze 1970) and other means of rearranging social and cosmological order (Okoro 1988). Precolonial Igbo jurisprudence often distinguished the subtle differences between *omenala* 'custom', *iwu* 'law', and *eziomume* 'good morals or admirable conduct' (Brecht 1941, p. 811). The genius of precolonial Igbo jurisprudence lay in the fact that these gradations of norms were allowed to influence, prefigure, and humanize the structure and process of Igbo jurisprudence in a holistic state.

Because individual responsibility was contingent on social status, age, gender, and mental fitness, the titled members of society were often weighed down by a litany of social restraints and codes of behavior, whereas the young and infirm all but escaped the normative strictures. The higher a person climbed the Igbo social ladder in terms of age, attainments, wealth, family, and other criteria, the tighter the restraints became. The various cumulative prescriptions of behavior were a check on both individual and group power and represented a system of behavioral codes (Holmes 1870, p. 215), signals, and complementary expectations.

The signals and mutual directions of expected conduct and behavior were deemed to have dual relevance (spiritual and temporal). It was incumbent on the *nmuo* 'evil and good spirits' as well as the *ndi di ndu* 'the living' to keep their end of the bargain and avoid disturbing the cosmological balance. Life on earth was seen as a dress rehearsal for both the afterlife and *ilo uwa* 'reincarnation'. By this logic Igbos showed particular regard for righteous life on earth and had fastidious ceremonies and religious rites for cleansing the earth and keeping one's hands straight (Odi 1994).

The Normative Character of Ceremonies
Among Precolonial Igbos

The precolonial Igbo attitude to ceremonies was a paradox. Although extremely pragmatic (Echeruo 1979) and often frugal, Igbos nonetheless acknowledged that ceremonies took the rough edge off communal interaction and allowed practical-minded people a measure of respectable frivolity. Hence, "it is to these details of ceremony that we have to go for concrete evidence of the life styles and values of any given society" (p. 10).

Among precolonial Igbos, ceremonies varied in elaborateness and intricacy. Some ceremonies lasted months; others could last only a few seconds.[7] Whether simple or elaborate, precolonial Igbo ceremonies were ritualistic and profoundly religious. Ceremonies in precolonial Igboland were often accompanied with dances, merry-making, and festivities. This observation may explain why the mistaken view held sway in some quarters that Africans were idlers who spent the better part of their lives making merry. Another mistaken view that proselytizing Christians propagated about precolonial Igbo ceremonies was that such ceremonies were devilish and barbaric rites devoid of value or normative significance.

Regardless of the nature of ceremonies in Igboland, one fruit has long been pivotal in all of them—the *oji* 'kola nut'. No ceremony proceeds among Igbos without this fruit. The kola nut is the supreme symbol of hospitality in Igbo culture. No serious subject or matter is discussed among the Igbo without a ceremony over the kola nut (Achebe 1958). The kola nut is the ultimate sign of benevolence, love, and common purpose and has paramount ritualistic meaning among Igbos. Igbos distinguish between *cola acuminata* (*oji Igbo*) and other species of kola nut such as *cola nitida* (*gworo*) (Uchendu 1964). The latter is cheaper and more commonly available but is not valued as much as the former. There is also a distinction between the white kola nut, known as *oji ugo* 'eagle's kola nut', and the ordinary reddish-brown kola nut, which is described as *oji efu* 'ordinary kola nut' by cultural snobs. The former is the rarest type of kola nut and commands more respect even though all kinds of kola nuts may be offered in greetings and hospitality, along with the *ofo* (the Igbo cult symbol of strength, authority, and justice). The *ofo* is a stick of the *ogilisi* tree, a flower tree used to demarcate borders in Igbo land (Nwoga 1984).

In Igbo mysticism and philosophy the number four is central. Hence, a four-cotyledon kola nut has the greatest symbolic value among the Igbo and is coveted for ritualistic purposes. The reason is that

> the number four is sacred among Igbo. Igbo have a four-day week. In divination, the number four count is auspicious; *ofo* is struck four times on the ground in any ritual in which it

---

[7] An example of a simple ceremony is the tying or knotting of a tender palm frond on a disputed property. This is known as *itu omu* and is a direct function of the Igbo belief that oath-taking is a sacred and serious business. Tender palm fronds indicate danger in Igboland. Fruit-bearing trees may be dedicated to the gods with palm fronds, and only the owner may access the tree. This simple ceremony speaks volumes about the unshakeable belief of the person in question that the property in dispute belongs to her and that judgment thereon has been submitted to the gods.

is needed and in most sacrifices the four-path road is an important sacrificial center. It is not surprising, therefore, that Igbo should give *oji aka ano*—a four cotyledon kola nut—a very high ritual status. (Uchendu 1964, p. 48)

There is considerable normative value in the ceremonies surrounding the celebration of kola nuts. The first step is the presentation of the kola nut. The presentation of kola nuts is often a male privilege. In offering the kola nut, the host touches the kola with his lips to convey that he welcomes his guests in good faith. The kola nut is broken by the eldest person in the group (or someone the eldest person delegates to perform that task), and the resulting smaller nuts are distributed for consumption.

Another focal ceremony is *omumu nwa* (child birth) and postnatal confinement (*o'mugwo*). Igbos place a high premium on children. The birth of a child is usually announced by the mother of the new mother. The proud father fires a gunshot and brings out palm wine to honor his manhood. An infertile man is reckoned as a half-man (Umeasiegbu 1969). The celebration then moves to the house of the grandfather (if he is still living), who fires gunshots. Merriment follows, and dancers receive gifts of money. Most villagers converge on the market square, which is the typical celebration spot. Two days after the birth, close family members are allowed to see the new baby. No one visits the baby empty-handed; gifts must be made to the child.

Immediately after the birth of a child, the new mother is confined to her house or hut in the family compound. Only the woman's husband is allowed in on the day of the child's birth. The wife's mother visits as soon as practicable and then moves in and stays with her daughter to perform postnatal services required of her. The normative implication is that mothers can thereby sanction their disagreeable daughters by refusing to make the postnatal visit. This disciplinary act would, of course, shame the recalcitrant daughters among their peers. It is taboo for the new mother to work or engage in any physical activity of any kind. She must stay in bed and be washed there. The visit of the grandmother lasts 28 days. During the period of postnatal confinement, it is taboo for the husband to have sex with the new mother. During this period, the husband also shows his love for his wife and displays his social status by spending heavily on food and delicacies. While the family engages in this "conspicuous consumption" (Veblen 1899, Chap. 4), financial prudence and frugality are unseemly (Uchendu 1965).

Another ceremony involves the planting of the *ogbu chi* (birth tree), the burial of the umbilical cord (*ili alo*), or both. Every man, on having his first child, plants a ritual tree in front of his house. This *ogbu chi* is chosen for the newborn child by the mother, who selects the tree as the most fruitful oil palm of the many indicated by her husband. (The oil palm, *elias guineensis*, is indigenous to West Africa, and no other tree may be used for this ceremony.) The *ogbu chi* symbolizes procreation and represents the child's *chi*—an agent of the Creator and a guardian agent of sorts. The tree belongs to the child and is cut down the day she or he dies.

Sometimes the *ogbu chi* is planted on the baby's umbilical cord, which is tended with care until it detaches about 4 days after the newborn's birth. The father digs a hole, places the shriveled organ at the bottom of it, and positions the oil palm seedling above it. As the young tree grows, the child invests enormous emotion and sentiment in it and fondly refers to this tree of status as *nkwo alom* (the palm for my navel cord). This palm belongs to the child and cannot be alienated.

The burial of the child's umbilical cord is delayed until the child is named. The act used to be a simple, key ritual in precolonial Igbo laws on citizenship and full participation in certain activities of the mini-Igbo states, and Igbo citizenship norms still distinguish between persons described as *nwadiala* (the freeborn) and other classes of inhabitants of Igbo societies. In precolonial Igbo society naturalized Igbo citizens could under certain circumstances attain freeborn status, and the planting of the *ogbu chi* often evidenced this change in their citizenship status (Dike 1985).

The ceremony itself is very bare but profound in its implications because, among other things, it often gave rise to what Uchendu (1965) has aptly termed the "navel complex" (p. 56). In precolonial Igbo jurisprudence, a male Igbo "who cannot point to the burial place of his navel cord was not a *diala*—freeborn" (p. 35). The import of this incapacity was that the *ogbu chi* directly indicated the citizenship and legitimacy of the male child's birth and afforded the foundation of his future ambition and status among freeborn males in the society. Only an irresponsible parent would deny a child this crucial basis for social legitimacy and citizenship. Conversely, it disempowered children born out of wedlock. In addition, the location of the tree conferred on the child an insuperable claim to the land in question and assured his position in disputes pertaining to succession to real property, particularly the portion of land where the navel was buried. Land on which a person's navel was buried could not be granted or gifted to another person.

The next most notable ceremony of childhood is *ibie ukwu* 'circumcision of the male child', which takes place on the eighth day after birth and marks the first rite of passage into adulthood. All male Igbos must be circumcised. Female circumcision is optional and depends on the specific locality. The naming of the child also takes place on the eighth day. The parents usually select the names of the child. However, the child's grandparents often give a name of their own. The giving of names is a privilege reserved for elders. Shortly after being named, the child is taken out of the closed room and introduced to the lineage group. This presentation normally takes place with the new mother and child dressed in their best clothes and accompanied to the market, where they will be given presents. Songs celebrating the joys, pain, and rewards of motherhood will be sung. This occasion constitutes the child's formal admission into the family and lineage. In view of the ultimate importance of personal names in Igbo societies, Igbo names are carefully crafted and often loaded with meanings. Igbo names tend to be shorthand accounts of the fortunes of the family, of its joys, sorrows, hopes, and fears.

A second rite of passage is the *ipu eze* 'teething ceremony'. It is a taboo for a child to be born with natal teeth. It is also a minor taboo for a child to cut his or her upper teeth first. The first person privileged to notice the teeth is obliged to give the child a hen. The eggs and chicks produced by the hen are shared between the child and the benefactor and establishes friendship between them. Later, the teething ceremony serves the purpose of notifying the larger family that the child may now eat some types of adult food. The ceremony therefore signals to the community what acceptable types of food they may give the child.

Of all rites of passage, the *iwa akwa* 'cloth-wearing' ceremony is probably the most dramatic. It is the one that initiates Igbo males into adulthood. The *iwa akwa* was the rite of passage of Igbo males into adulthood and signified the right of the new adult to partake in all activities reserved for adults. The activities included enlistment for warfare, granting of farmlands to the new adult male, and preparation for marriage. All children born in the same year invariably belong to what is commonly referred to as an 'age grade', a cohort of male children born within the same period of about 1 year, and should all undertake their cloth-wearing ceremony at the same time. This ceremony involves elaborate rituals that place the entire community on notice that the young adult male has attained maturity and is now a sane, full-fledged member of the society. After the *iwa akwa*, the young male is expected to move out of his parents' home and create his own household, an act that extinguishes, or at least radically curtails, certain legal claims that he could otherwise make on his parents or their estate.

The initiation into adulthood may also take the form of another rite of passage known as *itu anya* 'opening of the eyes'. It is a moment of privation, challenges, and entrance into manhood. The *itu anya* is often conducted by a *dibia* 'medicine man'. The young adult male passes through many stages designed to equip him with strength, knowledge, and courage. This process is the gateway to manhood, for the Igbo male is largely supposed to possess knowledge of the relationships between the spiritual and the mundane. The *itu anya* also recreates the death–rebirth cycle, and a new name may be given the reborn male.

A general ceremony among Igbo males is *igba mgba* 'wrestling'. Precolonial Igbo clans often bragged about the wrestling prowess of their kinsmen, ancestors, and friends. In Igbo wrestling, to be thrown on one's back or buttocks meant defeat. Wrestling was accompanied by music and martial songs designed to inspire the competitors. The normative relevance of wrestling ceremonies stemmed from the Igbo belief that life itself is a struggle and that a man should thus fight aggressors, human and spiritual, to the best of his ability. Wrestling in Igbo culture was a metaphor for life's struggles. It was not a statement that the winner was necessarily the biggest or the strongest but rather that he was skillful and courageous. Igbo tales are often full of wrestling encounters won by the wiliest and cleverest wrestlers who defeated stronger and bigger rivals.

For Igbo women the analogous ceremony calling attention to the passage into adulthood was the *ihu nso* 'the ceremony of the first menstruation'. This ceremony announced the commencement of female adulthood and signaled to society that the woman was ripe for marriage. After this transition the Igbo female may no longer play freely with members of the opposite sex. Prior to the *ihu nso*, no serious talk of marriage can ever be entertained between the girl's family and any interested parties. The *ihu nso* usually used to be rounded out by a noticeable change in the attire and hairstyle of the newly menstruating woman. The hair would be specially done in a style designed to attract the attention of both her peers and potential suitors. For the most part, precolonial Igbos regarded long hair on the female as a sign of fertility. After marriage, however, the flamboyant hairstyle would yield to a modest hairstyle befitting her dignified status as a wife.

The second arm of the rite of passage into adulthood is the ceremony of *iru mgbede*, the fattening of a woman before marriage. Under the sponsorship of the proposed groom, the bride is separated from her usual household or abode in order to hone her home-making abilities. The normative implication of this ceremony is that it epitomized the Igbo belief that healthy offspring physically fit enough to cope with life can be born only to healthy, physically fit mothers. The *iru mgbede* also reiterated the Igbo belief that an impecunious bachelor has no business with marriage. This message may be inferred from the fact that the cost of keeping the young bride in *iru mgbede* is borne exclusively by the proposed groom. At the *iru mgbede* the bride lives in luxury and relative idleness, treating herself to expensive delicacies at the expense of the groom. It is not an overstatement to assert that the *iru mgbede* ceremony was designed partly to stretch the financial capabilities of the proposed groom.

Among precolonial Igbos, marriage was at its roots an alliance between two families rather than a contract between two individuals. The marriage process was long and expensive. Precolonial Igbos placed more emphasis on the compatibility and social standing of couples. Still today, the search for marriage partners is infused with pragmatic considerations. For example, screening for hereditary illnesses, insanity, and consanguinity precede marriage ceremonies (Bosah 1977). The overriding criteria for choosing a marriage partner are fertility, socialization, economic support, collective responsibility, and status.

A traditional marriage ceremony proceeds in several phases. First, the girl's consent is sought through an *onye aka ebe* 'witness'. The witness probes the parties' background and character. Generally speaking, no parent or groom may approach a girl's family for discussion of marriage without first securing the "tentative consent of the girl" (Agbasiere 2000, p. 43). This approach is usually a tactful mission undertaken with the agreement of the girl's mother. It is, however, not unusual for the girl to be pressured to marry someone she does not really have affection for.

The next step is the *mmanyi ajuju* 'wine for inquiries', whereby the groom, accompanied by at least two close friends, publicly declares his interest and initiates open inquiry as to whether the girl would publicly accept him as a husband. It is at this stage that the families mutually probe their genealogy to determine whether a marriage of the couple would be precluded on grounds of consanguinity. Family reputation and the health history of the family are also highly relevant. Families with a record of insanity, epilepsy, and genetic diseases are often avoided.

After the stage of *ajuju* 'inquiry', the real negotiations for the marriage begin. They usually lead to the actual marriage, called *igba nkwu* 'wine carrying': a day of grandiose merriment and excessive consumption of palm wine, food, and other delicacies. Participation in the *igba nkwu* is public. The wide publicity surrounding such ceremonies is the legal equivalent of a marriage certificate or license issued by modern-day authorities (Fuller 1975, p. 89). The *aku isi nwanyi* 'bride price' is also paid on the day of *igba nkwu*. Contrary to popular misconception, this transaction is not a wife purchase but rather the part of the marriage contract that cements the groom's right to any offspring from the marriage. A groom who carries the palm wine but fails to pay the bride price is not legally entitled to any children from the marriage.

Yet another Igbo ceremony is the festival known as *otite*, a celebration of the Igbo penchant for itinerancy. Every itinerant person must return in the seventh month for a head count of the living and the dead. The *otite* was not just a celebration of the rigors and education inherent in travels, but a primitive form of census by which precolonial Igbo societies regularly checked on their populations.

As unusual as the *otite* may appear to Western commentators, no Igbo ceremony is as misunderstood as the *igbu ichi or ichi* 'facial scarification' (Jeffreys 1951), in which the individual's face was marked with lines drawn deep into the flesh without painkillers. Drawings and early pictures of Igbo nobles show facial scarification as a common feature among precolonial Igbos. Some of the available oral traditions suggest that

> the significance of scarification as well as scarification as the means of enhancing the status and intrinsic quality of the children sacrificed creates a precedent for all specially marked people who bear responsibility for the welfare of the people. The myth defines the minimal expectation to which persons vested with public office are held: they should wear those scars or *ichi* on their faces 'which would always remind them that they are something of sacrificial victims'—dedicated to service just as the mythic children were sacrificed for the benefit of humanity. (pp. 93–94)

It was unseemly for the Igbo noble, male or female, to flinch or show pain while the scarification ritual was in progress. In effect, facial scarification entrenched Igbo respect for and deference to courage and perseverance. A child or adult who submitted to facial scarification was not allowed to wince or whimper or utter a sound during the ceremony. Of necessity, the *Ozo* title was bestowed on every person who underwent *ichi* facial scarification, but not every one of its bearers had the special distinction of having his face scarified. *Ichi* facial scarifications were for Igbo noble men, their first and seventh sons, and distinguished women who could endure the requisite ascetic stoicism. Beyond physical courage, the eligibility of the aspirant is determined by her or his truthfulness and honesty. It was taboo to kidnap or assault a person whose face had undergone ritual facial scarification.

Precolonial Igbos had elaborate ceremonies for the planting and harvesting of crops, chief of which was the festival yam. The discovery of yam cultivation was a fundamental and definitive epoch in Igbo life and culture. As suggested by the West African origin of some species of yams, Igbos were the most enthusiastic cultivators of yam. Yam is still the king of crops among Igbos, and the eating of *iri ji ohuu* 'new yams' commences with religious ceremonies. No self-respecting Igbo eats new yams before this religious ceremony occurs. Yams sometimes have the status of deities in Igbo culture and jurisprudence. Oaths can be taken with yams (Jordan 1949/1971). Indeed, in the pantheon of Igbo gods, *Ahiajoku*, the god of yams,— occupies a prominent position. It is not unusual for people to curse sociopaths by flinging yams at them.

Precolonial Igbos had ceremonies on warfare as well. Prior to the commencement of any armed conflict, the parties, through trusted and sagacious intermediaries, especially elders and titled personages, had to conduct negotiations honestly and diligently to prevent warfare. If negotiations failed, a formal declaration of war followed. The declaration of war had to contain a notice of the nature of weapons to be used; usually long sticks. This type of armed conflict was known as *ogu osisi* 'the war of sticks'. It was taboo to commence wars with destructive or excessively lethal weapons. The negotiators would also agree on other rules of engagement, such as the days on which the battles might be fought. Bloodshed was permitted only in o*gu egbe* 'war with guns' or *ogu nma* 'war of machetes' (Isichei 1973).

In all wars women and children were exempt. Markets and the people visiting them were left in peace. Similarly, shrines and places of religious importance could not be attacked. Wars were never fought during religious festivals, such as the New Yam festival. Given the high rate of intermarriage between villages, the pursuit of warfare was a tricky, awkward business. Deliberate attempts were made to keep the number of casualties low and to ensure that the vanquished were only defeated, not humiliated.

Wars were commenced or resumed when the bearer of the *opa igbe ogu* 'war ark' of each village entered or left the war zone. It was unlawful to engage in combat if the war ark was not placed in the field of battle. It was not unusual to have wars of attrition or wars that lasted several years. This type of war was known as *ogu mkp-uru oka* 'war of the maize cob' (referring to the numerous grains on the maize cob). Wars were concluded with ceremonies and peace treaties. According to Isichei (1973),

> when a community desired to make peace, it sent a delegation carrying palm fronds, which symbolize peace. Typically, women played a key role as a pressure group for peace, because of their unique role as daughters and sisters in one community, and wives and mothers in another. (p. 97)

Precolonial Igbo jurisprudence rested squarely on the fundamental norm or principle of *pacta sunt servanda* (agreements must be kept), as was apparent in Igbo treaties and pacts sealed in blood and otherwise known as *igbandu-blood pacts*. The blood pact was a traditional and religious ceremony in which treaties or important agreements were consummated by "mixing drops of participants' blood and eating kola nuts dipped therein. Once the ceremony had been concluded, the parties thereto regarded themselves as "clansmen" (Ejidike 1999, p. 90). These blood pacts were treaties in their own rights, and no one was permitted to shirk obligations created by duly appointed representatives of the participating ministates or clans. It is equally instructive that most blood pacts or treaties were concluded in the presence of great oracles or at *ihu ala* 'face of the earth', which Igbos revered, even dreaded.

Igbo jurisprudence sometimes conferred legal personality on trees and forests. Elaborate rituals analogous to the burial of humans or the establishment of a shrine often preceded the felling of big trees. As Nwoga (1984) observed,

> when villagers want to cut down a big and aged tree, they take four sticks of *ogilisi* (a flower tree used to demarcate borders in Igboland) and plant them in a small square, put the neck of a pot in between, put some stones around the pot-neck and call upon the tree to move to that place and not be angry with them for removing it from its earlier habitation. (p. 18)

The normative significance of this practice was the intrinsic respect for nonhuman life forms. Precolonial Igbos largely considered the earth force—*ala*—to be the fundamental source of norms. Hence, cleansing ceremonies had to be performed to atone for misconduct or offences against the earth force. There were two types of such offences: *nso ala* 'conduct forbidden by the earth' and *alu* 'conduct forbidden by the community'. The former referred to abominations against the earth force that had to be cleansed (Agukoronye 2001). Examples included murder, adultery, incest, torture, and cruelty to animals (Onwuejeogwu 1987, p. 52). These wrongs were punishable by banishment, death and/or[8] very expensive earth-cleansing ceremonies known as *ikpu alu* 'to drag abomination away'. This ceremony was intended to revalidate the society's desecrated source of norms and to sanitize the earth force.

Given the belief that precolonial Igbos had in the continuity of life, they expressed much anxiety in their lifetimes as to how they were to be buried. Wealthy members of that society were buried with pomp and pageantry, and the poor buried their dead with as much festivity as they could mount. The demise of an eminent person was often announced by the intermittent beating of the *ikoro*, a gigantic wooden drum whose sound can travel about 10 miles. The variations of the music told people whether the deceased died poor or rich and whether that person had successfully trained his or her children.

The *ese*, a type of music played by professional musicians on the death of an elderly and important person, could also be played to announce the death of such an individual. Family members or relations of the deceased were not permitted to weep or cry in public unless and until the *ikoro* or *ese* had formally announced the passing of their elder relation. The corpse was washed, powdered, scented, and dressed in the deceased's best clothes. The body was then laid in a coffin and buried. Burial took place in front of the house or a place designated by the deceased in his lifetime. A tree or shrub was planted on the burial spot in remembrance. There was a second burial, the *ikwa ozu*, which was actually the real burial for titled men. It was fundamental to the Igbo because it marked the real celebration of the life of the deceased and functioned as a means of wealth redistribution. As Njoku (1990) explained regarding the second burial, "a chief on earth would be happy only as a chief after death. He would not be happy if his dignity was lowered" (p. 187). The second burial may be likened to an estate tax because heavy expense was imposed on the estate of the deceased. An Igbo family would lose face and reap social ridicule if members of the family, particularly the *umuada* 'the daughters of the clan' failed to appear at the second burial ceremonies. The *umuada* usually threatened refractory family members in their lifetimes with absence at their second funerals. In most cases, no funerals could take place if lineage daughters were not present (Amadiume 1987b, p. 59). This practice had a way of empowering female members of Igbo society (Nzekwu 1994).

---

[8] Igbo jurisprudence seldom imposed the death penalty. Instead, a murderer was expected to commit suicide or make adequate reparations.

Conclusion

The traditional knowledge and practices of precolonial Igbo as manifested and encoded in their jurisprudence were a complex and sophisticated web of norms, practices, and prescriptions of social and cosmic ordering (Obi 1963). Igbo ecological and traditional knowledge was often enmeshed in a body of law that asserted the multiple linkage of mankind in a complex chain and interconnection of other parts. It was believed that changes in this delicate web of relations would disturb the intangible and tangible forces that unite society. These beliefs were narrated and validated in Igbo law and life through ceremonies and religious observations. Those ceremonies also afforded an avenue for modifying the worldview and dramatizing the people's beliefs and value systems. Virtually all members of the society participated in these ceremonies in various ways. Since the age of colonialism, however, Igbo cultural practices have been subjected to Eurocentric sanitization and cleansing processes designed to remake native law in the image of English common law. In the colonial encounter, Igbo culture, though somewhat displaced, has not been totally vanquished.

Some of the ceremonies had major ecological and environmental implications. For example, the planting of trees at the birth of children and the taboos on felling those trees ensured the protection of trees. Similarly, the taboos on killing young animals were protective measures designed to ensure the sustainability of animal species. The elaborate ceremonies on adulthood and maturation ceremonies were equally seminal in teaching teenagers the importance of self-reliance and respect for constituted authority.

Regrettably, the colonial attitude to indigenous normative prescriptions ignored the commendable attributes of local norms, such as certainty, legal reasonableness, immemorial origin, and continuity. The ceremonies amounted to a normative chart of how members of society lived together in their social and environmental space. The larger meaning of ceremonies is that they embedded, transmitted, and modified normative prescriptions. Those ceremonies were not ritualistic and pointless drama but invaluable methods of communicating and transferring signals of meaning, knowledge, and information among the citizens themselves and in their relationships with the larger universe (Odi 1996). They had implications for both the environment and the ecology. Consequently, studying such ceremonies is useful and should no longer remain a subject of arid scholarly debate or anthropological curiosity (Fuller 1975).

References

Achebe, C. (1958). *Things fall apart*. London: Anchor.
Adler, M. (1931). Legal certainty. *Columbia Law Review, 31*, 91–108.
Afigbo, A. E. (1981a). *The age of innocence: The Igbo and their neighbours in pre-colonial times* (Ahiajoku lectures, Vol. 3). Owerri: Ministry of Information, Culture Division.

Afigbo, A. E. (1981b). *Ropes of sand: Studies in Igbo history and culture*. London: University Press.

Agbasiere, J.-T. (2000). *Women in Igbo life and thought*. London: Routledge.

Agukoronye, O. C. (2001). Landscape practices in traditional Igbo society, Nigeria. *Landscape Research, 26*, 85–98.

Amadiume, I. (1987a). *Male daughters, female husbands: Gender and sex in an African society*. London: Palgrave Macmillan.

Amadiume, I. (1987b). *African matriarchal foundations: The Igbo case*. London: Karnak House.

Animalu, A. O. E. (1990). *Ucheakonam: A way of life in the modern scientific age* (Ahiajoku lectures, Vol. 11). Owerri: Ministry of Information, Culture Division.

Anya, A. O. (1982). *The environment of isolation or the ecology and sociobiology of Igbo cultural and political development* (Ahiajoku lectures, Vol. 4). Owerri: Ministry of Information, Culture Division.

Anyanwu, U. D., & Aguwa, J. C. U. (1993). *The Igbo and the tradition of politics*. Enugu: Fourth Dimension Publishing.

Arinze, F. A. (1970). *Sacrifice in Igbo traditional religion*. Ibadan: Ibadan University Press.

Austin, J., & Austin, A. (1885). *Lectures on jurisprudence, or, the philosophy of positive law: Vol. 1*. London: John Murray. Balogun v. Oshodi, 10 Nigeria Law Report 50 (1929).

Bosah, S. I. (1977). *Groundwork of the history and culture of Onitsha*. Ibadan: University of Ibadan.

Brecht, A. (1941). The myth of *is* and *ought*. *Harvard Law Review, 54*, 811–831.

Chijioke, M. O. (1989). *Ugwumba: The greatness of a people* (Ahiajoku lectures, Vol. 10). Owerri: Ministry of Information, Culture Division.

Cohen, M. R. (1941). Should legal thought abandon clear distinctions? *Illinois Law Review, 36*, 727–746.

Dayrell, E. (1911). Further notes on the 'Nsibidi signs with their meanings from the Ikom District, Southern Nigeria. *Journal of the Royal Anthropological Institute of Great Britain and Ireland, 41*, 521–540.

Dike, A. A. (1985). *The resilience of Igbo culture: A case study of Awka Town*. Enugu: Fourth Dimension.

Dike, K. O., & Ekejiuba, F. I. (1990). *The Aro of south-eastern Nigeria, 1650–1980: A study of socio-economic formation and transformation in Nigeria*. Ibadan: University Press.

Donnelly, J. (1989). *Universal human rights in theory and practice*. Ithaca: Cornell University Press.

Echeruo, M. J. C. (1979). *Afamefula: A matter of identity* (Ahiajoku lectures, Vol. 1). Owerri: Ministry of Information, Culture Division.

Ejidike, M. (1999). Human rights in the cultural traditions and social practice of the Igbo of south-eastern Nigeria. *Journal of African Law, 43*, 71–98. doi:10.1017/S0021855300008731.

Emenanjo, E. N. (1996). *Olumefula: Asusu Igbo na ndu ndi Igbo* [My voice shall endure: Language and life of the Igbos]. Owerri: Archdayosis Katolik nke Owerri.

Emenanjo, E. N. (2001). *Igbo or igboid: Asusu n'agburu ndi* [Igbo language in Igbo civilization] (Ahiajoku lectures, Vol. 17). Owerri: Ministry of Information, Culture Division.

Equiano, O. (1794). *The interesting narrative of the life of Olaudah Equiano, or Gustavus Vassa, the African*. London: O. Equiano. Eshugbayi Eleko v. Government of Nigeria, Appeal Cases 662 673 (1931).

Finnis, J. (1980). *Natural law & natural rights*. Oxford, UK: Oxford University Press.

Frank, J. (1932). Mr. Justice Holmes and non-Euclidean legal thinking. *Cornell Law Quarterly, 17*, 568–603.

Fuller, L. L. (1969). Human interaction and the law. *American Journal of Jurisprudence, 14*, 1–36.

Fuller, L. L. (1975). Law as an instrument of social control and law as a facilitation of human interaction. *Brigham Young University Law Review, 1*, 89–96.

Greenberg, J. H. (1966). *The languages of Africa*. Bloomington: Indiana University Press.

Hart, H. L. A. (1961). *The concept of law*. Oxford, UK: Oxford University Press.

Hartle, D. D. (1967). Archaeology in eastern Nigeria. *Nigeria Magazine, 93*, 134–143.

Henderson, R. N. (2000). *The king in every man: Evolutionary trends in Onitsha Ibo society and culture*. New Haven: Richard Henderson.

Herskovits, M. J. (1937). *Life in a Haitian valley*. New York: Doubleday and Company.

High Court Law of Lagos State. (1994). Vol. 3. Johannesberg: n.p.

Holmes, O. W., Jr. (1870). Codes and the arrangement of the law. *American Law Review, 5*, 212–221.

Holmes, O. W., Jr. (1897). The path of the law. *Harvard Law Review, 10*, 457–478.

Horton, J. A. B. (1969). *West African countries and peoples*. Edinburgh: Edinburgh University Press.

Isichei, E. A. (1973). *The Ibo people and the Europeans: The genesis of a relationship—to 1906*. London: Faber and Faber.

Isichei, E. A. (1976). *A history of the Igbo people*. London: Macmillan.

Jeffreys, M. D. W. (1951). The winged solar disk, or Ibo Ichi scarification. *Journal of the International African Institute, 21*, 93–111.

Jordan, J. P. (1971). *Bishop Shanahan of southern Nigeria*. Dublin: Elo Press. (Original work published 1949).

Llewellyn, K. N. (1960). *The bramble bush: On our law and its study*. New York: Oceana.

MacGregor, J. K. (1909). Some notes on Nsibidi. *Journal of the Royal Anthropological Institute of Great Britain and Ireland, 39*, 209–219.

Morel, E. D. (1911). *Nigeria, its peoples and its problems*. London: Smithe, Elder.

Ndem, E. B. E. (1961). *Ibos in contemporary Nigerian politics: A study in group politics*. Onitsha: Etudo Limited.

Njoku, J. E. E. (1990). *The Igbos of Nigeria: Ancient rites, changes, and survival*. New York: E. Mellen Press.

Nwabueze, R. N. (1985). *The Igbos in the context of modern government and politics in Nigeria: A call for self-examination and self-correction* (Ahiajoku lectures, Vol. 6). Owerri: Ministry of Information, Culture Division.

Nwabueze, R. N. (2002). The dynamics and genius of Nigeria's indigenous legal order. *Indigenous Law Journal, 1*, 153–199.

Nwoga, D. I. (1984). *Nka na nzere: The focus of Igbo worldview* (Ahiajoku lectures, Vol. 5). Owerri: Ministry of Information, Culture Division.

Nzekwu, N. (1994). Gender equality in a dual sex system: The case of Onitsha. *Canadian Journal of Law and Jurisprudence, 7*, 73–95.

Obi, S. N. C. (1963). *The Ibo law of property*. London: Butterworths.

Obiechina, E. N. (1994). *Nchetaka: The story, memory and continuity of Igbo culture* (Ahiajoku lectures, Vol. 14). Owerri: Ministry of Information, Culture Division.

Odi, A. (1994). Library and information dissemination in a traditional society: The Igbo of eastern Nigeria. *International Information and Library Review, 26*, 1–9. doi:10.1006/iilr.1994.1001.

Odi, A. (1996). As it was in the beginning: The influence of traditional elements in Igbo information systems. *The International Information and Library Review, 28*, 193–202.

Ohadike, D. C. (1994). *Anioma: A social history of the western Igbo people*. Athens: Ohio University Press.

Ohuche, R. O. (1991). *Ibu anya danda (No load defeats the ant): The centrality of education in Igbo culture* (Ahiajoku lectures, Vol. 12). Owerri: Ministry of Information, Culture Division.

Okigbo, P. N. C. (1986). *Towards a reconstruction of the political economy of Igbo civilization* (Ahiajoku lectures, Vol. 7). Owerri: Ministry of Information, Culture Division.

Okoro, A. N. (1988). *Chukwu ka dibia* [God is greater than the medicine man] (Ahiajoku lectures, Vol. 9). Owerri: Ministry of Information, Culture Division.

Olivecrona, K. (1971). *Law as fact*. London: Stevens.

Onwuejeogwu, M. A. (1972). *A brief survey of an Anambra civilization in the Igbo culture area about A. D. 850–1937*. Onitsha: Odinani Museum.

Onwuejeogwu, M. A. (1981). *An Igbo civilization: Nri kingdom and hegemony*. London: Ethnographica.

Onwuejeogwu, M. A. (1987). *Evolutionary trends in the history of the development of the Igbo civilization: In the culture theatre of Igboland in southern Nigeria* (Ahiajoku lectures, Vol. 8). Owerri: Ministry of Information, Culture Division.

Oriji, J. N. (1998). *Ngwa history: A study of social and economic changes in Igbo mini-states in time perspective*. New York: Peter Lang.

Uchendu, V. C. (1964). 'Kola hospitality' and the Igbo lineage structure. *Man, 64*, 47–50.

Uchendu, V. C. (1965). *The Igbo of southeast Nigeria*. Austin: Holt, Rinehart and Winston.

Umeasiegbu, R. N. (1969). *The way we lived: Ibo customs and stories*. London: Heinemann Educational.

Umeh, J. A. (1999). *Igbo people: Their origin and culture area*. Enugu: Gostak Print.

Umezuruike, G. M. (1992). *The hub of Igbo renaissance in the scientific age* (Ahiajoku lectures, Vol. 13). Owerri: Ministry of Information, Culture Division.

Veblen, T. (1899). *The theory of the leisure class*. New York: Macmillan.

Chapter 13
Knowledge, Behavior, and Culture: HIV/AIDS in Sub-Saharan Africa

William T.S. Gould

The Knowledge/Behavior Gap

As in other subdisciplines across the social sciences, concern for understanding "knowledge" is growing within development studies. A key marker in refocusing this emphasis was the 1999 World Development Report, *Knowledge for Development* (World Bank 1999). During the previous half century, developmental thinking and activities had stressed "education for development" as an idea to parallel "knowledge for development"—with the expansion and improvement of formal education systems from primary school to the university as a necessary condition of economic and social improvement in the less-developed countries. However, formal school systems increasingly proved insufficient in themselves to generate an increasing human resource base of sufficient breadth to allow genuine human development to bring people out of poverty and into more successful and sustainable livelihoods. As recognition of this fact spread, attention inevitably switched from the acquisition of technical knowledge by relatively few people in formal schooling to the acquisition of the knowledge required by the population as a whole. What knowledge was needed to promote social and economic development, and how could it be accessed? As the modernist project for formal knowledge acquisition faltered, new awareness of the need to understand and build on existing indigenous knowledge systems grew. From the 1990s on, the revolution in information and communication technologies, with the World Wide Web and the Internet, clearly renewed concentration on *how*: How could the digital divide for knowledge acquisition between North and South be reduced, and with what effect? What knowledge was needed, and how could its acquisition be facilitated by the new technologies (see Gould 2015, especially Chap. 8)?

W.T.S. Gould (✉)
Department of Geography and Planning, University of Liverpool,
29 Beryl Road, Prenton, Merseyside CH 49 3RS, UK
e-mail: wtsg@liv.ac.uk

© Springer International Publishing Switzerland 2016
P. Meusburger et al. (eds.), *Ethnic and Cultural Dimensions of Knowledge*,
Knowledge and Space 8, DOI 10.1007/978-3-319-21900-4_13

Underpinning both the modernist project on formal schooling and the new approach to knowledge acquisition is the assumption that people need access to more and better information than they currently have if they are to enhance their livelihoods and life chances. Because the acquisition and application of knowledge are generally acknowledged to have been seminal factors in western development, it is assumed that knowledge enhancement by the local population must also be a necessary, but not sufficient, condition for development in the currently rather poor countries. Their knowledge base will affect how people can become effective in the development process: how they behave, how they respond to economic opportunities, how they respond to lifestyle opportunities such as those for improving their health prospects, and how they cooperate with others locally and with society at large, including their responses to government initiatives. Knowledge will affect their attitudes as well as their behavior and practices, a point recognized, for example, in knowledge, attitude, and practice (KAP) studies, which commonly inform in medical research.

However, knowledge alone is not sufficient to influence behavior. Because the path from knowledge to behavior is filtered through attitudes and cultures, there is always a knowledge/behavior gap. People do not always react directly or immediately to new knowledge. Their behavior may or may not be affected by that new knowledge, and a sizable knowledge/behavior gap may survive for a variety of economic and cultural reasons. Not all cigarette smokers respond positively, immediately, or even at all to medical evidence that smoking affects health and reduces life expectancy! Governments may seek to narrow the knowledge/behavior gap through legislation (e.g., prevention or restriction of sales of some smoking materials), pricing (e.g., heavy taxation of cigarettes), or persuasion (e.g., public information campaigns), but a gap of choice continues.

One of the serious contemporary threats to livelihoods in sub-Saharan Africa is human immunodeficiency virus/acquired immune deficiency syndrome (HIV/AIDS), a disease of the human immune system caused by infection with the human immunodeficiency virus. The disease is essentially a behavioral condition and spread by human agency, mostly through unprotected heterosexual sex in Africa (Gould 2009b). Infection with the HIV virus can be avoided, and governments are anxious to promote knowledge about the disease in order to bring about the behavioral changes needed to reduce its prevalence. Although knowledge about how to avoid HIV infection is ubiquitous and shared, that increasing level of awareness has coincided with high and, in some countries, still rising infection rates (Bongaarts et al. 2008). A conspicuous knowledge/behavior gap for HIV/AIDS remains. The purpose of this chapter is to examine the nature and extent of that gap in sub-Saharan Africa.

HIV/AIDS as an Exceptional Disease

Any appreciation of the nature of the knowledge/behavior gap for HIV/AIDS needs to start with the exceptional nature of this devastating disease. AIDS is an extreme condition caused by the HIV virus in the blood stream (Gould 2009b; Gould and

Woods 2003). The world still lacks a cure or vaccine for AIDS, but antiretroviral (ARV) therapies can suppress the levels of HIV infection sufficiently so that being HIV+ no longer need be a barrier to normal living. Yet the expense of ARV therapies continues to put them beyond the reach of the vast majority of people who need them in sub-Saharan Africa (Posse et al. 2008). Therefore, preventing infection in the first place and managing the condition when it does arise depends on individual behavior, basically meaning the avoidance of sexual contact, rather than on a solution rooted in modern or even traditional therapies.

The HIV/AIDS condition is unparalleled in three main respects.

- Its demography: Unlike most other epidemic diseases, HIV/AIDS disproportion-ately affects young adults, that is, the sexually active, with women normally experiencing higher prevalence of the infection than men do. Sexual behavior thus becomes a pivotal aspect of policy and prevention initiatives (Gould 2005, 2015, pp. 111–113, Box 4.3).
- Its local geography: The global geography of the disease is well known. Sub-Saharan Africa accounts for over three-quarters of the world's more than 30 million affected people, with southern Africa having much higher prevalence rates (typically about 20 % of adults being HIV+) than West Africa (typically less than 5 %) and with most countries of eastern Africa ranging somewhere between these extremes (UNAIDS 2009). Less well known is that rates throughout sub-Saharan Africa are generally higher in urban areas than in rural ones, typically by a factor of two (Dyson 2003). The finding that urban rates can be twice as high as rural rates is certainly unusual for any disease, a divergence attributable, again, to the dissimilarity of sexual behaviors. Urban dwellers typically have more sexual partners and wider sexual networks than the inhabitants of rural areas do and are thus more exposed to infection.
- Its social prevalence: From the beginning of the epidemic in the 1980s, the mobile, the educated and the wealthy seemed to be at much greater risk than the poor (on Uganda, for example, see Kirunga and Ntozi 1997). Meanwhile, the Demographic and Health Surveys (DHSs) now available for most countries in sub-Saharan Africa have confirmed that HIV/AIDS has affected the wealthier groups in the population more than the poorer, the educated more than the uned-ucated, and the mobile rather more than those who have not become migrants (Gillespie et al. 2007; Glick and Sahn 2008; Gould 2004, 2009a). This pattern is the reverse of that for all other epidemic diseases, which are associated normally with the poor and marginalized and with people living in difficult and polluted environments. This evidence further consolidates the rural–urban differential in the distribution of the disease, a differential that seems to persist in most African countries. Some evidence does show that growing proportions of the uneducated and poor in those societies are becoming affected, as in Tanzania (Hargreaves and Howe 2010) and Zimbabwe (Lopman et al. 2007), but the broad ecological relationship between wealth and HIV status holds.

Whereas demography, geography, and social distribution are the key structural or background identifiers of the disease's extraordinary nature, the fundamental proximate, or immediate, determinant of HIV status in African populations is the number of sexual partners a person has. The immediate causes of infection in this case are neither environmental nor economic, as they are with classic epidemic diseases like malaria or measles, but behavioral. The causes of HIV infection are therefore directly influenced by individual knowledge, lifestyles, and life choices in sexual behavior within and outside marriage or other forms of consensual union. As a behavioral condition, HIV/AIDS seems likely to respond directly to improved knowledge about the disease.

Enhancing the Knowledge Base

National programs for AIDS treatment, care, and prevention exist in all high-prevalence African countries. Treatment currently therefore centers on the roll out of ARVs, but that approach has been neither easy nor uncontroversial in countries where the potential demand far outstrips the availability of and resources for therapies (Posse et al. 2008). In terms of care, communities (including churches, as in Botswana; see Klaits 2010) and families have focused intensely on problems confronting children and the older generation and are called on to assume much more active roles as caregivers and direct breadwinners than is normal (Robson et al. 2006). In the matter of prevention, efforts have aimed to prevent initial infection, particularly at ensuring and managing behavioral change. These endeavors have encompassed two broad dimensions of behavioral change: reducing the number of sexual partners and promoting safe, protected sexual intercourse.

The leading mechanisms by which governments have sought to achieve these two behavioral changes, which are generally considered necessary to prevent infection, have been information, education, and communication (IEC) programs. They have taken several forms, featuring locally specific mixes of components. Some countries have emphasized public information campaigns through the media—TV, radio, newspapers, billboards, and even mobile phones, YouTube, and Facebook (e.g., " 'Side Dishes'," 2010). Other countries, such as Kenya (Ferguson and Morris 2007), have mounted occupation-specific programs targeting the most vulnerable groups, including sex-workers and truckers, and have changed educational curricula and school-specific programs for young, often still school-age people beginning their sexual activity. In all these components the implicit, if not explicit, message to the target audiences for changing their behavior has been generally recognized as the ABC approach: Abstain, Be faithful, use a Condom. Abstinence is clearly much more of an option for young people before marriage. Within marriage or other consensual union it has little practical value. In such circumstances fidelity (i.e., minimization of the number of sexual partners) is obviously a possible option. And in all cases, using a condom will prevent cross-infection during intercourse.

Nowhere has the argument for increasing the knowledge base through public information programs been more apparent than in Uganda (Iliffe 2006, pp. 126–130). By the late 1980s, Uganda had the world's highest rate of HIV/AIDS infection, over

12 %, by far the most serious and best known national case. By 1995 the national prevalence rate had declined by more than half, to 5 %, a unique experience at that time and not replicated since, though more recent national prevalence rates indicate that they are beginning to fall again slightly. The steep reduction in prevalence in Uganda is attributed to behavioral change, with both the number of sexual partners and the amount of unprotected sex plunging in urban and rural areas alike for both women and men (Low-Beer and Stoneburner 2003; Ntozi et al. 2003; Stoneburner and Low-Beer 2004). The government of Uganda had embarked on information-and-awareness campaigns through the national media and interventions in school curricula, but the critical difference in thought came from President Yoweri Museveni and the strong political credibility he lent the campaign in his own national speeches and extensive travel at the grass-roots level. He exhorted people in earthy and direct language to reduce the number of sexual partners and encouraged the use of condoms for safe sex. Such overt political engagement in Uganda stands in spectacular contrast to the response in South Africa. During the growth of that country's HIV/ AIDS prevalence in the late 1990s and early 2000s, the national President, Thabo Mbeki, was in apparent denial about the severity and cause of the epidemic, despite a relatively well-educated population, a solid informational infrastructure, media massages, and nongovernmental agencies urging behavioral change (De Waal 2006; Iliffe 2006; Patterson 2006). The HIV prevalence rates remain high in South Africa, which has Africa's richest and best educated population but also far and away the continent's largest number of HIV+ inhabitants.

Knowledge, Attitudes, and Behavior: DHS Information

Understanding of the knowledge/behavior gap for HIV/AIDS in sub-Saharan Africa has advanced rapidly in recent years through the collection of appropriate data in the national DHSs. These internationally comparable national projects are characteristic biosocial surveys with similar questions and output formats. The DHSs were first completed in the 1980s with nationally representative samples of between 5,000 and 10,000 women. The program has continued ever since, covering many countries several times, and in most cases it now also includes surveys of men. The primary achievements of the earliest surveys were the identification and measurement of the main parameters of fertility and mortality changes (Gould and Brown 1996).

A significant breakthrough in eliciting systematic information on AIDS came in 2001. In that year the DHS, under direct instructions from the President of United States and the USAID (its principal funders), began to include modules with items that both inquired about the incidence of AIDS—with AIDS testing under strict protocols—and gathered associated KAP information from the sample of informants. Since beginning with Mali in 2001, the surveys have included self-reported questions on HIV status, verified by a blood test. The responses have vastly improved the calculation of prevalence rates because the indices are derived from samples of whole populations, not from hospital-based, and therefore biased, blood samples.

The differences between the methods of calculation stem largely from urban and female bias in recording. The DHS data have revolutionized knowledge about the patterns of prevalence, facilitating the detection of HIV by age, gender, wealth, and education. More important in this chapter, though, is that the DHS modules have included questions on sexual behavior, knowledge of and attitudes to AIDS, and the sexual practices of the women and men included in the samples, the very people at greatest risk (the population 15–50 years old). The modules provide a broad range of indicators on various aspects: a massive data base from 219 surveys, over half of which come from sub-Saharan African countries. This resource, *The HIV/AIDS Survey Indicators Data Base*, is publicly available in detail on the DHS web site, MeasureDHS, at http://www.measuredhs.com/hivdata/ and, more specifically, under AIDS knowledge, attitudes, and behavior at http://www.measuredhs.com/What-We-Do/Survey-types/AIS.cfm

The data base allows comparative identification of knowledge parameters through a number of basic variables and facilitates the generation of cross-tabulations and associated mapping. I illustrate these tabulations with examples based on recent DHS survey evidence from three countries: Kenya (Central Bureau of Statistics et al. 2004), Tanzania (TACAIDS 2005), and Uganda (Ministry of Health & ORC Macro 2006). In all three cases the data are cross-tabulated by rural–urban location and major educational subgroups.

Knowledge

The data base addresses three aspects of knowledge about HIV/AIDS: awareness of the existence of the disease, of ways in which it is transmitted, and of methods for preventing it. The first of these three facets is the most rudimentary and is addressed with a basic question on whether the respondent had or had not heard of HIV/AIDS. The high proportions of people who have ever heard of HIV/AIDS are evident from Table 13.1. In sub-Saharan Africa, as represented by the three countries covered there, that knowledge unsurprisingly proves to be effectively universal now, even in rural areas and among the least educated people. The name itself, and therefore its importance, is recognized. Many national surveys have additional detailed questions for 15–30-year-olds, the most sexually active group and, hence, the most critical one for HIV transmission.

A second set of questions inquires about knowledge of how the disease is transmitted, specifically, whether the respondents know about transmission during sexual intercourse. The data base also encompasses responses to questions about knowledge of mother-to-child transmission among lactating women, an issue critical for pregnant women and mothers.

A third set of questions addresses knowledge of prevention methods, concentrating on three of them: the use of condoms, abstention from sex, and confinement of sexual contact to only one partner. These three methods are consistent with the ABC approach in IEC programs and are used to calculate a composite knowledge vari-

Table 13.1 Interviewees who have ever heard of HIV/AIDS (in percentages)

Population-based sub-Saharan Africa	Residence (all ages surveyed)				Education (all ages surveyed)					
	Urban		Rural		No education		Primary		Secondary and higher	
	Male	Female	Male	Female	Male	Female	Male	Female	Male	Female
Kenya Demographic and Health Survey, 2003	99.8	99.2	99.3	98.3	94.8	92.6	99.6	99.2	99.9	99.8
Tanzania Demographic and Health Survey, 2004–2005	99.6	99.8	99.0	98.6	95.9	96.2	99.5	99.7	100.0	100.0
Uganda HIV/AIDS Sero-Behavioural Survey, 2004–2005	99.9	99.8	99.0	98.4	94.8	96.0	99.4	99.3	100.0	99.9

Compiled from USAID, UNAIDS, UNICEF, WHO, CDC, U.S. Census Bureau, & DHS (2009)

Table 13.2 Knowledge of HIV prevention methods—composite of three components (in percentages)

Population-based sub-Saharan Africa	Residence (all ages surveyed)				Education (all ages surveyed)					
	Urban		Rural		No education		Primary		Secondary and higher	
	Male	Female	Male	Female	Male	Female	Male	Female	Male	Female
Kenya Demographic and Health Survey 2003	74.4	63.0	64.9	51.8	34.6	25.1	63.8	54.7	78.4	67.3
Tanzania Demographic and Health Survey 2004–2005	69.3	77.3	66.9	70.0	52.7	60.0	69.8	75.3	67.1	80.8
Uganda HIV/AIDS Sero-Behavioural Survey 2004–2005	74.7	72.1	62.5	55.1	45.2	38.5	63.1	60.4	72.5	75.0

Compiled from USAID, UNAIDS, UNICEF, WHO, CDC, U.S. Census Bureau, & DHS (2009)

able (see Table 13.2). This part of the data base also contains questions for lactating and pregnant women about how mother-to-child transmission can be reduced if the mother takes special medication, notably ARVs, during pregnancy. Unlike the levels of knowledge about the existence of HIV/AIDS, those for the three cardinal prevention methods fall sharply for all subgroups (see Table 13.2). Yet clear differentials do emerge. The level of urban knowledge is generally higher than that of rural knowledge, as is to be expected, not only because towns offer better access to IEC messages than rural areas do but also because urban areas have more cases of HIV/AIDS to be aware of (a neighborhood effect). There are also gender differences in knowledge. It is somewhat surprising that men are not always better informed than women (e.g., in Tanzania for both urban and rural areas), but the gender differences are less striking than the differences by educational status. As expected, the difference between people with no education and those with some education is fairly steep, but the curve flattens out at the top end for people with the most education. For any given level of education, women and men exhibit small differences in knowledge, and these do not always go in the same direction. Tanzanian women seem better informed then their men; rather the opposite is true in Kenya.

The DHS data on knowledge of HIV/AIDS also reflect an array of beliefs and misconceptions about the disease, its transmission, and its prevention. These misconceptions can be pervasive and nationally or even locally specific. They include a fairly common belief among some people that—

- Having sex with a virgin will prevent AIDS (a rather gender-biased excuse for sexual license!);
- AIDS is spread by mosquito bites;
- AIDS can be transmitted by supernatural means (a fallacy that leads to frequent recourse to the traditional therapies and indigenous medical belief systems);
- People can be infected by food shared with someone who has AIDS; and
- A healthy-looking person cannot have AIDS (which is often called *slim* in East Africa, given the disease's wasting of the human body).

All these beliefs are identified through further questions in national survey modules. Information gathered from the IEC sources is far from complete, and personal knowledge remains subject to personal biases in its source and how it is understood and acted upon.

Attitudes

The DHSs also have direct questions designed to educe attitudes toward HIV/AIDS. These items are alternative (yes–no) questions, the responses to which indicate how accepting the respondent is of a person infected with the virus. Is the respondent willing to—

- Care at home for a family member sick with the AIDS virus?
- Buy fresh vegetables from an HIV-positive shopkeeper?
- Allow an HIV-positive female teacher who is not sick to continue teaching?
- Reveal the HIV-positive status of a family member?

Attitudes vary enormously from country to country, by gender, and from one educational subgroup to the next. In general, HIV+ people and AIDS sufferers across sub-Saharan Africa live under a substantial stigma that stems chiefly from incomplete awareness of the nature of the disease. The broad national and regional patterns of attitudes and stigma can be illustrated by the mapping facility of the HIV/AIDS Survey Indicators Data Base.

Behavior

Although gathering reliable information on sexual behavior is difficult in any situation, it is plainly necessary in HIV/AIDS work. Questions on behavior, however imperfectly answered, have therefore been intrinsic to the DHS modules, which focus on the respondent's knowledge about avoiding the disease: use of condoms, awareness of when and with whom one has sexual intercourse (regular partners and/or casual encounters), the number of sexual partners over a given period, and the nature of these partnerships (e.g., consensual/commercial).

Table 13.3 summarizes the main findings in Uganda, Tanzania, and Kenya on behavior during high-risk sex (defined as sexual intercourse without a condom with a noncohabiting partner in the 12 months preceding the survey). Knowledge of sexual partnerships is seen in the percentages of 15- to 49-year-old men and women who had sex with more than one partner in the 12 months preceding their participation in the survey. Evidence of knowledge about condom use and high-risk sex is derived from the percentages of 15- to 49-year-old men and women who used a condom the last time they had sex with a nonmarital, noncohabiting partner among those who have had sex with such a partner in the 12 months preceding the survey.

Table 13.3 High-risk sex in the previous 12 months (in percentages)

Population-based sub-Saharan Africa	Residence (all ages surveyed)				Education (all ages surveyed)					
	Urban		Rural		No Education		Primary		Secondary and higher	
	Male	Female	Male	Female	Male	Female	Male	Female	Male	Female
Kenya Demographic and Health Survey 2003	40.9	23.8	36.0	15.6	21.8	10.3	40.4	18.0	35.7	20.1
Tanzania Demographic and Health Survey 2004–2005	49.9	28.5	44.1	15.3	36.4	12.5	46.4	20.2	50.8	31.7
Uganda HIV/AIDS Sero-Behavioural Survey 2004–2005	51.2	28.6	31.3	12.7	17.2	6.0	32.3	14.0	44.4	32.4

Compiled from USAID, UNAIDS, UNICEF, WHO, CDC, U.S. Census Bureau, & DHS (2009)

The rates of high-risk sex are slightly higher in urban areas than in rural ones, but the rates for men are roughly double those for women. The crux of this discussion is that the proportions for women in all three countries are highest among the most educated. In Kenya the group most vulnerable to high-risk sex consists of men with only a primary school education. The gap between men and women is widest among people with primary schooling only and narrows for secondary schooling. Between 33 % and 50 % of people with secondary education and above have had high-risk sex in the 12 months preceding their participation in the survey. The corresponding percentages for people with no education range from less than one third to less than one half those rates. Counterintuitively, the survey results thus state that the most knowledgeable people take the highest risks! Nowhere is this finding more apparent than among university students. Fully exposed to IEC programs and well equipped with scientific understanding of the disease, they were still reluctant in 2009 to "wrap up" (use condoms) during casual sex, despite a generation of devastation by HIV/AIDS ("Kenya: University," 2009).

These and other DHS data make it evident that knowledge itself does not seem to be providing the means to change attitudes or behavior in Kenya, Tanzania, and Uganda. Knowledge does increase with the level of education, for education will improve objective understanding, but it proves to be anything but a magic solution for changing sexual behavior. Increased knowledge from IEC programs does not necessarily translate into changed behavior immediately or even in the medium term. Vazquez-Alvarez and Froelich (2009) cite detailed survey evidence from Kenya that IEC programs do indeed have some positive effect on behavior. Yet the effect is limited overall and is rather more noticeable for women than for men. The disjunction between knowledge and behavior is especially conspicuous for men. In male-dominated societies and male-dominated gender relationships, even within the family, this differential has been a central cause of the weakness in behavioral response to the HIV/AIDS epidemic in sub-Saharan Africa.

The Cultural Limits to Narrowing the Knowledge/Behavior Gap

Obviously, the knowledge/behavior gap for HIV/AIDS is still wide. Whereas much remains to be understood on the knowledge side of the relationship, it is exceptionally important not to consider knowledge in isolation from behavior. Unmistakably, any deepened understanding of the disease and improved attitudes toward people who are either directly or indirectly afflicted by it must include examination of the kinds of knowledge that can affect behavior. A more critical, but more elusive, question is therefore what knowledge is needed to facilitate appropriate behavioral change?

One way to explore this question is to examine the analysis of what is thought to be effective in behavioral programs, as done in a major policy review in the influential journal *Science* a number of years ago. It identified the span of approaches that have been used globally, differentiating those having only weak evidence of their effectiveness from those for which the evidence is stronger (Potts et al. 2008). The authors concluded that many popular ones fall into the former category. They included three pillars of policy: condom promotion and distribution, voluntary counseling and testing, and treatment of other sexually transmitted infections. By contrast, it was found that two practices in particular do seem to reduce HIV prevalence rates: male circumcision and reduction in the number of sexual partners. What they have in common is that they impinge on culture. Both the acceptance of male circumcision and the compass of individual sexual networks are firmly rooted cultural realities and relationships that are often specific to subgroups within national populations. Just as with mortality and especially with fertility, it is widely recognized that the prevalence of HIV/AIDS is culturally variable and that cultural factors explain much of the substantial variation in the rates and timing of African fertility decline in particular (Newman and Lura 1983).

Male Circumcision

Medical evidence distinctly indicates that circumcision substantially reduces a man's chances of becoming HIV+. The conclusions by the World Health Organization and The Joint United Nations Programme on AIDS (WHO & UNAIDS 2007) are stridently argued in favor of encouraging male circumcision as a public health measure to directly reduce HIV prevalence. Three individual randomized control studies in Africa have shown that circumcision can reduce HIV incidence in men by 50–60 %. The technical literature indisputably suggests that male circumcision is cost effective in southern Africa, at least if all men are targeted (Fieno 2008; White et al. 2008). WHO therefore argues that education about the medical advantages of circumcision needs to be increased, as does training, access, and circumcision itself—above all, safe circumcision provided in public health facilities and meant for young men, including boys before sexual debut.

 Compared to the rest of the world, sub-Saharan Africa already has fairly high male circumcision rates, not only as performed in Moslem Africa but as traditionally practiced among many ethnicities across the continent. There tends to be a cultural link between male and female circumcision, but the two practices have very different medical and social effects. Whereas female circumcision has been broadly reviled as a physically dangerous and socially oppressive practice, male circumcision has certainly not been branded in those terms (though hygiene during the circumcision ceremonies is a concern). Nevertheless, one finds a well-known belt of noncircumcising peoples in eastern and southern Africa. The spatial link with HIV/AIDS in Africa was identified early in the epidemic, most prominently by one of the maps in Caldwell and Caldwell (1993). The mechanism of reduced transmission may not have been known then, but the spatial correlation seemed highly indicative. The scientific community is now aware that the foreskin is especially vulnerable to opportunistic infections during sexual intercourse and that its removal offers partial protection.

The central and southern African countries with the highest HIV prevalence tend to be those with the lowest circumcision rates. The most intense public interest in male circumcision campaigns as a public health measure is expressed in South Africa. There seems to be a general sense that demand for male circumcision is increasing, encouraged by the circulation of knowledge about the supporting medical evidence. However, information campaigns cautiously avoid painting circumcision as the magic bullet, the opportunity to continue high-risk behavior without the risk ("Male circumcision," 2009).

Given the biomedical evidence and increasing knowledge about this evidence, why does noncircumcision persist? The technical literature, epitomized by WHO and UNAIDS (2007), rests on the assumption that the principal causes of low demand are associated with poor hygiene and unsafe circumcisions. Thus, there is a sense that sanitizing and institutionalizing the event—seeing it primarily as a medical procedure rather than a cultural rite of passage—is necessary for increasing acceptability. However, in discussion of the primary barriers to acceptability, biomedical determinants such as pain, safety, costs, and surgical complications need to be juxtaposed with culturally unacceptable ones (WHO & UNAIDS 2007, p. 24, chiefly Table 4). Male circumcision is practiced not only, or even primarily, for health reasons. It is not a mere medical procedure but rather a significant cultural event, a rite of passage, and a marker of identity, and it is often done in groups. It serves as a marker of the circumcised and uncircumcised alike, differentiating peoples and declaring allegiance to some ritual authority and thereby to the economy (land) and traditional authority structures. To see it merely as a cultural fashion that can be easily altered with better knowledge or safer implementation is to grossly underestimate the strength of cultural identities and how they are marked.

This tension between health and culture is abundantly clear in Kenya, where the two largest ethnic groups are the Kikuyu, who traditionally circumcise men as well as women, and the Luo, who have not practiced male circumcision in the past. The 2003 Kenyan DHS recorded a national male circumcision rate of 84 %, which was 94 % among the Kikuyu but only 16 % among the Luo (Central Bureau of Statistics et al. 2004). For the Kikuyu, male circumcision, traditionally performed on boys aged 8–14 years with great public ceremony and rejoicing for the whole community, brings full manhood. The Luo, too, practice elaborate male initiation rites, but they do not include circumcision. This difference in cultural practice between the two major ethnic groups in Kenya did not used to be seen as a political issue, but it has acquired substantial political importance in recent years. In the 2009 presidential elections, some Kikuyu politicians and others argued that having a noncircumcised male as president would be like having a child as president and thus implied that the Luo were a tribe of children and that a Luo president or prime minister could not be taken seriously. Such attitudes can be symptomatic of the rationalizations implicit in the current and recurrent political standoff between the Kikuyu and Luo, where a Kikuyu President and a Luo Prime Minister now sit uneasily together to manage the range of political and economic issues facing that country.

The Kenyan state is anxious to reduce HIV prevalence through behavioral change that, in theory, will come from better knowledge of the medical advantages of male circumcision. Such a change is held to be particularly necessary among the Luo, whose HIV prevalence rates are the highest in Kenya (over 15 % in Nyanza Province compared with 7 % nationally). Despite heavy pressure, one finds mixed messages in the Luo community. The prime minister and the minister of health, the latter of whom is also a Luo, have been promoting the national male circumcision plan that began in November 2008 with a target of circumcising one million additional men by 2013. The government recorded 40,000 circumcisions of men in the first year of the plan at the 124 sites opened with trained assistants and appropriate equipment ("Kenya: Million," 2009). Nonetheless, organized resistance to this program exists among the Luo. The traditional Luo Council of Elders in 2007 refused to endorse the plan. They would be satisfied if the circumcision were to be voluntary (as it is), but it must not be imposed on cultural traditions. They argue for greater condom use rather than circumcision for controlling infection, despite the general lack of enthusiasm about condoms in Kenya, where rates of their use are especially low in Nyanza (0.3 %) as compared to Central (3.3 %) and the country as a whole (1.2 %).

Reducing the Number of Sexual Partners

The DHS data, associated tabulations, and the exceptional, but important, case of Uganda in the early 1990s demonstrate that the policy preference for encouraging behavioral change has tended to focus on reducing the number of sexual partners, that is, to reduce sexual networks to an ideal of faithful monogamy (or to a faithfully closed network in polygamous marriages). Substantial behavioral change has undoubtedly occurred as a result of better knowledge, especially where that information has come in politically direct and readily accepted messages. Regardless, fairly open sexual networks are still the cultural norm, even for formally married and notionally monogamous men in urban areas. A widespread culture of male identity and status is being enhanced, if not sustained, by sexual activity often characterized by sugar daddy relationships between a relatively rich older man and younger girls, or *deuxième bureau* arrangements for supporting mistresses outside the family home. The term *zero grazing*, meaning faithful adherence to a monogamous relationship, is prevalent in the media and in private conversation, but the practice is not always sustained (see Table 13.3). The practice of having extramarital sexual partners, inelegantly termed "side dishes" in recent Ugandan campaigns ("Side Dishes," 2010), is more common in urban areas than in rural ones, even among the better educated.

Fairly open sexual networks, however, are associated with many traditional behaviors associated with African marriage, including the phenomenon of sexual cleansing. This practice involves a widow's remarriage after her husband's death to cleanse the woman from unwanted spirits associated with death. The widow has sexual intercourse with another male relative, normally a brother of the deceased man, where endogenous marriage is the expectation. It can be a one-off event or a spiritual cleansing. More commonly in southern Africa, the widow becomes an additional wife with continuing sexual contact, a relationship known as a *levirate arrangement*. In both cases the relationship has a definite economic function in that it cements inheritance within the group. In pre-AIDS times, given the likelihood of a groom being considerably older than his bride, particularly in polygamous societies, the probability of the husband dying before the wife was quite high, so levirate arrangements were frequent. In contemporary high-prevalence HIV societies, early male death is even more likely, mainly because of AIDS. In cases when the husband has died of AIDS, it is more likely that his widow, too, is HIV positive than if he has died of other causes. She can carry the infection to her new husband, even if he had not been infected. With AIDS mortality and the associated economic stress within households, though, the need to cement relationships over land and property by traditional behaviors is even stronger.

Thus, sexual cleansing is seen in the literature as a barrier to reducing HIV prevalence. Concerted attempts to change this behavior have included suggestions of alternative cleansing rituals, which can have limited success according to some evidence (e.g., in Zambia; see Malungo 2001). Yet the need for sexual cleansing seems to be deeply entrenched in many ethnic cultures. Among the Luo in Kenya, for example, sexual intercourse has a symbolic function, and not just at death. It has an important role in festivals for the fertility of people and land, for the harvest, and for prosperity in general (Ayikukwei et al. 2008). The role of the community is also important. Widows are expected to uphold tradition and inheritance systems, so there is clan pressure, if not compulsion, to comply with sexual cleansing practices. Widows are commonly ostracized if they do not behave as expected. Furthermore, they may need to engage in risky sexual behavior to keep themselves and their families out of extreme poverty and starvation (see Bryceson 2006, on Malawi). The cultural and economic imperatives for women to continue maintaining extensive sexual networks raise larger issues about women's empowerment, even for educated women, and create room for negotiation about sexual behavior in such a context.

Both resistance to male circumcision and reluctance to abandon sexual cleansing rituals are two specific examples of cultural attitudes that many people regard as barriers to reducing HIV prevalence and AIDS deaths. Despite growing public and private knowledge, despite official policies and media messages, the behaviors endure, though they are somewhat weakening at the edges. They are inconsistent with a biomedical model of behavior that seeks to minimize risk by reducing the opportunities for infection. The technical opportunities are known and can readily be accepted, but the actual behaviors can be construed, at best, as evidence of denial rather than irrationality. Modernist assumptions about rationality need to be juxtaposed with assumptions about how people are to lead their everyday lives, live in a community, relate to the fundamental institutions of their society, and relate to the political leadership and the land from which they derive their livelihoods. They have their own part in the material and social culture of their group. This culture can change, and development from above is intended to put mechanisms in place to bring about this change, with the objective of narrowing the gap between biomedical and traditional perspectives on causes of illness and on the ways in which behavior will affect and, if necessary, prevent infection.

Cultural change—change in institutions and local power relations, including gender relations—has to be seen as a long-term project, above all where there is a tension between a modernizing state and traditional society. A short-term fix based on political messages, as in Museveni's Uganda; initiatives within the health system that increase the accessibility and acceptability of condoms or circumcision, as in South Africa; and an overall narrowing of the gap between knowledge and behavior are more likely to be effective with grass-root political support and local opinion leaders who are in sympathy with national political leaderships.

In Kenya this eventuality is much more likely among the Kikuyu than among the Luo, so the policy goal of reducing the high HIV prevalence rates among the Luo will remain elusive. As long as ethnicity and regional difference figure prominently in politics and economic differentiation, ethnicity will remain as a differentiating factor in the risk of infection. DHS data in Kenya technically confirm this ethnic factor as a significant variable in logistic regression analysis:

> Kenya has a mix of cultural beliefs and practices emanating from the existence of over 41 different ethnic groups. The diversity in culture has been a major hindrance in the successful implementation of intervention programs since no single program has been found to suit all the different communities. (Akwara et al. 2003, p. 407)

A similar argument has been applied in many other multiethnic African states, but neither the international community nor international policy programs are structured to deal with anything other than national programs. It seems uncommon for a state and its population to have similar perspectives or to reach similar conclusions on what is needed, especially in a political environment of failed and failing states and lack of accountability. The yawning knowledge/behavior gap then becomes part of a much broader context about political legitimacy and local populations and national authority.

Conclusion

The wide knowledge/behavior gap for HIV/AIDS in Africa remains. Despite increased knowledge shared by more and more people with ever greater access to objective knowledge–based program initiatives for initiating behavioral change, there are still daunting barriers to accepting, internalizing, and acting on modern, bioscientific understanding. The behavioral change necessary to reduce HIV prevalence is constrained by the cultural realities of the continent, which are often associated with poverty and the need to avoid risks with relationships involving land and livelihood. Behaviors rooted in traditional values, such as male circumcision and multiple partnerships, are not based on ignorance. Nor are they irrational in the social and cultural contexts in which they must be elaborated, though they may conflict with biomedical knowledge and policy assumptions and the social imperatives of modernization. Modern IEC policies and programs for increasing the access to and use of knowledge have been technically narrow and directed to achieving a short-term fix rather than the relatively long-term change in values and cultural understandings needed to change sexual behavior.

The persistence of this knowledge/behavior gap for HIV/AIDS, despite the strenuous efforts of the international community, is as much a failure of the modernizing state as it is a failure of knowledge-based policies. The state has all too often failed to engage with local cultural realities, preferring the safety of international respectability and funding as the basis for HIV/AIDS programs. There needs to be a wider appreciation of knowledge, an understanding that extends beyond the basics of biomedical information to a cultural grasp of what the biomedical knowledge might mean to those who feel the threat of HIV/AIDS in their daily lives. Those directly affected need to deal with AIDS within livelihoods of chronic poverty and struggle. They must also do so within socioeconomic contexts of modernization where the educated segments of the population and people with a degree of biomedical expertise need to negotiate their own behaviors with ethnic kin and urban neighbors.

References

Akwara, P. A., Madise, N. J., & Hinde, A. (2003). Perception of risk of HIV/AIDS and sexual behaviour in Kenya. *Journal of Biosocial Science, 35*, 385–411. doi:10.1017/S0021932003003857.

Ayikukwei, R., Ngare, D., Sidle, J., Ayuku, D., Baliddawa, J., & Green, J. (2008). HIV/AIDS and cultural practices in western Kenya: The impact of sexual cleansing rituals on sexual behaviors. *Culture, Health and Sexuality, 10*, 587–599. doi:10.1080/13691050802012601.

Bongaarts, J., Buettner, T., Heilig, G., & Pelletier, F. (2008). Has the HIV epidemic peaked? *Population and Development Review, 34*, 199–224.

Bryceson, D. F. (2006). *Ganyu* casual labour, famine and HIV/AIDS in rural Malawi: Causality and casualty. *The Journal of Modern African Studies, 44*, 173–202. doi:10.1017/S0022278X06001595.

Caldwell, J. C., & Caldwell, P. (1993). The nature and limits of the sub-Saharan African AIDS epidemic: Evidence from geographic and other patterns. *Population and Development Review, 19*, 817–848.

Central Bureau of Statistics, Ministry of Health, Kenya Medical Research Institute, National Council for Population and Development, Centers for Disease Control and Prevention, & ORC Macro. (2004). Kenya: Demographic and health survey, 2003. Retrieved from http://www.measuredhs.com/pubs/pdf/FR151/FR151.pdf

De Waal, A. (2006). *AIDS and power: Why there is no political crisis—yet.* London: Zed Press.

Dyson, T. (2003). HIV/AIDS and urbanization. *Population and Development Review, 29*, 427–442.

Ferguson, A. G., & Morris, C. N. (2007). Mapping transactional sex on the Northern Corridor highway in Kenya. *Health and Place, 13*, 504–519.

Fieno, J. V. (2008). Costing adult male circumcision in high HIV prevalence, low circumcision rate countries. *AIDS Care, 20*, 515–520.

Gillespie, S., Kadiyala, S., & Greener, R. (2007). Is poverty or wealth driving HIV transmission? *AIDS, 21*(Suppl. 7), S5–S16. doi:10.1097/01.aids.0000300531.74730.72.

Glick, P., & Sahn, D. E. (2008). Are Africans practicing safer sex: Evidence from demographic and health surveys for eight countries. *Economic Development and Cultural Change, 56*, 397–439. doi:10.1086/522893.

Gould, W. T. S. (2004). Rural/urban interactions and HIV/AIDS in Eastern Africa. In L.-N. Hsu (Ed.), *Development, spatial mobility and HIV/AIDS: A workshop on interrelations and programmatic responses* (pp. 35–50). Bangkok: UNDP.

Gould, W. T. S. (2005). Vulnerability and HIV/AIDS in Africa: From demography to development. *Population, Space and Place, 11*, 473–484.

Gould, W. T. S. (2009a). Exploring the anomalous positive relationship between AIDS and poverty in Africa. *Geography Compass, 3*, 1449–1464.

Gould, W. T. S. (2009b). HIV/AIDS in developing countries. In R. Kitchin & N. Thrift (Eds.), *International encyclopedia of human geography* (Vol. 5, pp. 173–179). Oxford, UK: Elsevier.

Gould, W. T. S. (2015). *Population and development*. 2nd edition. London: Routledge.

Gould, W. T. S., & Brown, M. S. (1996). A fertility transition in sub-Saharan Africa? *International Journal of Population Geography, 2*, 1–22.

Gould, W. T. S., & Woods, R. I. (2003). Population geography and HIV/AIDS: The challenge of a 'wholly exceptional disease'. *Scottish Geographical Journal, 119*, 265–281. doi:10.1080/00369220318737178.

Hargreaves, J. R., & Howe, L. D. (2010). Changes in HIV prevalence among differently educated groups in Tanzania between 2003 and 2007. *AIDS, 24*, 755–761. doi:10.1097/QAD.0b013e328336672e.

Iliffe, J. (2006). *The African AIDS epidemic: A history*. Oxford, UK: James Currey.

Kenya: Million Man Cut, The. (2009, November 17). IRIN humanitarian news and analysis service. Retrieved from http://www.irinnews.org/report/87074/kenya-the-million-man-cut

Kenya: University Students Don't Always Wrap-Up. (2009, June 25). IRIN humanitarian news and analysis service. Retrieved from http://www.irinnews.org/report/85019/kenya-university-students-don-t-always-wrap-up-study

Kirunga, C. T., & Ntozi, J. P. (1997). Socio-economic determinants of HIV serostatus: A study of Rakai District, Uganda. *Health Transition Review, 7*(Suppl.), 175–188.

Klaits, F. (2010). *Death in a church of life: Moral passion during Botswana's time of AIDS*. Berkeley: University of California Press.

Lopman, B., Lewis, J., Nyamukapa, C., Mushati, P., Chandiwana, S., & Gregson, S. (2007). HIV incidence and poverty in Manicaland, Zimbabwe: Is HIV becoming a disease of the poor? *AIDS, 21*(Suppl. 7), S57–S66. doi:10.1097/01.aids.0000300536.82354.52.

Low-Beer, D., & Stoneburner, R. L. (2003). Behavior and communication change in reducing HIV: Is Uganda unique? *African Journal of AIDS Research, 2*, 9–21.

Male Circumcision Slowly Taking Off. (2009, July 23). IRIN humanitarian news and analysis. Retrieved from http://www.irinnews.org/report/85405/africa-male-circumcision-slowly-taking-off

Malungo, J. R. S. (2001). Sexual cleansing (*kusalazya*) and levirate marriage (*kunjilila mung'anda*) in the era of AIDS: Changes in perceptions and practices in Zambia. *Social Science and Medicine, 53*, 371–382.

Ministry of Health of Uganda, & ORC Macro. (2006, March). *Uganda HIV/AIDS sero-behavioral survey, 2004–05*. Calverton: Ministry of Health and ORC Macro. Retrieved from http://www.measuredhs.com/pubs/pdf/AIS2/AIS2.pdf

Newman, J. L., & Lura, R. P. (1983). Fertility control in Africa. *Geographical Review, 73*, 396–406.

Ntozi, J. P. M., Najjumba Mulwinda, I., Ahimbisibwe, F., Ayiga, N., & Odwee, J. (2003). Has the HIV/AIDS epidemic changed sexual behaviour of high risk groups in Uganda? *African Health Sciences, 3*, 107–116.

Patterson, A. S. (2006). *The politics of AIDS in Africa*. Boulder: Lynne Rienner.

Posse, M., Meheus, F., van Asten, H., van der Ven, A., & Baltussen, R. (2008). Barriers to access to antiretroviral treatment in developing countries: A review. *Tropical Medicine and International Health, 13*, 904–913.

Potts, M., Halperin, D. T., Kirby, D., Swidler, A., Marseille, E., Klausner, J. D.,…, Walsh, J. (2008). Public health: Reassessing HIV prevention. *Science, 320*, 749–750. ISSN No. 1095-9203.

Robson, E., Ansell, N., Huber, U. S., Gould, W. T. S., & van Blerk, L. (2006). Young caregivers in the context of the HIV/AIDS pandemic in sub-Saharan Africa. *Population, Space and Place, 12*, 93–111.

"Side dishes" Campaign Gets People Talking. (2010, March 24). IRIN humanitarian news and analysis service. Retrieved from http://www.irinnews.org/report/88543/side-dishes-campaign-gets-people-talking

Stoneburner, R. L., & Low-Beer, D. (2004). Population-level HIV declines and behavioral risk avoidance in Uganda. *Science, 304*, 714–718. ISSN No. 1095–9203.

TACAIDS (Tanzania Commission for AIDS), NBS (National Bureau of Statistics), & Opinion Research Corporation (ORC) Macro. (2005). *Tanzania: HIV/AIDS indicator survey 2003/04.* Calverton: TACAIDS, NBS, and ORC Macro. Retrieved from http://www.measuredhs.com/pubs/pdf/AIS1/AIS1.pdf

UNAIDS. (2009). *2009 Report on the global AIDS epidemic.* Geneva: UNAIDS.

USAID, UNAIDS, UNICEF, WHO, CDC, U.S. Census Bureau, & DHS. (2009). HIV/AIDS survey indicators database. Retrieved from http://www.measuredhs.com/hivdata

Vazquez-Alvarez, R., & Froelich, M. (2009). HIV/AIDS knowledge and behavior: Have information campaigns reduced HIV infection? The case of Kenya. *African Development Review, 21*, 86–146.

White, R. G., Glynn, J. R., Orroth, K. K., Freeman, E. E., Bakker, R., Weiss, H. A.,…, Hayes, R. J. (2008). Male circumcision for HIV prevention in sub-Saharan Africa: Who, what and when? *AIDS, 22*, 1841–1850. doi:10.1097/QAD.0b013e32830e0137

WHO (World Health Organization) & UNAIDS (Joint United Nations Programme on HIV/AIDS). (2007). *Male circumcision: Global trends and determinants of prevalence, safety and acceptability.* Geneva: WHO and UNAIDS.

World Bank. (1999). *The World Development Report, 1998–1999: Knowledge for development.* Washington, DC: The World Bank.

The Klaus Tschira Stiftung

Physicist Dr. h.c. Dr.-Ing. E. h. Klaus Tschira established the German foundation Klaus Tschira Stiftung in 1995. The Klaus Tschira Stiftung is one of Europe's largest privately funded non-profit foundations. It promotes the advancement of natural sciences, mathematics, and computer science and strives to raise appreciation for these fields. The focal points of the Foundation are "Natural Science – Right from the Beginning", "Research" and "Science Communication". The Klaus Tschira Stiftung commitments begin in the kindergartens and continue at primary and secondary schools, universities and research facilities. The Foundation champions new methods of scientific knowledge transfer, and supports both development and intelligible presentation of research findings. The Klaus Tschira Stiftung pursues its objectives by conducting projects of its own, but also awards subsidies upon application and positive assessment. The Stiftung has also founded its own affiliations that promote sustainability among the selected topics. Klaus Tschira's commitment to this objective was honored in 1999 with the "Deutscher Stifterpreis", the prize awarded by the National Association of German Foundations.

The Klaus Tschira Stiftung is located in Heidelberg with its head office in the Villa Bosch (Fig. A.1), which used to be the residence of the Nobel laureate in chemistry Carl Bosch.

© Springer International Publishing Switzerland 2016
P. Meusburger et al. (eds.), *Ethnic and Cultural Dimensions of Knowledge*,
Knowledge and Space 8, DOI 10.1007/978-3-319-21900-4

Fig. A.1 The Villa Bosch (Source: Peter Meusburger)

Fig. A.2 Participants of the symposium "Ethnic and Cultural Dimensions of Knowledge" at the studio of the Villa Bosch in Heidelberg (Source and copyright: Elisabeth Militz, Heidelberg)

In Memory of Klaus Tschira

On March 31, 2015, the world lost a splendid human being and great benefactor for science and research. We mourn the passing of Dr. h.c. Dr.-Ing. E.h. Klaus Tschira, one of the world's greatest benefactors of German-speaking geography. For two decades, Klaus Tschira and his foundation have sponsored students and young academic professionals in a most generous and uncomplicated way. Being a physicist himself, Klaus Tschira pursued his funding of research primarily in the natural sciences, mathematics, and computer science. Among the outstanding initiatives his foundation Klaus Tschira Stiftung has supported in human geography were the Hettner Lectures between 1997 and 2006 and the Knowledge and Space symposia since 2006, both of which have enabled a whole generation of students and young researchers from various disciplines and countries to engage intellectually with internationally renowned scientists. These events have contributed significantly to the creation of a global network of research cooperation.

Fig. A.3 Klaus Tschira (1940–2015) (Source and copyright: Klaus Tschira Stiftung)

© Springer International Publishing Switzerland 2016
P. Meusburger et al. (eds.), *Ethnic and Cultural Dimensions of Knowledge*,
Knowledge and Space 8, DOI 10.1007/978-3-319-21900-4

Index

A
Academic degree, 96, 97
Acculturation, 6
Acephalous community, 12
Achievement
    gap, 66
    test, 31
Adult, 40, 81, 82, 104, 120, 148, 149, 258, 264, 265, 267, 277
Adulthood, 258, 264, 265
    ceremony, 16, 266, 270
Afghanistan, 14, 15, 116, 211, 222–225
Africa, 16, 46, 216, 217, 219, 220, 263, 276, 277, 279–287, 289
African Americans, 10, 33, 59, 70–75, 77, 84–87
African-Canadians, 115
Agency, 219
Agropastoralism, 212, 216
Alabama, 33, 60, 75, 86
Albuquerque, 10, 94, 99, 104
Alfa grass, 215
Algeria, 212–216, 219
Algerian Forest Code, 215
Algerian Forest Service, 214
Alternative
    development, 151, 152, 155, 156
    knowledge, 129–157, 192, 197
American community survey, 95, 96
American Indian, 26, 28, 34, 41, 45, 95, 99
Andrew Carnegie, 60
Anglicans, 79
Anthropological, 173–175
Anthropology, 133, 145

Anticolonial
    discursive framework, 111
    perspective, 11, 109
    theory, 110, 111
Anti-culturalist, 7
Apache, 36, 41
Artificial glaciers, 203
Asian Americans, 86, 87
Assimilation, 44–46
Asylum seekers, 28
Asymmetric power relations, 3, 12, 24, 30
Austinian positivism, 258
Australia, 12, 133
Austria, 23, 36, 37, 44, 46
Authentic experience, 110
Authority, 147–151
Autochthonous minorities, 12, 26, 28

B
Baltimore, 57, 85
Basic veterinary worker (BVW) program, 223–225
Behavior/behaviour, 275–276, 279–289
Belief, 136, 166–170
    system, 1, 168, 169, 282
Bible Belt, 56
Bilingual
    area, 28, 36, 38, 43
    region, 28, 36, 38, 39
    school, 38, 39, 43
    teacher, 36–38
Biomedical knowledge, 16, 289, 290
Birmingham, Alabama, 59

© Springer International Publishing Switzerland 2016
P. Meusburger et al. (eds.), *Ethnic and Cultural Dimensions of Knowledge*,
Knowledge and Space 8, DOI 10.1007/978-3-319-21900-4

Black
    community, 61–63, 65, 113
    student, 9, 11, 63, 66, 113, 115, 123
    teacher, 9, 56, 58, 62, 64, 118
Boarding schools, 10, 34, 87
Bohemian, 36
Booker T. Washington, 60
Botswana, 135, 278
Brown v. Board of Education, 9, 64
Bureaucracy, 12, 71, 183
Bureau of Indian Affairs (BIA), 26, 28,
    34, 36
Byelorussian, 38

C
California, 56, 66, 75, 86, 100
Camel, 212, 213, 218, 220
Canada, 113, 115, 116, 119, 121, 126
Canadian, 4, 11, 35, 109, 110, 112,
    113, 115, 116, 118–120,
    122, 126
Canoe, 15, 16, 229–254
Career, 101
Caribbean, 70, 118
Carinthia, 28, 38, 39
Carnegie Foundation, 60
Catacomb schools, 45, 46
Categorical knowledge, 168
Catholic, 46, 55, 79, 104
Catholicism, 71, 150
Center, 214
Centralized, 42, 43, 55, 60
Ceremony, 15, 16, 249, 250, 257–270,
    285, 286
Cheder, 45
Chicago, 25, 81
China, 86, 195, 197
Chinese, 86, 121
Christian, 34, 148
Circumcision, 285–287
Civil Rights, 70, 88
Civil War, 9, 57–59, 63
Clandestine schools, 37, 45
Class, 120–123
Classroom, 64
Climate change, 2, 14, 63, 148, 191–193, 199,
    201–204, 226
Cognition, 173–174
Cognitive, 173–175, 184
Cold War, 63
Collective, 29–31

Colonial, 5, 11–13, 15, 16, 24, 70, 79, 95, 111,
    133, 135, 152–154, 211, 213–216,
    218–220, 225, 270
Colonialism, 213–218
Communal, 15, 125, 214, 230, 233, 235–237,
    240, 242–248, 250, 262
Commune, 43
Communication, 4, 42, 47, 130, 138, 143, 150,
    152, 172, 250, 260, 275
Communist, 65
Community, 61–63, 95, 96, 125
Comparative analysis, 13
Compulsory
    ignorance, 33, 45, 84
    school, 25, 42
    schooling, 23, 104
Conflict
    cultural, 7, 8
    ethnic, 3, 5, 7, 8, 17, 23–48
    political, 3, 7, 8, 24, 30
Contestable matter(s), 2
Context
    cultural, 2, 12, 13, 17, 27, 32, 36,
        151, 165, 174, 289
    spatial, 12, 17, 32, 47
Craftsmanship, 233, 243
Croatia, 38
Crop cultivation, 196
Cultural, 1, 4–7, 14–17, 24, 259–261
    barriers, 87
    change, 133, 250, 252, 288, 289
    context, 2, 12, 13, 17, 27, 32, 36, 151,
        165, 174
    diversity, 7, 46, 119, 259
    environment, 1, 2, 34, 42
    framing, 198, 203
    geography, 4, 5, 94, 96, 99, 106, 107
    homogeneity, 7, 24
    identity, 6–8, 10, 11, 13, 16, 24, 26, 32, 34,
        35, 40, 44, 45, 47, 100, 103, 106, 107,
        109, 110, 112, 113, 117–120, 122, 252
    knowledge, 42, 111, 120, 122, 125
    landscape, 193–194
    memory, 2, 8, 23, 25, 203
    practices, 2, 4, 6, 7, 9, 12, 16, 24, 25, 42, 45
    product, 13, 166
    studies, 5, 6, 47, 136
    traditions, 1, 4, 7, 13, 287
Culturally-centered pedagogy, 125
Curriculum, 115–117
Customs, 134, 249, 257–270
Czarist Russia, 45

D

Decentralized, 43, 149, 172
Declarative, 168
Democracy, 10, 23, 46, 63, 80
Democratic Party, 38, 58, 63
Demography, 277
Desegregation, 27, 32, 64, 88
Development
    intervention, 129, 151, 191–205
    program, 15, 130, 217, 224, 225
    studies, 3, 12, 275
Dinghies, 248, 249
Discrimination, 72, 84
Distributed knowledge, 16, 246, 249
Dmowski, Roman, 38
Dropout rates, 77
Dropouts, 10, 76–78, 113, 122, 123
DuBois, W.E.B., 86
Dugout, 231, 234, 235–237, 239, 240, 243,
    244, 246, 247, 252, 253
Duration of the school year, 10, 86

E

Eastern Europe, 26
Economic development, 60, 94, 153,
    155, 275
Education, system, 36, 47, 104, 105, 110, 275
Educational
    achievement, 8, 9, 12, 25, 29–31, 47, 56,
        69, 72, 75, 77, 78, 84
    attainment, 5, 10–12, 17, 25, 29–31,
        75–78, 84, 94–96, 98, 100, 106, 107,
        114, 124
    inequalities, 6, 10, 11, 29–33, 93–107
    institution(s), 4, 5, 7, 10, 24, 25, 34, 39, 45,
        93, 94, 96, 100, 103, 106, 107
    success, 11, 93, 95, 96, 102, 106, 107
Educator, 11, 46, 109, 111–120, 124, 125
Egalitarian, 10, 56, 58, 148, 153–155, 157
Egalitarianism, 56, 69
Elementary school, 8, 9, 27, 36, 37, 39, 40,
    42–45, 59
Embeddedness, 96–106
Emic, 174, 178, 182, 183
Empowerment, 11, 111, 288
Environment, 2–4, 11, 13, 34, 41, 62, 65, 99,
    103, 106, 113, 119, 122, 125, 136, 149,
    167, 169, 176, 178, 184, 193, 194, 198,
    201–205, 211, 213, 215, 216, 218, 220,
    225, 226, 257, 270, 289
Environmental, 175, 176

Ethnic
    awareness, 9, 25, 27, 29, 42
    conflicts, 3, 5, 7, 9, 23–48
    identity, 8, 28, 40, 42, 46
    minorities, 3, 5, 6, 8–11, 24–31, 42, 44–46,
        72, 84, 87, 88, 105–107, 114, 125, 133
    self-esteem, 9, 29, 40, 44
    undereducation, 84–89
Ethnicity, 6–7, 71, 95
Ethnocide, 12, 133
Ethnographic method, 173
Ethnologists, 145, 173, 174
Ethnoveterinary, 222
Eurocentric, 16, 112, 116, 257, 270
Europe, 8, 9, 26, 28, 37, 46, 55, 69–71, 205
Expert, 2, 12–15, 76, 144, 147, 148, 152, 153,
    156, 157, 167, 183, 211, 212, 221, 224,
    225, 229, 230, 233–250
Expertise, 170
Expulsion rates, 31
External, 12, 14, 27, 28, 154, 192–194, 198,
    202, 204

F

Facial scarification, 267
Factual knowledge, 2
Ferguson, Missouri, 70, 278
Fieldwork, 173
Financial endowment, 10, 80, 84
First World War, 32, 37, 38, 61, 70, 85
Floating identity, 28
Food-distribution ceremonies, 15, 250
Forced assimilation, 44–46
Fort Apache Indian reservation, 36
Foucault, 7, 111, 138
Foundationalism, 137
Fourteenth Amendment, 63
Franklin Roosevelt, 33, 45, 62, 85, 140
Free Blacks, 57
Free-born Blacks, 57
Freedmen's Bureau, 60
French
    Forest Code, 214
    occupation, 213
Funding inequalities, 65

G

Galicia, 37
Gatherings "under the trees", 45, 85
Gender, 120–123

General Education Board (GEB), 60, 61
Genocide, 12, 13, 34, 133
Gentile, 7, 38
Geographical embeddedness, 93–107
Geography, 5, 39
Geography of education, 47, 76
Georgia, 33, 59
German, 30, 36–38, 41, 43–45, 70, 96
Germany, 41, 46, 70, 85, 121, 126, 205
Giertych, Jędrze, 38
Giovanni, 7
Globalization, 13, 14, 41, 42, 71, 134,
    136, 249
Governance, 16, 130, 191, 257
Grabski, Stanisław, 38, 43
Graduate education, 63, 102
Great depression, 62
Great Society program, 79

H
Habsburg Empire, 36, 37, 45
Health survey, 277, 281, 283
Hesse, 70
Higher education, 10, 59, 60, 62, 74, 79,
    93–107
High school, 55, 56, 59, 61, 65, 66, 74, 77, 78,
    85, 95, 106, 109, 113, 117, 118, 120
Himalaya, 14, 191
Himalayan, 191–194, 200, 203, 205
Hispanic(s), 8, 10, 28, 56, 76, 77, 84, 86, 87,
    94–96, 98–100, 102–107
History, 103–106, 115–117, 119
HIV/AIDS, 16, 17, 276–285, 289, 290
Holism, 151, 172
Horace Mann, 80
Hualapai, 41
Human capital, 93
Human geography, 30

I
Identification, 82, 217
Identity, 40
Ideology, 2, 10, 32, 62, 83, 111, 145, 198
Idiom, 155, 156, 182, 183
Igbo
    jurisprudence, 261, 264, 268
    law, 16, 264, 270
    names, 264
    societies, 16, 264, 267
Ignorance, 33, 166–170
Illiteracy, 86
Illiterate, 12, 30, 31, 57

Immigrants, 76
Inclusive, 114, 116, 121, 122, 174
    school(ing), 110, 114, 116, 120–123
India, 192, 195, 197, 200
Indigeneity, 109, 111, 112
Indigenous, 12–17, 109, 111, 112, 124, 125,
    129, 135, 211–225
Indisputable matter(s), 2
Indonesia, 13, 165–167, 176
Industrial education, 61
Inequalities, 96–106
Informal learning, 172
Information, 279–284
Institution, 4, 5, 7, 10, 13, 24, 25, 34, 39,
    42, 45, 47, 56, 59, 60, 70, 79, 81–84,
    93–107, 138, 156, 192, 200, 214, 224,
    258, 259, 288, 289
Institutional, 84, 124, 191, 193
Integration, 63–66
Intellectuals, 7, 13, 133
Intersectionality, 93, 107
Interview(s)/interviewing, 98–100, 102, 103,
    106, 178–181
Irish, 71, 132
Iron Curtain, 46
Irrigation, 14, 193, 197, 199–202, 212
Islamic, 148, 224
Italian-Americans, 70
Italy, 7, 27, 38, 41, 44

J
Jewish, 45, 132

K
Kenya, 16, 17, 278, 280–284, 286–289
Kikuyu, 286, 289
Kindergarten, 38, 45, 55
Knowledge
    base, 119, 221, 276, 278
    as a cultural/social product., 13, 166
    and culture, 1–17
    for development, 275
    ecological, 15, 16
    and power, 2, 4–7, 9, 14, 29–31, 33,
        42, 43, 46, 47, 82, 110, 111, 115,
        118, 120, 122
    production, 7, 10, 17, 110–112, 114, 117,
        118, 123, 211
    and space, 7, 12, 42, 47, 192, 198
    system, 14, 15, 17, 111, 123,
        166, 169, 170, 192, 198, 204,
        211, 225, 259, 275

technical, 15, 80, 130, 142, 151, 275
traditional, 2, 3, 5, 8, 12, 13, 16, 17,
    132–134, 136, 151, 157, 165, 168
Koochi, 15, 211, 222–225
K–12 education, 55
Ku Klux Klan, 58
Kula canoes, 231
Kven, 34, 36

L

Ladakh, 14, 192–200, 202, 203
Land management, 15, 218, 219
Land Ordinance, 80
Land-use practices, 14, 192
Language, 34–39, 44, 120–123, 155
Language of instruction, 44
Las Vegas, 104
Latino, 28, 94, 95
Law, 36, 258
Learning, 116, 174, 175
Learning environment, 125
Learning process, 2, 4, 5, 8, 26, 40, 46, 125
Legitimacy, 16, 257, 264, 289
Legitimate, 11, 26, 40, 109–112, 115,
    120–123, 131, 133
Leo von Thun und Hohenstein, 36
Lineage, 70, 164, 169, 261
Literacy, 5, 23, 57, 73
Literate, 57
Lithuanians, 38
Little Rock, 64
Livelihood, 14, 131, 154, 155, 191–193,
    198, 203–205, 212, 214, 275, 276,
    288–290
Livestock, 216
Local
    knowledge, 2, 3, 8, 11–14, 17, 165–184
    practices, 130, 202
Loss of knowledge, 16, 250
Louisiana, 33, 64, 80
Luo, 17, 286–289
Lvov, 26

M

Maghreb, 14, 15, 211–222, 225
Magic, 230, 233–238, 241, 243–246, 248–251,
    284, 286
Magical, 15, 198, 203, 233, 235, 237–241,
    243–246, 248–251
Magician, 230, 235, 236, 238–241, 244–246
Makassar, 167, 176, 177, 180
Marginalization, 111, 113, 115, 212, 226

Marginalized, 123–126
Marriage, 168, 170, 258, 265–267,
    278, 287
Martin Luther King, 64
Masawa canoe, 229–254
Maturation, 16, 270
Medicine, 15, 130, 170, 214, 216, 217, 219,
    221, 224, 225, 264
Melting pot, 10, 71
Memory, 2, 8, 23, 26, 203, 251
Meritocracy, 12, 104
Methods, 173–175
Michael Gamper, 46
Migration, 219
Minority, 27–29, 31–44, 115–123, 135
    education, 11, 113, 120, 123
    research, 28–30
    rights, 8, 12, 24, 28, 32, 42, 135
    youth, 11, 109, 110, 113, 115, 118, 123,
        124
Misinari(ies), 233, 248–251
Mississippi, 33, 63, 64, 85
Mobile, 57, 218
Modernization, 104, 135, 136, 202, 289, 290
Mormon, 76
Morocco, 212, 216–221
Mother-tongue, 28, 29, 45
Multiethnic, 25–27
Multilingual, 8, 25, 31, 34, 36, 38
Muslim, 15, 211, 224, 225
Mysticism, 262

N

Narratives, 115–123
Nation, 23–25, 70, 72, 73
National Assessment of Educational
    Progress, 31
National Association for the Advancement of
    Colored People (NAACP), 63, 65
National Center for Education Statistics
    (NCES), 72, 73, 76, 78
Nationalism, 26
Nationalist movements, 9, 26
Nation-state, 7, 23, 26, 38, 41, 46, 70, 156
Native Americans, 10, 71, 84, 86, 87
Native knowledge, 12
Native people, 13, 26
Nat Turner, 57
Natural environment, 13, 169, 170, 184, 194
Natural hazards, 192
Natural resource management, 191
Navajo, 34, 36
Neo-Liberalism, 220

New Deal, 62
New England, 79, 86
New Mexico, 10, 86, 94–97, 99–107
New Mexico Institute of Mining and
    Technology, 104
New Orleans, 57, 59, 64, 66, 85
New York, 56, 66, 70, 79, 203
Nigeria, 16, 257, 258, 260
Nomad(s), 214, 216, 221–225
Normative, 259, 261–269
Normativity, 258, 259
Norms, 1, 5, 7, 24, 36, 103, 105, 107, 121,
    122, 147, 224, 261, 264, 269, 270
North Tyrol, 44
Norway, 27, 32, 34, 35

O

Occupational aspirations, 73, 74
Ohio, 33, 104
Old South, 10, 75
One-class schools, 39, 44
Ontario, 110, 114, 126
Oral history, 251
Orientation, 1, 2, 5, 7, 25, 35, 42, 79, 104,
    105, 172
Orientation knowledge, 1, 2, 7, 25, 42
Ötztal, 46
Outrigger canoe, 230, 231, 234, 246

P

Pakistan, 195, 197, 200, 203, 223
Palatinate, 70
Papua New Guinea, 15, 129, 133, 134, 137,
    145, 155
Participatory approaches, 174, 175
Participatory learning approach (PLA),
    174, 177
Participatory rapid appraisal (PRA), 174, 175,
    177, 178
Particular knowledge, 1–4, 117, 129, 146, 171
Pastoralism, 212
Pastoralist, 212–222
Peach springs, 41
Periphery, 3
Persistence, 14, 31, 75, 192, 290
Personal experience, 13, 102, 138, 149
Petrol, 248
Philanthropy/philanthropic, 60–61
Philosophy, 125, 147, 262
Physical disability, 120–123
Physical environment, 1, 2, 11, 99
Place-based knowledge, 2, 3, 192
Pledge of Allegiance, 81

Poland, 37, 38, 43, 45
Polish, 37, 38, 41, 43, 45
Polish National Democratic Party, 38
Political ecology, 191
Political economy, 152, 211–226
Politics, 138, 218–222
Polygamous society, 288
Population, 94–96
Portugal, 94
Post-colonialism, 13
Post-development, 197, 198
Postmodern, 138, 144, 149
Poverty, 77
Power, 6–7
Poznań (Posen), 37, 41
Practices, 135, 175
Precolonial, 212–222, 261–269
Preindustrial, 12
President Barak Obama, 8, 72–75
President George W. Bush, 72
President Lyndon B. Johnson, 79, 82, 83
President Yoweri Museveni, 279, 289
Primary school, 25, 27, 37–39, 41–44, 106,
    275, 284
Procedural knowledge, 172, 175
Protestant, 55, 79
Prussia, 37, 60, 70
Public education, 56, 69, 83
Public schools, 55, 56, 59, 78, 79
Public sector, 59
Pupil Level Annual School Census, 31
Purdah, 223
Puritans, 79

Q

Quechua, 150, 151

R

Race, 8, 95, 120–123
Race theory, 8
Racial discrimination, 63
Racialization, 5, 8
Racial segregation, 32, 55, 56, 58, 61, 63
Racial solidarity, 62, 65, 122
Racial studies, 8
Range ecology, 211
Range management, 15, 211, 218–221,
    225, 226
Rapid Rural Appraisal (RRA), 174, 175, 177
Rationality, 12, 129, 166, 167, 176, 288
Reading, 72, 73, 75
Refugees, 28, 46
Religion, 36

Repertory grid, 175–179, 181, 182
Representation, 6, 11, 75, 109–126, 139, 168, 182
Republican Party, 58, 60
Republicans, 58, 80, 260
Research, 27–29, 110, 113, 115–123, 126, 182, 184
Residential, 176–177
Responsibility, 11, 79, 102, 111, 112, 124, 125, 144, 233, 260, 261, 266, 267
Ritual, 2, 17, 42, 81, 125, 149, 168, 172, 247, 249–251, 260, 263, 265, 267, 268, 286, 288
Rockefeller, John D., 60
Rockwell, Norman, 87–89
Rules, 4, 30, 36, 45, 46, 65, 104, 147, 168, 169, 175, 183, 211, 224, 225, 232, 249, 258, 259, 261, 268
Rural schools, 9, 59, 82, 86
Russia, 37, 65
Russian, 37, 45
Russification, 45

S
Sail, 229–234, 238, 239, 241–243, 246, 247, 249, 252, 253
Sami, 26, 34, 35
Sapir-Whorf hypothesis, 131
School
  architecture, 80
  attendance, 31, 38, 45, 59, 79, 85
  authority, 16, 40, 66, 99, 111, 135, 144, 146, 147, 152–153, 233, 262, 286
  district, 43, 55, 56, 64, 77, 80, 84
  enrollment quota, 31
  funding, 84
  life expectancy, 31
  system
Schooling, 8–11, 23–25, 29, 117, 120–123
Scientific knowledge, 14, 148, 151, 154, 166, 168, 192, 204
Scripts, 111, 168, 169, 175, 198
Secondary School, 8, 9, 37–39, 59, 106, 284
Second Morrill Act, 60
Second World War, 30, 36, 62, 71, 82, 83
Secondary School, 8, 9, 37–39, 59, 106, 284
Segregated school, 65
Segregation, 62
Seminomads, 212, 216
Serb, 31
Serbian, 30
Sexual Behavior, 16, 277, 278, 280, 283, 284, 288, 289
Sexual cleansing practices, 288

Sexuality, 120–123
Sharecropper, 9, 59, 62, 216
Site-specific knowledge, 14
Situated knowledge, 2
Size of elementary schools, 8, 43, 44
Skills, 4, 10, 13, 15, 36, 38–41, 45, 57, 72, 75, 100, 146, 149, 170–173, 184, 213, 219, 225, 237, 248
Slavery, 9, 57, 58, 60, 70, 85
Slave(s), 33
Slovenia, 38
Slovenian, 30, 38, 39
Small schools, 38, 42, 43, 82, 87
Social
  activity, 16
  cognition, 167
  cohesion, 2, 9, 40, 42, 82, 93
  construction, 6, 7, 15, 16, 98, 107, 230, 247
  cooperation, 14, 193
  Darwinism, 34
  development, 113
  difference(s), 6, 11, 109, 114, 120–122
  distribution of knowledge, 243
  environment, 47, 167, 184
  events, 16, 45, 85, 250
  governance, 16
  identification, 117, 118
  interaction, 258
  justice, 11, 16, 80, 111, 126, 257
  mobility, 9, 25, 29, 70, 93, 100, 114
  network(s), 40, 125
  status, 5, 9–12, 15, 25, 27, 29–31, 121, 149, 168, 180, 260, 263
  values, 80, 125
Society, 83, 147
Socio-cultural, 7, 10, 24, 34, 77, 93–107, 112, 130, 131
Socioeconomic disability, 120–123
Socio-hydrological system, 193–197
Sociology, 126
Sociology of education, 47
Solidarity, 7, 27, 42, 46, 47, 62, 65, 82, 122, 125
South Carinthia, 28, 38, 39
South Carolina, 33, 75
Southeast Asia, 167, 173
Southern United States, 9, 55–66
South Tyrol, 37, 38, 41, 43–45
Spain, 94
Spanish, 77, 86, 94, 95, 98, 99, 103, 150
Spatial inequalities, 30, 32, 94
State, 35, 81, 85, 99, 100, 104, 194, 258
Static traditionalism, 13, 112
Stonewall, Texas, 82, 83
St. Petersburg, 37

Student enrollment, 105
Sub-Saharan Africa, 277
Supreme Court, 9, 32, 55, 56, 63–65
Suspension rates, 31
Sustainability, 151, 194, 270
Systematic inequality, 63

T
Taboos, 261, 263, 264, 267, 268, 270
Tacit knowledge, 145, 146, 166, 172
Taliban, 33, 224, 225
Tanzania, 16, 277, 280–284
Tauwema, 229–232, 235–240, 246, 248,
    250, 251
Teacher, 104
Teaching staff, 41, 94, 100–103, 105–107, 113
Technical knowledge, 15, 275
Territory, 23, 24, 26–28, 31, 37, 41, 42, 47, 87,
    94, 103, 104, 170, 259, 260
Texas, 59, 81, 82, 86
Texas Hill Country, 83
Textbook, 8, 23, 26, 27, 32, 37, 41, 46, 81,
    174, 181, 182, 184
Theory of culture, 6
Tibetan plateau, 194
Toliwaga, 15, 230, 233–248
Transculturality, 41
Transculturation, 24
Transformation, 5, 11, 14, 29, 98, 100, 173,
    192, 197, 249
Tribal, 28, 34, 41, 133–135, 219, 220, 259
Tribe, 28, 41, 45, 213–215, 223, 286
Trobriand Islands, 16
Truancy rates, 31
Trust, 152
Tunisia, 212, 216, 217, 219
Tyrol, 43–45

U
Uganda, 16, 116, 277, 278, 280, 281, 283,
    284, 287, 289
Ukraine, 37
Ukrainian, 26
Uncertainty, 2, 138
Undereducation, 84–89
United Kingdom, 31, 148
United States, 8–10, 12, 28, 31, 32, 34, 36, 45,
    55–56, 69–75, 77, 78, 80, 81, 85, 94,

95, 100, 102–104, 107, 133, 148, 169,
    219, 279
Universal knowledge, 1–4, 13, 112
University, 10, 26, 59, 85, 94, 97–105, 113,
    126, 205, 284
Urban knowledge, 13, 165, 167, 182, 281
USAID, 15, 219, 223, 279, 281, 283
U.S. census data, 94–96
U.S. Civil Rights Movement, 26
U.S. Congress, 55
U.S. Department of Education, 55

V
Value, 1, 2, 4, 5, 7, 9–11, 16, 17, 25–30,
    35, 40–42, 70, 71, 74, 75, 79–81,
    83, 87, 94, 104, 105, 111, 118,
    125, 130, 133, 134, 148, 155, 169,
    171, 175, 177, 180, 192, 195, 197,
    224, 247, 248, 260, 262, 263,
    270, 278, 289
Veterinary
    knowledge, 15, 211, 216, 219, 225
    laws, 217, 218
    medicine, 15, 216, 217, 219, 221, 225
    services, 217–220
Vienna, 37
Virginia, 33, 59, 75, 79, 101

W
Warnier law(s), 214
Washington, D.C., 56, 75, 101
Water distribution technologies, 193
White-Anglo America, 10, 87
White flight, 9, 64
Whiteriver, 41
White supremacy, 61, 62, 65
WHO, 281, 283, 285, 286
William Lloyd Garrison, 57
Wola, 134, 136, 137, 139–150,
    152–155
Women, 222–225
World War I, 30, 32, 36–38, 61–63, 70, 71,
    83, 85
World War II, 30, 36, 62, 63, 71, 83

Y
Yam cultivation, 267